Semiconductor Device Modelling

Christopher M Snowden (Ed.)

Semiconductor Device Modelling

With 111 Figures

Springer-Verlag
London Berlin Heidelberg New York
Paris Tokyo

Christopher M. Snowden, BSc, MSc, PhD, CEng, MIEE, MIEEE
Department of Electrical and Electronic Engineering, University of Leeds, Leeds L52 9JT, UK

ISBN 3–540–19545–9 Springer-Verlag Berlin Heidelberg New York
ISBN 0–387–19545–9 Springer-Verlag New York Berlin Heidelberg

British Library Cataloguing in Publication Data
Semiconductor device modelling.
1. Semiconductors. Mathematical models
I. Snowden, M. (Christopher M.)
621.3815'2'0724
ISBN 3–540–19545–7

Library of Congress Cataloging-in-Publication Data
Semiconductor device modelling.
Bibliography: p.
Includes index.
1. Semiconductors—Mathematical models. I. Snowden, Christopher.
TK7871.85.S4676 1989 621.3815'2 89–4155
ISBN 0–387–19545–9 (U.S.)

Printed in Great Britain

Printed and bound by The Alden Press, Osney Mead, Oxford

2705/3916–543210 (Printed on acid-free paper)

Preface

Semiconductor device modelling has developed in recent years from being solely the domain of device physicists to span broader technological disciplines involved in device and electronic circuit design and development. The rapid emergence of very high speed, high density integrated circuit technology and the drive towards high speed communications has meant that extremely small-scale device structures are used in contemporary designs. The characterisation and analysis of these devices can no longer be satisfied by electrical measurements alone. Traditional equivalent circuit models and closed-form analytical models cannot always provide consistently accurate results for all modes of operation of these very small devices. Furthermore, the highly competitive nature of the semiconductor industry has led to the need to minimise development costs and lead-time associated with introducing new designs. This has meant that there has been a greater demand for models capable of increasing our understanding of how these devices operate and capable of predicting accurate quantitative results. The desire to move towards computer aided design and expert systems has reinforced the need for models capable of representing device operation under DC, small-signal, large-signal and high frequency operation. It is also desirable to relate the physical structure of the device to the electrical performance.

This demand for better models has led to the introduction of improved equivalent circuit models and a upsurge in interest in using physical models. The increasing interest in the area of semiconductor device modelling has spawned a number of conference series (such as SISDEP, NASCODE and IEEE/SIAM), which deal with recent developments in this area, particularly on the topic of physical device models. However, up to the present time there has been little effort devoted to introducing the wide range of available modelling techniques to the engineering and scientific community. It was with this in mind that a short course in semiconductor device modelling was initiated in 1987, to provide insight into semiconductor device modelling techniques and their application. Following the success of this course a second was organised for 1989, covering the material contained in this text.

This text aims to provide an insight into the many modelling techniques available for semiconductor devices. A review of basic device physics and carrier transport theory is followed by a discussion of classical and semi-classical modelling techniques. This is followed by a summary of essential numerical techniques used to implement many of the models presented in the text. Equivalent circuit modelling techniques for silicon and compound semiconductor technologies are dealt with in separate chapters. The practical aspects of semiconductor device modelling are discussed in the context of present trends in CAD. A detailed foundation in Monte Carlo particle simulation techniques is presented. This is followed by chapters on physical models for silicon VLSI and compound semiconductor devices, together with practical details of their implementation and examples. The important area of modelling noise processes is covered in detail. The rapidly evolving area of quantum mechanics and quantum transport modelling and their application in nano-electronics extends the text to state-of-the-art devices. The modelling of optoelectronic devices, including laser diodes and photodiodes, is addressed. The text is completed by a chapter describing the principles of computer simulation programs, with a number of examples together with simulation results.

I am very grateful to the authors who have contributed to this volume, making this a unique collection of monographs for use as an introduction by those new to the subject and as a helpful reference for the more experienced modeller.

University of Leeds, 1988 Christopher M. Snowden

Contents

List of Contributors

Roel Baets
Laboratory for Electromagnetism and Acoustics, University of Gent-IMEC, Sint-Pietersnieuwstraat 41, B-9000 Gent, Belgium

John Barker
Nanoelectronics Research Centre, Department of Electronics and Electrical Engineering, University of Glasgow, Glasgow, UK

Joseph A Barnard
Barnard Microsystems Limited, 13 Long Spring, St Albans, Hertfordshire, AL3 6EN, UK

Trevor M Barton
Microwave Solid State Group, Department of Electrical and Electronic Engineering, University of Leeds, Leeds, LS2 9JT, UK

Margaret E Clarke
Electrical Engineering Laboratories, The University, Manchester, M13 9PL, UK

Alain Cappy
Centre Hyperfrequences et Semiconducteurs, Universite des Sciences et Techniques de Lille, 59655 Villeneuve D'Ascq, France

Michael J Howes
Microwave Solid State Group, Department of Electrical and Electronic Engineering, University of Leeds, Leeds, LS2 9JT, UK

Derek B Ingham
Department of Applied Mathematical Studies, University of Leeds, Leeds LS2 9JT, UK

Robert E Miles
Microwave Solid State Group, Department of Electrical and Electronic Engineering, University of Leeds, Leeds, LS2 9JT, UK

Stephen D Mobbs
Department of Applied Mathematical Studies, University of Leeds, Leeds, LS2 9JT, UK

Mustafa Al-Mudares
Nanoelectronics Research Centre, Department of Electronic and Electrical Engineering, University of Glasgow, Glasgow, UK

Siegfried Selberherr
Institut fur Mikroelektronik, Technical University Vienna, Gußßhausstraße 27–29, A-1040 Wien, Austria

Michael Shur
Department of Electrical Engineering, University of Minnesota, 123 Church Street, Minneapolis, Minnesota, 55455, USA

Christopher M Snowden
Microwave Solid State Group, Department of Electrical and Electronic Engineering, University of Leeds, Leeds, LS2 9JT, UK

Review of Semiconductor Device Physics

Robert E Miles
University of Leeds

INTRODUCTION

As a basis for the work to be covered in subsequent lectures an outline of the physics of a number of common semiconductor devices will be presented. The purpose here is not only to familiarise the reader with the modes of operation but also to point out the limitations of the simple theories. Nevertheless, even with these limitations, the analytical models outlined in this chapter have played a vital role in the development of semiconductor devices. As the performance demanded from these devices becomes ever more exacting, current densities and electric fields approach their limits and many of the approximations necessary for an analytical description no longer apply. In these situations we must resort to a physical model based on semiconductor physics in order to understand the processes taking place. These models can account for the full non-linear behaviour of a device but they do require a large amount of computing power for a successful simulation. After a number of years of development in this area, device simulation is now giving meaningful results by including realistic models of both surface and bulk phenomena such as surface states, trapping levels and hot electron effects.

An alternative but less accurate approach is to model the device in the form of an equivalent circuit. By incorporating non-linear elements these models can be used to represent devices at high frequencies and large signal levels. The problem with an equivalent circuit model is that it is only loosely related to the material properties and device geometries and provides little insight into the physical processes taking place.

At the present time circuit design software such as SPICE, TOUCHSTONE and SUPER COMPACT use equivalent circuit models whereas physical simulation is more useful as an aid to understanding the operation of a device and hence as a tool in its development. Nevertheless with the ever increasing speed and power of modern computers and using some recent quasi two dimensional models, simulation is fast becoming a viable circuit design technique. This is particularly true for microwave integrated circuits where the number of active devices is usually quite small (often only 1). Such techniques are essential if the nonlinear behaviour of devices in oscillator, amplifier and mixer circuits is to be explored and exploited.

1. DEVICE OPERATION - GENERAL CONSIDERATIONS

1.1. Energy Bands, Electrons and Holes.

The behaviour of electrons in a periodic lattice is a problem in quantum mechanics which is beyond the scope of this introductory lecture. Detailed calculations lead to the now familiar picture of electrons in a crystal populating a set of energy bands, filling them from the lowest levels upwards. The electrical properties of the material are determined by the electrons in the highest filled levels. This is because for an electron to move in an electric field it must gain a small amount of kinetic energy and it is only the electrons near the top of the pile that have empty levels sufficiently close to them in energy to move into. The main interest in this course is in the class of materials called semiconductors which are characterised by having a completely filled energy band at the absolute zero of temperature but with an empty band separated from it by a forbidden energy gap E_G of about 1 eV. Thus at absolute zero semiconductors are insulators but as the temperature increases more and more electrons gain enough energy to be excited across the forbidden gap where they are free to move in the sparsely populated energy band. See Fig. 1.

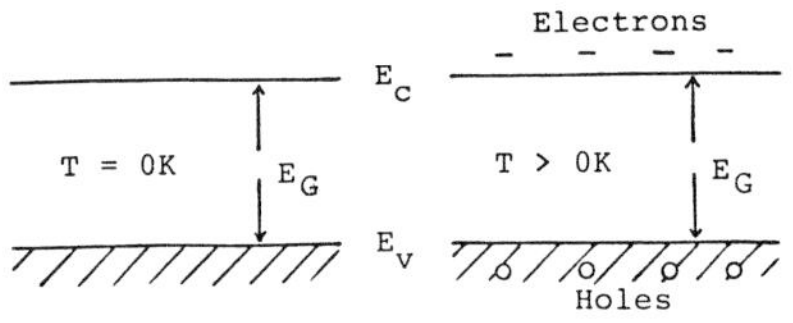

Figure 1: Conduction and Valence bands of a semiconductor at and above absolute zero.

The lower of these bands is called the valence band and the upper is the conduction band. Electrons excited into the conduction band leave empty levels or holes behind which in turn enables movement of charge in the valence band as electrons jump into these empty states. The effect is that of a flow of positive charges or "holes" in the valence band.

The condition that n = p is only true for a pure or intrinsic semiconductor. With the deliberate addition of a low density of impurities a semiconductor can be doped such that n > p or p > n but in all cases the n.p product is a constant at a given temperature.

The density of electrons in the conduction band at a particular temperature is the result of a dynamic equilibrium. Electrons are being generated steadily at a rate g(T) which depends on temperature. Initially this leads to a continuous increase of the electron density in the conduction band and of course an equal increase in hole density in the valence band. Electrons and holes recombine at a rate which is proportional to the n.p product so that eventually an equilibrium exists when

$$g\,(T) = r.n.p$$

where r is a proportionality constant. Variations in temperature change g(T) and a new equilibrium is achieved with different electron and hole densities. In doped material, where one carrier type predominates the rate of recombination is proportional to the excess minority carrier concentration because the majority carrier concentration is relatively unaffected and the recombination equations can be written as :-

$$\frac{dn}{dt} = -\frac{\delta n}{\tau_{rn}} \quad \textit{for a p-type semiconductor}$$

and

$$\frac{dp}{dt} = -\frac{\delta p}{\tau_{rp}} \quad \textit{for an n-type semiconductor}$$

where δn and δp are the excess carrier densities and τ_{rn} and τ_{rp} are the recombination times. The generation rate can also be changed considerably by illuminating the material with light of sufficient photon energy to excite electrons across the forbidden gap. A new equilibrium will then be achieved with excess electron and hole densities over and above those generated by the ambient temperature. The excess densities will depend on light intensity and it is this interaction (and its inverse) between light and semiconductors which gives rise to many applications such as photo-detectors, LEDS, LASERS and photoelectric-cells.

1.2. Charge Transport

When electrons in a crystal move under the influence of an electric field the quantum mechanical nature of the phenomenon is simply allowed for by the concept of effective mass. Because they are moving in the periodic electric field of the crystal the electrons (and holes) behave as though they have a mass which is not equal to rest the mass m_o of an electron. i.e.

$$m^* = \alpha m_o$$

where m^* is the effective mass and $\alpha \neq 1$. In most cases $\alpha < 1$ for both electrons and holes and in many semiconductors of practical interest the effective mass of electrons is less than that of holes. As a consequence for a given electric field electrons will be accelerated to a higher velocity than holes over a given distance. This has important implications for high speed (or high frequency devices) where carrier velocity is a critical parameter.

In an electric field electrons and holes do not move through a crystal with a uniform acceleration as they would through a vacuum. Their motion is of a stop/start nature as they continuously loose the energy gained from the electric field to the crystal lattice by collisions. The nature of these collisions is interesting because theory shows that electrons move without colliding in a perfectly regular lattice i.e. they do not bump into the atoms. (For GaAs the mean distance between collisions can be as much as 3 microns compared to an atomic spacing of 0.0003 microns.) The key here is the phrase 'perfectly regular'. Such a lattice is rare for a number of reasons. There may be impurities in the lattice which distort the periodic field and act as collision-centres (these impurities may even have been deliberately introduced to dope the material). Atoms may be missing or out of place in the lattice again leading to distortion of the periodic potential through which the electrons move. Finally even for a 100% pure and perfect lattice, if the temperature is raised above absolute zero the constituent atoms vibrate about their mean position, once more distorting the perfect periodicity. The amplitude of vibration increases with temperature making the effective distance between collision centres decrease with temperature. See Figs. 2 & 3.

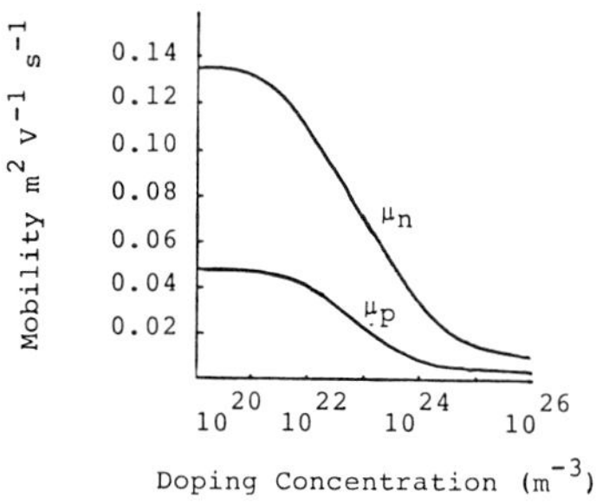

Figure 2: Variation of electron and hole mobility with doping concentration. Si at 300K

The net result of these collisions is that in an electric field E electrons and holes drift along with a velocity given by

$$v = \frac{q\tau}{m^*} E = \mu E$$

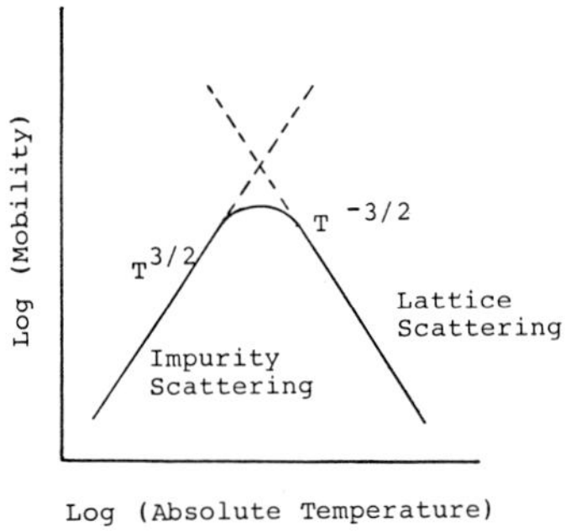

Figure 3: Temperature variation of mobility for impurity and lattice scattering.

where μ is the mobility and τ is the mean time between collisions which as described above depends on lattice perfection and temperature. (τ is not to be confused with the recombination times τ_r mentioned earlier.) This drift velocity is imposed upon the random thermal velocity giving

$$J_p = \sigma_p E \quad \textit{where} \quad \sigma_p = pq\mu_p$$

$$J_n = \sigma_n E \quad \textit{where} \quad \sigma_n = nq\mu_n$$

The above formulae are low field approximations where the drift velocity is much less than the random thermal velocity. As the field increases this approximation breaks down and leads to an effective reduction in τ and hence drift velocity. At even higher fields the energy gained by the carriers is transferred efficiently into elastic waves in the lattice and the drift velocity saturates. The resulting velocity field curve for Si is illustrated in Fig. 4. The saturation of the drift velocity at high electric fields plays an important part in device operation.

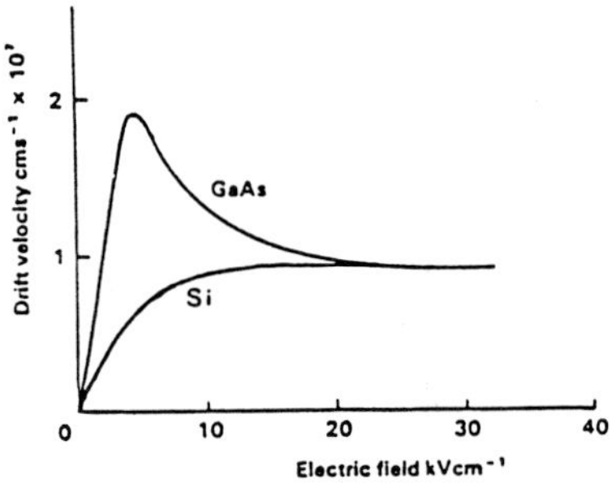

Figure 4: Room temperature velocity-field characteristics for Si and GaAs.

The velocity field curve for GaAs, also shown in Fig. 4, is somewhat different from that of Si. This results from the peculiarities of the band structure of the III-V semiconductors. There are in effect two overlapping conduction bands, one with its band edge at a somewhat higher energy than the other (0.3 eV in GaAs). Electrons in this second band have a higher effective mass than those in the lower band. If the electrons can gain enough energy from the electric field they will transfer to the heavy mass band. This accounts for the maximum in the velocity field curve as more and more electrons make the transfer. At high fields, electrons in GaAs behave in a similar way to those in Si. At low fields they have a lower effective mass and a correspondingly higher drift velocity at a given field. For high speed devices there are advantages in keeping electrons in the light mass band where "velocity overshoot" is thought to have a significant effect on device performance.

Carriers can also move by diffusion when concentration gradients exist. Simple one dimensional theory relates current density J to concentration gradient by

$$J_p = -qD_p \frac{dp}{dx} \qquad \textit{for holes}$$

$$\textit{and } J_n = +qD_n \frac{dn}{dx} \qquad \textit{for electrons}$$

where D_p and D_n are diffusion coefficients and $\frac{dp}{dx}$ and $\frac{dn}{dx}$ are the concentration gradients. Considerations of equilibrium yield the Einstein relationships between diffusion coefficients and mobilities:-

$$D_p = \frac{kT}{q}\mu_p \qquad D_n = \frac{kT}{q}\mu_n$$

Current in a semiconductor can therefore be carried by electrons and holes and be by drift and diffusion leading to the following general expressions for current density.

$$J_p = pq\mu_p E - qD_p \frac{dp}{dx}$$

$$J_n = nq\mu_n E + qD_n \frac{dn}{dx}$$

Furthermore for continuity of carriers in a region where there is both current flow and carrier recombination. The following continuity equations apply.

$$\frac{dp}{dt} = -\frac{1}{q}\frac{\partial J_p}{\partial x} - \frac{\delta p}{\tau_p}$$

$$\frac{dn}{dt} = \frac{1}{q}\frac{\partial J_n}{\partial x} - \frac{\delta n}{\tau_n}$$

1.3. Semiconductor Surfaces

The ideas of energy bands are based on an infinitely periodic crystal but in the real world crystals must terminate at an interface with another material. For semiconductor devices, terminations at vacuum, air, dielectric and metal interfaces are of particular importance. (The properties of interfaces between semiconductors of different energy gaps are increasingly used in modern devices. i.e. heterojunctions)

The termination of the lattice at whatever medium can give rise to energy levels at the surface in the forbidden gap known as "surface states". Surface states can also result from impurities or lattice re-arrangement at the surface. Whatever their source, if the states are in densities of greater than $10^{16}\ m^{-2}$, they completely dominate the nature of the surface. (It was the high densities of surface states on germanium and silicon that plagued the early development of field effect devices)

The charge occupation of the surface states is of course governed by the position of the Fermi Level. With a high density of states, even a small movement of the Fermi Level results in a large

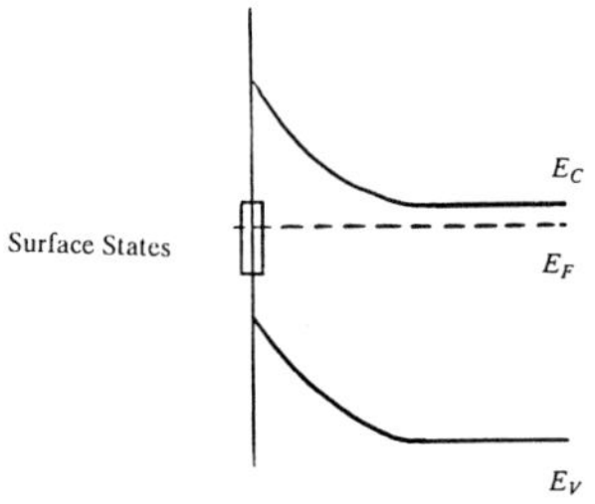

Figure 5: n-type semiconductor with the surface Fermi Level clamped near mid gap.

change in the surface charge. As a consequence the Fermi Level tends to remain in a relatively fixed position in the forbidden gap at the surface and the bulk of the semiconductor is effectively shielded from the outside world. The surface potential of the semiconductor, as measured by the position of the Fermi Level relative to the intrinsic Fermi Level at the surface, is "clamped". This is illustrated by a band diagram for an n-type semiconductor with clamping near mid gap in Fig. 5.

Silicon technology has developed to the point where the density of states at the silicon-silicon dioxide interface has been reduced to a negligible value but this is generally not the case for the III-V semiconductors.

The concepts outlined in Section 1 will now be applied to some common devices.

2. BIPOLAR DEVICES

Both electrons and holes play a part in the operation of these devices. The important processes take place in the bulk of the device and surface effects can be often be ignored.

2.1. The p-n junction diode

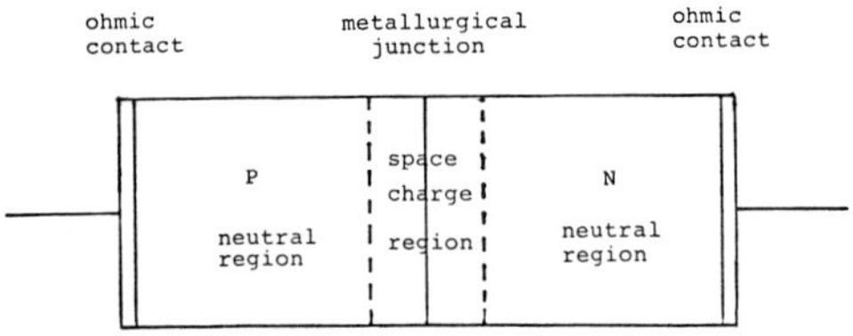

Figure 6: Schematic diagram of a p-n junction diode.

In the simple theory of a p-n junction all of the applied voltage is assumed to fall across the junction region and the current flow is determined by diffusion of minority carriers. This is considered to take place in the field free (or neutral) regions outside the space charge on each side of the metallurgical junction. (Fig. 6) The electric field in the the depletion region represents a potential barrier to the flow of majority carriers across the junction as shown in Fig. 7.

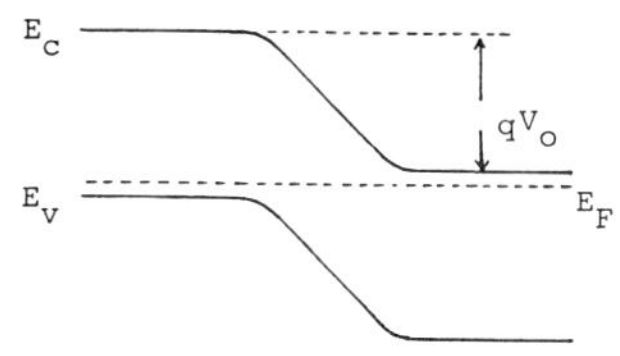

Figure 7: Band diagram of a p-n junction showing the potential barrier qV_0 to majority carrier flow across the junction.

Because the electron and hole diffusion lengths are much longer than the depletion length, minority carriers of both types are swept rapidly across the junction by the high electric field with no loss or gain by recombination or generation on the way. The excess density of minority carriers injected across the barrier is then governed by the Boltzman factor which determines the fraction of carriers having enough energy to surmount the potential barrier at the junction. So, remembering that at zero bias there is an equal and opposite flow of charge from each side of the junction we get the equation

$$I = I_o e^{qV/kT} - I_o$$

where V is the applied bias voltage and I_o is the reverse bias saturation current made up of the diffusion currents in the field free regions ($E = 0$). I_o is given by solution of the continuity equations in steady state:-

$$\text{i.e. with} \quad \frac{dp}{dt} = 0 \qquad D_p \frac{d^2\delta p}{dx^2} = \frac{\delta p}{D_p \tau_{rp}}$$

$$\text{and with} \quad \frac{dn}{dt} = 0 \qquad D_n \frac{d^2\delta n}{dx^2} = \frac{\delta n}{D_n \tau_{rn}}$$

Solution of these equations gives an exponential decay of excess carriers on both sides of the junction with the diffusion currents determined by the gradients at the space charge edges.

$$\text{ie} \qquad I_o = I_{po} + I_{no}$$

$$= qn_i^2 A \left\{ \frac{1}{p}\frac{D_n}{L_n} + \frac{1}{n}\frac{D_p}{L_p} \right\}$$

where p and n refer to the majority carrier concentrations on each side of the junction, $L_n = (D_n \tau_{rn})^{1/2}$ and $L_p = (D_p \tau_{rp})^{1/2}$ are the diffusion lengths for electrons and holes respectively. The approximations made in this elementary calculation are many and it is no wonder that the simple theory breaks down. Let us look at some of the other effects that must be included.

2.1.1. In Forward Bias

a) Heavy injection of minority carriers from both sides results in recombination in the junction region leading to a lowering of I_o in forward bias.

b) The large currents flowing in forward bias result in a potential drop across the contact and thus not all of the applied voltage falls across the junction.

c) The large current flow and lowered barrier also result in an electric field outside the junction region making the assumption of diffusion in a field free region invalid.

2.1.2. In Reverse Bias

a) The approximations work better in reverse bias because the current is usually small ($= I_o$) but there can be an effect if crystalline imperfections produce generation/recombination centres in the forbidden gap at the junction. As a result, instead of saturating, the reverse bias current increases with applied voltage as the depletion region extends over a larger volume at the junction.

b) Eventually at a sufficiently high reverse bias voltage the diode will break down either by quantum mechanical tunneling (Zenner breakdown) or avalanche effects. These breakdown mechanisms are high field effects and are always liable to occur when the electric field in a semiconductor exceeds a critical value.

2.1.3. At High Frequencies

The response time of a diode, particularly in forward bias is limited by the recombination times of excess carriers injected across the junction. This is known as the charge storage effect.

2.2. Diode Equivalent Circuit

Considered as a circuit element the diode is simply a voltage dependent capacitor C in parallel with a resistance R_p, both in combination with a series resistor R_s as shown in Fig. 8.

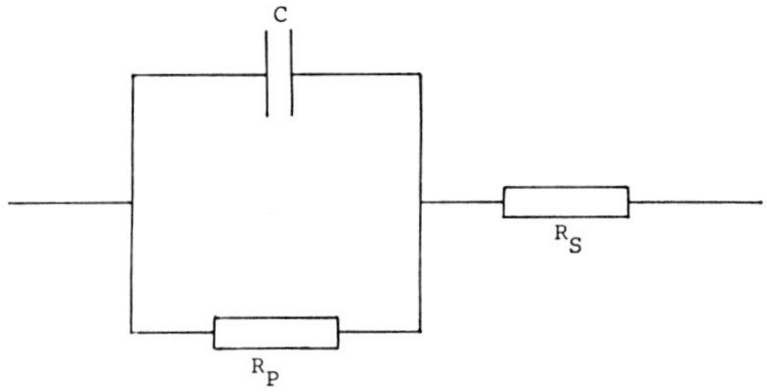

Figure 8: Equivalent circuit of a p-n junction diode.

In reverse bias this model works very well with C as the well known depletion capacitance. In a good diode R_p is very large and R_s is small.

In forward bias C becomes the charge storage capacitance of minority carriers injected across the junction and increases strongly with applied voltage. It also has a frequency dependence, especially at frequencies corresponding to carrier recombination times.

As can be seen from the above, the p - n junction is a bipolar device because its mode of operation is intimately connected with the interaction of electrons and holes across the forbidden gap.

2.3. Bipolar Junction Transistors(BJT)

These transistors are based on p-n junctions and as implied in their name are bipolar devices. Their construction is that of two p-n junctions located close together to give the familiar n-p-n or p-n-p structure as shown in Fig. 9

In operation the device is biased such that the emitter-base junction (see Fig. 9) is forward biased and the base-collector junction reverse biased. The important point is that the base region is much narrower than the diffusion length so that minority carriers injected from the emitter cross the base and are swept off into the collector. This is illustrated in the energy band diagram of a biased

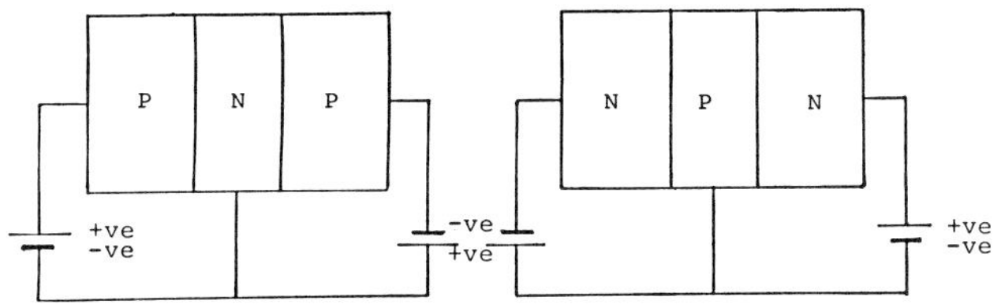

Figure 9: p-n-p and n-p-n transistors with bias (common base configuration).

n-p-n device shown in Fig. 10.

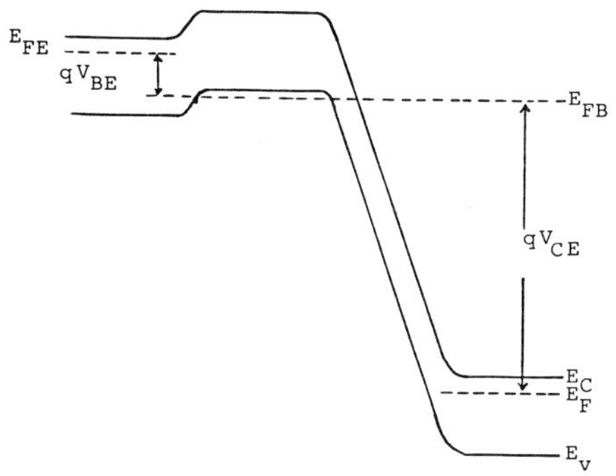

Figure 10: Band diagram of a biased n-p-n bipolar transistor.

The effect is to produce a current flow in the reverse biased base-emitter junction whose magnitude is more characteristic of a forward biased junction. Because this current flows through the relatively large potential difference of the reverse bias, power gain is achieved.

In order to get current gain the emitter current must be controlled by a small base current. This is accomplished by making the base doping much less than the emitter doping such that the current across this junction is predominantly that of the carrier type in the emitter.

This device has been analysed extensively in the literature and many analytical models have been developed to account for its observed behaviour under all manor of conditions.

BJTs can be well represented by equivalent circuit models such as that due to Ebers & Moll (Fig. 11) in circuit design software. The double diode nature of the device is reflected in the equivalent circuit.

Transistor performance has been improved many times over the original design both in power handling and frequency response and has been largely responsible for the electronics revolution witnessed over the last 40 years.

For high frequency operation there are similar considerations as in the p-n diode where capacitance effects at the junctions must be minimised. In addition performance is limited by the transit time of minority carriers across the base. The base must therefore be very thin and it is also necessary to take into account the field dependence of carrier velocity in what turns out to be a quite complicated electric field region.

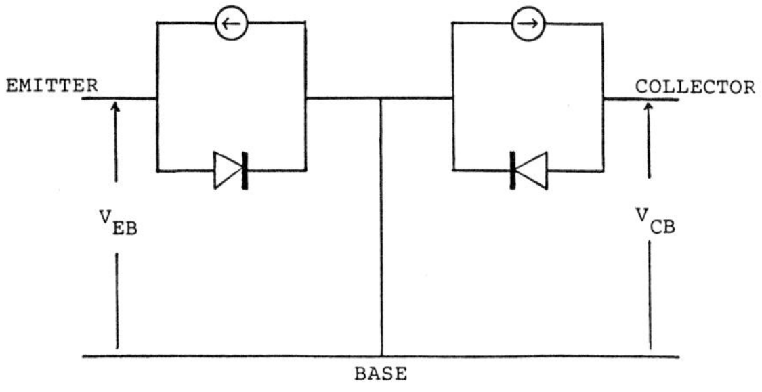

Figure 11: Ebers-Moll model of a p-n-p transistor.

A recent development of the BJT made possible by modern technology is the heterojunction bipolar transistor (or HBT) which will be discussed in the section on heterostructure devices.

3. MAJORITY CARRIER OR MONOPOLAR DEVICES

In these devices only one carrier type is involved - usually electrons because they travel at higher velocities than holes.

3.1. Schottky Barrier Diodes

An equivalent structure to the p-n junction diode is the Schottky barrier diode formed by a metal/semiconductor contact as represented in Fig. 12.

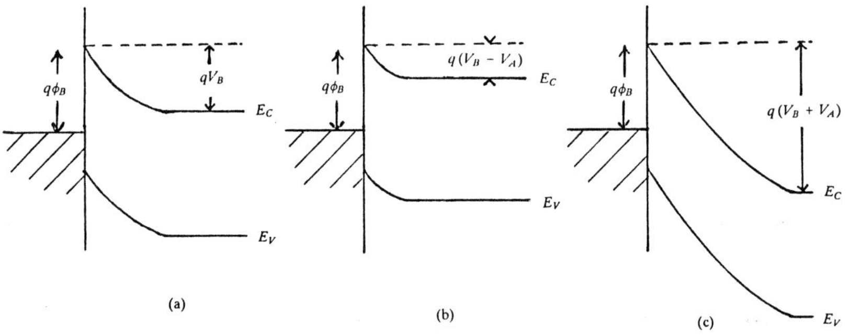

Figure 12: Schottky barrier in a) zero bias b) forward bias c) reverse bias

Current flow in these diodes is entirely by majority carriers, i.e. electrons if, as is usually the case, the semiconductor is n-type. There are therefore no minority carrier storage effects and Schottky diodes have excellent performance up to frequencies > 100 GHz. It is of interest to note that metal/semiconductor contacts were amongst the first solid state devices used in the form of the 'cats whisker' in crystal radios and are still used today up to the highest frequencies of operation.

The essential mode of operation of Schottky barrier devices is adequately explained by a very simple theory that ignores many of the detailed features of the barrier. This theory illustrated in Fig. 12 assumes that the Schottky barrier behaves very much like a $p^+ - n$ junction where the metal plays the part of the p^+ semiconductor. The applied voltage V_A either adds to or subtracts from the built in barrier voltage V_B to reverse or forward bias the device respectively. In reverse bias the width of the depletion region increases owing to the increased voltage drop across it and conversely decreases in width in forward bias. In practice Schottky diodes are mostly used in reverse bias. (For mixer applications they are used near zero bias where the curvature of the I-V characteristic is greatest)

The current-voltage relationship of a Schottky diode is usually written as

$$I = I_o \left(e^{(qV/nkT)} + 1\right)$$

The term I_0 is is determined by the mechanism of electron flow over the barrier ϕ_B as seen from the metal. In its simplest form thermionic emission dominates and

$$I_o = device\ area \times A^* T^2 exp\left(-q\phi_B/kT\right)$$

where A^* is Richardsons constant. If the semiconductor is highly doped the barrier becomes thin enough, especially near the top, for electrons to tunnel through. The emission is then controlled by both thermionic and field effects with a consequent modification to the above equation. The barrier is also lowered by image force effects. The quantity n in the diode equation is known as the ideality factor and as its name implies covers a multitude of effects. Its ideal value is unity but the presence of recombination and generation in the depletion can increase this. Good diodes have n values of about 1.01 but in extreme cases it can take on values as high as 6 or 7 where its interpretation is open to much speculation.

A major assumption is that the space charge is continuously distributed giving a uniform barrier across the junction area but simple calculations throw an interesting light on this idea. For the doping densities commonly used i.e. $\approx 10^{23}\ m^{-3}$ the donor impurities have a separation of about 0.02 microns which is about one half of the corresponding zero bias depletion length. The barrier must therefore be lumpy on this scale with a thickness varying considerably from place to place depending on the exact location of the charges. This effect is usually ignored.

3.2. Field Effect Transistors

There is a basic family of three field effect transistors illustrated in Fig. 13. They are the JFET, MISFET (or MOSFET) and MESFET.

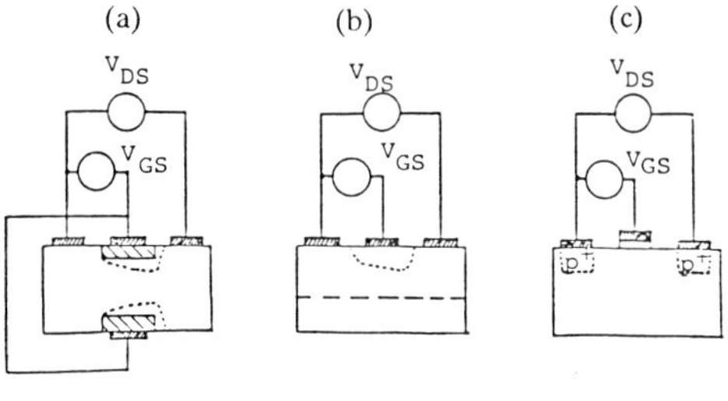

Figure 13: The field effect transistor family
a) JFET b) MESFET c) MOSFET

All three have the same basic operational principle where the resistance of a conducting channel is modulated by applying a signal voltage V_{GS} between gate and source. This effectively controls the width of the space charge region under the gate and hence the cross sectional area available for conduction. JFET's use conventional p-n junction technology. MOSFETS based on silicon are used extensively in integrated circuits, the excellent insulating properties of silicon dioxide playing an important part here. MESFETS are invariably used in GaAs technology for analogue and digital circuits. A considerable effort has been put into the simulation and modelling of GaAs MESFETS at high frequencies and for this reason the rest of this section will concentrate on these devices.

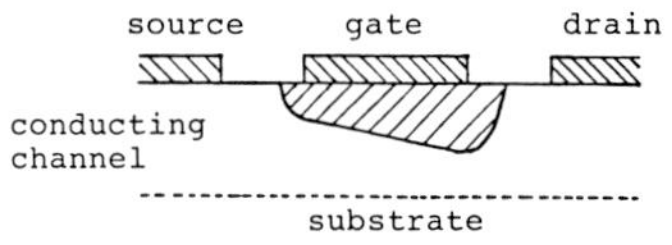

Figure 14: Typical conducting channel in a MESFET.

The simple theory is essentially that of Shockley's 'gradual channel' approximation which describes conduction in a channel of tapering cross sectional area as shown in Fig. 14 (the channel tapers because the reverse bias voltage across the depletion region increases from the source to the drain end of the gate due to the source-drain voltage V_{DS}). The gradual channel approximation assumes that the field at right angles to current flow between source and drain is negligible and is really only applicable to relatively long devices. Even then an analytical solution can only be achieved up to the point where saturation occurs in the output characteristics (see Fig. 15). Beyond this point different approximations must be made in different regions determined by the relevant part of the velocity-field curve (Fig. 4).

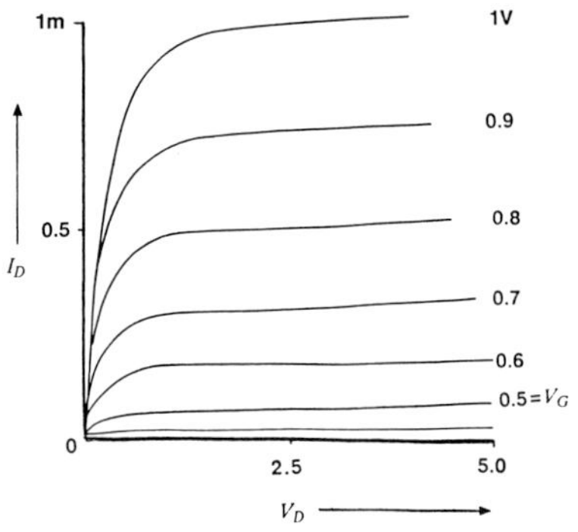

Figure 15: Output characteristics of an enhancement mode GaAs MESFET showing current saturation.

For short channel (< 1 micron) GaAs devices there is no adequate analytical treatment because of the complex interactions of electrons with the high electric fields set up under the gate. A much deeper understanding of the physical processes taking place has been gained from simulation work in this area.

It is not uncommon for modern MESFETS to have gate lengths of 0.25 microns or less. As for the Schottky barriers mentioned above there will only be 5 or 10 donors under the gate making for a very non-uniform situation. These effects are expected to become more important as device size shrinks and quantum mechanical ideas are required to explain the operation. It is therefore somewhat surprising to find that semi-classical ideas appear to adequately account for device operation for gate lengths even as small as 0.05 microns.

4. HETEROJUNCTION DEVICES

Modern techniques of semiconductor crystal growth have resulted in what is known as bandgap engineering. Based on the III-V semiconductors Molecular Beam Epitaxy (MBE) and Metal Organic Chemical Vapour Deposition (MOCVD) allow us to grow very thin layers of semiconductor materials of controlled energy gap. Furthermore these layers can be grown epitaxially on top of one another providing there is a reasonable match of the crystal lattice constant between layers. This latter limitation is becoming less restrictive with for example reports of successful growth of GaAs on Si substrates. (MBE is increasingly used in Si technology - an interesting example of III-V technology leading the way.)

There are many new device structures using the principles of band gap engineering currently under investigation but his section will concentrate on just two. They are the Heterojunction Bipolar Transistor (HBT) and the High Electron Mobility Transistor (HEMT) which are the Heterostructure equivalents of the Bipolar Junction Transistor and the MESFET. These two devices illustrate how band gap engineering can be used to overcome some of the limitations of conventional devices.

4.1. Heterojunction Bipolar Transistor

It has already been mentioned that in a successful BJT the current flow from emitter to base must be predominantly of the carrier type that is in a majority in the emitter (i.e. electrons in an n-p-n transistor). Another way of expressing this is that the emitter injection efficiency γ defined as the ratio of the majority carrier current into the base to the total current is close to 100%. Normally this is achieved by making the carrier density in the emitter at least 100 times greater than in the base. This condition leads to problems in high frequency devices because a consequence of this base doping is a comparatively high base resistance which leads in turn to an inordinately large RC charging time of the junction capacitance.

This problem can be overcome by arranging for the emitter to have a larger band gap than the base as shown schematically in Fig. 16.

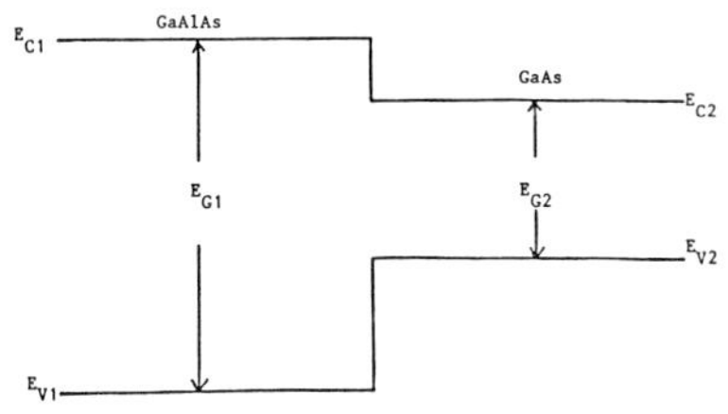

Figure 16: Simplified band structure diagram of the base emitter junction in a HBT.

Figure 17 shows a more realistic band diagram for a biased device. (compare this with Fig. 10 for the conventional BJT) For the n-p-n structure represented in the diagram the unwanted hole current from base to emitter is virtually eliminated by the potential barrier that the holes must surmount to cross from the narrow gap base material to the wide gap emitter. In theory this can lead to very high current gain devices but in practice this high injection efficiency is traded off against base doping to improve the high frequency characteristics.

It is interesting to note that the standard workhorse of semiconductor electronics - the silicon BJT - almost doesn't work. The high emitter doping actually narrows the band gap which has the reverse effect to that in the HBT and reduces the emitter injection efficiency.

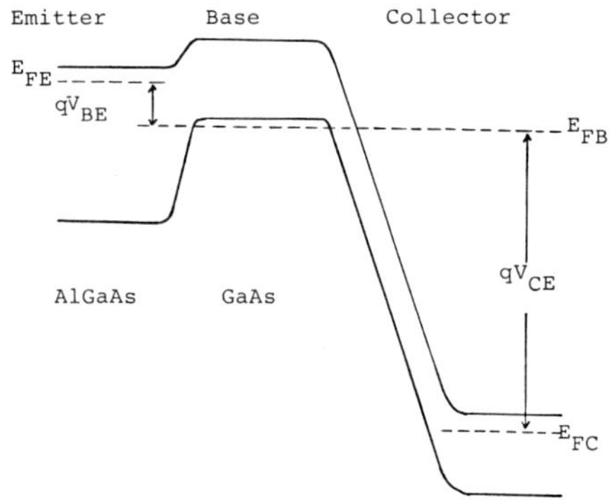

Figure 17: Energy band diagram of a n-p-n heterojunction transistor.

4.2. The HEMT

The HEMT is another method of overcoming the inherent limitations of conventional field effect devices. The optimum carrier density in the channel of a GaAs MESFET is about $10^{23}\ m^{-3}$. The donor impurities responsible for the carriers also have the effect of lowering the carrier mobility in the channel and thus degrading the performance of the device. The HEMT contrives to produce mobile carriers by conventional doping but then to constrain them to move in a very pure high mobility layer.

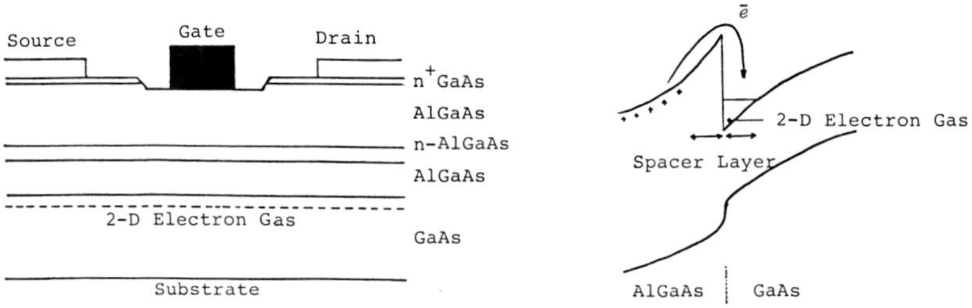

Figure 18: a) Structure of a HEMT b) Band diagram at the AlGaAs/GaAs interface showing the 2D electron gas.

A typical device structure is illustrated in Fig. 18 together with an energy band diagram. It can be seen that the electrons are constrained to move in the potential well formed in the high mobility GaAs layer due to the difference in energy gaps of the two materials. These devices have been shown to have superior properties to GaAs MESFETS of comparable dimensions.

The HEMT is also an interesting device from the modelling and theoretical point of view. The electrons move in a narrow potential well which is essentially two dimensional (giving rise to the alterative name of TEGFET or Two Dimensional Electron Gas FET). This problem was first encountered in inversion layers on silicon surfaces and requires the application of quantum transport in two dimensions for a proper treatment. Device engineers must now add quantum mechanics to their already long list of skills. We look forward to the rest of the course for enlightenment.

Classical and Semiclassical Models

Christopher M Snowden
University of Leeds

The traditional view of semiconductor device operation is based around a carrier transport model described by a set of semiconductor equations, which are often regarded as being phenomenological in origin. In a broader sense, carrier transport in semiconductor devices can be characterised in terms of either classical, semiclassical or quantum physical models. The majority of contemporary devices can be adequately characterised using the generalised classical approach based on the Boltzmann transport equation. A set of semiconductor equations derived from the Boltzmann transport equation form the basis of the majority of current device models and a widely used in both closed-form analytical models and numerical simulations. A full solution of the Boltzmann transport equation would be formidable and the final equations used in the models are arrived at after a series of simplifying approximations and depart significantly from the original form of the Boltzmann expression.

THE BOLTZMANN TRANSPORT EQUATION

In order to derive a set of semiconductor equations to describe carrier transport, a useful starting point is to establish the background to the Boltzmann Transport Equation (BTE). Charge particles can be characterised in terms of their position in space $\mathbf{r}$ and momentum $\mathbf{k}$ at time t. The density of particles $n(\mathbf{k},\mathbf{r},t)$ can then be described in terms of a distribution function $f(\mathbf{k},\mathbf{r},t)$, which is itself a function of phase space, momentum space and time,

$$n(\mathbf{r},t) = \int f(\mathbf{k},\mathbf{r},t)\, d\mathbf{k} \tag{1}$$

The derivative of the distribution function $f(\mathbf{k},\mathbf{r},t)$ with respect to time along a particle trajectory $\mathbf{r}$, $\mathbf{k}$ vanishes,

$$\frac{d}{dt} f\,(\mathbf{k},\mathbf{r},t) = 0 \tag{2}$$

This is the implicit form of the Boltzmann transport equation, which can be expanded as,

$$\frac{df}{dt} = \frac{\partial f}{\partial t} + \frac{\partial f}{\partial \mathbf{k}}.\frac{\partial \mathbf{k}}{\partial t} + \frac{\partial f}{\partial \mathbf{r}}.\frac{\partial \mathbf{r}}{\partial t} = 0 \tag{3}$$

The net force $\mathbf{F}$ acting on the particles can be expressed as

$$\mathbf{F} = \hbar \frac{d\mathbf{k}}{dt} \tag{4}$$

where $\hbar$ is the reduced Planck's constant ($\hbar = h/2\pi$). The total force $\mathbf{F}$ can be attributed to forces due to the internal crystal lattice $\mathbf{F}_I$ and external electromagnetic fields $\mathbf{F}_E$,

$$\mathbf{F} = \mathbf{F}_I + \mathbf{F}_E \tag{5}$$

It is convenient to express the effect of internal forces acting on the distribution function statistically by obtaining an expression for the internal collision mechanism as a function of a scattering probability,

$$\frac{\mathbf{F}_i}{\hbar}\frac{\partial f}{\partial \mathbf{k}} = \int d\mathbf{k}'[f(\mathbf{k}')\mathbf{P}(\mathbf{k}',\mathbf{k}) - f(\mathbf{k})\mathbf{P}(\mathbf{k},\mathbf{k}')] \tag{6}$$

where $\mathbf{P}(\mathbf{k},\mathbf{k}')$ is the probability that a carrier will be scattered, changing its wave vector from $\mathbf{k}$ to $\mathbf{k}'$. Using equations (3) to (6), the Boltzmann Transport Equation can be re-expressed as,

$$\frac{\partial f}{\partial t} + \mathbf{v}.\frac{\partial f}{\partial \mathbf{r}} + \frac{\mathbf{F}}{\hbar}.\frac{\partial f}{\partial \mathbf{k}} = \int d\mathbf{k}'[f(\mathbf{k}')\mathbf{P}(\mathbf{k}',\mathbf{k}) - f(\mathbf{k})\mathbf{P}(\mathbf{k},\mathbf{k}')] \tag{7}$$

where $\mathbf{v}$ is the group velocity of the particles,

$$\mathbf{v} = \frac{d\mathbf{r}}{dt} \tag{8}$$

It is extremely difficult to attempt a rigourous solution of equation (3) or (7) and there are no closed-form solutions available. In order to obtain a tractable model it is necessary to make a number of approximations. It is normally assumed that the scattering probability is independent of external forces and that all scattering processes are elastic. The duration of collisions is assumed to be much shorter than the average time between collisions. This is usually further simplified by assuming that the collisons are instantaneous. It is assumed that the external forces $\mathbf{F}_E$ are constant over a distance comparable with the dimensions of the wave packet describing the motion of a carrier. Particle interaction is assumed to be negligible. The Lorentz force due to magnetic fields is usually neglected, which means that the external force $\mathbf{F}_E$ is due to the electric field ($q\mathbf{E}$ for positive particles and $-q\mathbf{E}$ for negative particles). Finally, the distribution function is usually assumed to be symmetrical in momentum space. The implications of some of these approximations are discussed later in the text. The Boltzmann approximation obtained from these assumptions becomes,

$$\frac{\partial f}{\partial t} + \frac{F_E}{\hbar}.\frac{\partial f}{\partial \mathbf{k}} + \mathbf{v}.\frac{\partial f}{\partial \mathbf{r}} = \left(\frac{\partial f}{\partial t}\right)_{coll.} \tag{9}$$

The collison term is more fully expressed as,

$$\left(\frac{\partial f}{\partial t}\right)_{coll.} = \left(\frac{dn}{dt}\right)_{coll} + \left(\frac{d\mathbf{v}}{dt}\right)_{coll} + \left(\frac{d\xi}{dt}\right)_{coll} \tag{10}$$

where n is the carrier density, $\mathbf{v}$ the carrier velocity and ξ the carrier energy. The collision term can conveniently be expressed using the relaxation time approximation,

$$\left(\frac{\partial f}{\partial t}\right)_{coll.} = -\frac{f - f_o}{\tau} \tag{11}$$

where f_o is the spherically symmetric solution and τ is the relaxation time. The collision terms for the individual contributions due to generation-recombination, momentum and energy relaxation are given as,

$$\left(\frac{dn}{dt}\right)_{coll} = -R \tag{12}$$

$$\left(\frac{d\mathbf{v}}{dt}\right)_{coll} = -\frac{\mathbf{v}}{\tau_p(\xi)} \tag{13}$$

$$\left(\frac{d\xi}{dt}\right)_{coll} = -\frac{\xi - \xi_0}{\tau_e(\xi)} \tag{14}$$

where R is generation-recombination rate. R is positive for recombination and negative for generation. τ_p is the momentum relaxation time and τ_e is the energy relaxation time, both of which are functions of average electron energy.

Carrier, momentum and energy conservation equations can be obtained from the simplified Boltzmann equation by taking the first three moments of the BTE, by carrying out a series of integrals or utlilising the general moment equation (Snowden 1988). This is a fairly lengthy affair by either route. The particle continuity equation derived from the Boltzmann equation is,

$$\frac{\partial n}{\partial t} + \nabla.(n\mathbf{v}) = \left(\frac{\partial n}{\partial t}\right)_{coll} \tag{15}$$

The momentum conservation equation for electrons is obtained as,

$$\frac{\partial \mathbf{v}}{\partial t} + \mathbf{v}.\nabla\mathbf{v} + \frac{q\mathbf{E}}{m^*} + \frac{1}{m^*n}\nabla(nkT_e) = \left(\frac{\partial \mathbf{v}}{\partial t}\right)_{coll} \tag{16}$$

where m^* is the effective mass, k Boltzmann's constant, T_e electron temperature and $\mathbf{E}$ electric field.

The energy conservation equation obtained by assuming that the distribution function has a displaced Maxwellian distribution is,

$$\frac{\partial \xi}{\partial t} + \mathbf{v}.\nabla\xi + q\mathbf{v}.E + \frac{1}{n}\nabla.(n\mathbf{v}kT_e) + \nabla.\mathbf{s} = \left(\frac{d\xi}{dt}\right)_{coll} \tag{17}$$

where $\mathbf{s}$ is the energy flux. The total electron energy ξ is the sum of the kinetic and thermal energy,

$$\xi = 1/2m^*v^2 + \frac{3}{2}kT_e \tag{18}$$

The current continuity and Poisson equations can be extracted from Maxwell's equations. Maxwell's equations which govern the motion of charged particles in semiconductors are,

$$\nabla\times\mathbf{H} = \mathbf{J} + \frac{\partial \mathbf{D}}{\partial t} \tag{19}$$

$$\nabla\times\mathbf{E} = -\frac{\partial \mathbf{B}}{\partial t} \tag{20}$$

$$\nabla.\mathbf{D} = \rho \tag{21}$$

$$\nabla.\mathbf{B} = 0 \tag{22}$$

where $\mathbf{H}$ and $\mathbf{E}$ are the magnetic and electric field strengths respectively and $\mathbf{D}$ and $\mathbf{B}$ are the electric and magnetic flux densities respectively. $\mathbf{J}$ is the total conduction current density and ρ is the electric charge density.

The continuity equations may be derived from equations (19) and (21) by application of the divergence operator ($\nabla.(\nabla\times\mathbf{a}) = 0$),

$$\nabla.(\nabla\times\mathbf{H}) = \nabla.\mathbf{J} + \frac{\partial \rho}{\partial t} = 0 \tag{23}$$

The electric charge density ρ consists of the positive charge due to hole density p, negative charge due to the electron density n and the charge due to donor and acceptor densities N_D and N_A (surface and interface charges may also be included here),

$$\rho = q(p - n + N_D - N_A) \tag{24}$$

where q is the magnitude of the charge on the electron.

The total conduction current $\mathbf{J}$ is defined in terms of electron and hole current densities $\mathbf{J}_n$ and $\mathbf{J}_p$ respectively as

$$\mathbf{J} = \mathbf{J}_n + \mathbf{J}_p \tag{25}$$

Using equations (23), (24) and (25) two continuity equations for electrons and holes may be conveniently defined as

$$\nabla.\mathbf{J}_n - q.\frac{\partial n}{\partial t} = q.R \tag{26}$$

$$\nabla.\mathbf{J}_p + q.\frac{\partial p}{\partial t} = -q.R \tag{27}$$

where R is defined as the generation-recombination rate. By defining electron and hole current densities $\mathbf{J}_n$ and $\mathbf{J}_p$ in terms of the electron and hole velocities,

$$\mathbf{J}_n = -qn\mathbf{v}_n \tag{28}$$

$$\mathbf{J}_p = qn\mathbf{v}_p \tag{29}$$

The continuity equations (26) and (27) are in the same form as those derived from the Boltzmann equation above (equation (15)).

Further consideration of equation (23) allows us to define the total electric current density $\mathbf{J}_{tot}$ where,

$$\nabla.(\mathbf{J}_{tot}) = \nabla.\left(\mathbf{J} + \frac{\partial \rho}{\partial t}\right) = 0 \tag{30}$$

Assuming that the semiconductor has a time-independent permittivity and that polarization due to mechanical forces is negligible the electric flux density can be directly related to the electric field intensity,

$$\mathbf{D} = \varepsilon_o \varepsilon_r.\mathbf{E} \tag{31}$$

where ε_o is the permittivity of vacuum and ε_r is the relative permittivity of the semiconductor. Using equations (21) and (31), it is possible to define the total current density as

$$\mathbf{J}_{tot} = \mathbf{J}_n + \mathbf{J}_p + \varepsilon_o \varepsilon_r.\frac{\partial \mathbf{E}}{\partial t} \tag{32}$$

The third term on the right-hand side of the equation is known as the displacement current.

In practical semiconductor device models the time dependence in equation (20) is neglected. This allows the Poisson equation and electric field to be simply derived for the case of static fields (the electrostatic case), where from equation (20),

$$\nabla \times \mathbf{E} = 0 \tag{33}$$

Utilising the concept of a magnetic vector potential, it may be deduced from equation (33) that the electric field may be expressed in terms of the gradient of a scalar electric potential ψ (Snowden 1988),

$$\mathbf{E} = -\nabla\psi \tag{34}$$

Substituting for $\mathbf{E}$ in equation (31) yields,

$$\mathbf{D} = -\varepsilon_o \varepsilon_r.\nabla\psi \tag{35}$$

Substituting this expression for $\mathbf{D}$ in equation (21) yields

$$\nabla.(\nabla\psi) = -\frac{\rho}{\varepsilon_o \varepsilon_r} \tag{36}$$

In the case where the permittivity is homogeneous this equation reduces to the well known Poisson equation

$$\nabla.\nabla\psi = -\frac{\rho}{\varepsilon_o\varepsilon_r} = -\frac{q}{\varepsilon_o\varepsilon_r}(p - n + N_D - N_A) \tag{37}$$

CLASSICAL SEMICONDUCTOR EQUATIONS

The majority of current semiconductor device models are based on the classical semiconductor equations, derived from approximate solutions for the first two moments of the Boltzmann Transport Equation. This assumes that the carriers move under equilibrium conditions and that the velocity is an instantaneous function of the localised electric field.

The Drift-Diffusion Approximation

The drift-diffusion approximation utilises a simplified form of the momentum conservation equation (16). This assumes that the electron temperature T_e is equal to the lattice temeperature T, and that electron temperature gradient is zero. It is also assumed that the term $\mathbf{v}.\nabla\mathbf{v}$ is small compared with other terms in the equation. The time scales of interest to modellers also allow further simplifications. In the majority of simulations there is at least an order of magnitude difference in time between the device and circuit responses, which implies that a quasi-steady-state model is adequate for most purposes and hence it is convenient to assume that $\partial\mathbf{v}/\partial t = 0$. Under these circumstances, equation (16) for electrons reduces to the form,

$$\mathbf{v}_n = -(\mu_n\mathbf{E} + \frac{D_n}{n}\nabla n) \tag{38}$$

and the current density expression for electrons becomes,

$$\mathbf{J}_n = -qnv_n = qn\mu_n\mathbf{E} + qD_n\nabla n \tag{39}$$

This formulation treats electrons as negatively charged particles of charge $-q$ Coulombs. This form of the momentum equation requires the mobility μ_n and diffusion D_n parameters to be defined in the following familiar forms,

$$\mu_n = \frac{q\tau_p}{m*} \tag{40}$$

$$D_n = \frac{kT\mu_n}{q} \tag{41}$$

where k is Boltzmann's constant and T is the lattice temperature. Equation (41) is also known as the Einstein relationship for diffusion. A common variation of equations (38) and (39) is to replace the mobility-electric field product by drift velocity term $\mathbf{v}_d = -\mu\mathbf{E}$. This velocity term should not be confused with the total velocity v_n due to both the field and diffusion components.

This representation of carrier transport is often referred to as the 'drift-diffusion' model. The exact form of the diffusion term in equations (38) and (39) has been the subject of considerable debate. The diffusion coefficent D is shown outside the derivative term in these equations, which is the usual form. However, within the context of the Boltzmann model it is often argued that the diffusion coefficient D should be included inside the gradient operator, ∇Dn.

Summarising, the classical carrier transport equations (the *semiconductor equations*) are,

Continuity equations

$$\frac{\partial n}{\partial t} = \frac{1}{q}\nabla.\mathbf{J}_n - R \quad \textit{for electrons} \tag{42}$$

$$\frac{\partial p}{\partial t} = -\frac{1}{q}\nabla.\mathbf{J}_p - R \quad \textit{for holes} \tag{43}$$

where $\mathbf{J}_n$ and $\mathbf{J}_p$ are the electron and hole current densities respectively and R is the generation-recombination rate.

Current densities

$$\mathbf{J}_n = qn\mu_n\mathbf{E} + qD_n\nabla n \quad \textit{for electrons} \tag{44}$$

$$\mathbf{J}_p = qp\mu_p\mathbf{E} - qD_p\nabla p \quad \textit{for holes} \tag{45}$$

The electric field $\mathbf{E}$,

$$\mathbf{E} = -\nabla\psi \tag{46}$$

$$-\nabla^2\psi = \frac{q}{\varepsilon_o\varepsilon_r}\left(N_D - n + p - N_A\right) \tag{47}$$

where μ_n and μ_p are the electron and hole mobilities, D_n and D_p are the carrier diffusion coefficents, ψ is the electrostatic potential, q electronic charge, $\varepsilon_o\varepsilon_r$ permittivity, N_D donor doping density, N_A acceptor doping density, n electron concentration and p hole concentration.

The parameters of the semiconductor equations ψ, n and p have a very large number range (spanning 0 to 10^{24} in magnitude). This is undesirable from a computational aspect for solving these equations. The semiconductor equations are often scaled to reduce the range of magnitudes and simplify the expressions. The scaled semiconductor equations may be expressed in the forms,

$$\mathbf{J}_n = -n\mu_n\nabla\psi + D_n\nabla n \tag{48}$$

$$\mathbf{J}_p = -p\mu_p\nabla\psi - D_p\nabla p \tag{49}$$

$$\frac{\partial n}{\partial t} = \nabla.\mathbf{J}_n - R \tag{50}$$

$$\frac{\partial p}{\partial t} = -\nabla\mathbf{J}_p - R \tag{51}$$

$$\lambda^2\nabla^2\psi = n - p - N_D + N_A \tag{52}$$

All the parameters are normalised (scaled). There are a number of schemes available, the most notable of which are attributable to De Mari, Markowich and Polak (De Mari 1968; Markowich 1986; Polak et al 1987). λ is a scaling parameter related to the scheme chosen (unity for De Mari). The expressions for current densities $\mathbf{J}_n$ and $\mathbf{J}_p$ assume the Einstein relationship for the diffusion coefficients.

Many simulations use algorithms which are based on the electron and hole quasi-Fermi levels ϕ_n and ϕ_p,

$$\phi_n = \psi - \frac{kT}{q}\ln\left(\frac{n}{n_i}\right) \tag{53}$$

$$\phi_p = \psi + \ln\left(\frac{p}{n_i}\right) \tag{54}$$

where n_i is the intrinsic carrier concentration. Quasi-Fermi level models are often used for simulating bipolar devices and for modelling semiconductor devices which have complex energy band structures such as heterojunction devices.

A non-linear Poisson equation may be derived by substituting for n and p in equation (47),

$$-\nabla^2\psi = \frac{q}{\varepsilon_o\varepsilon_r}\left[-n_i\exp\left(\frac{q(\psi - \phi_n)}{kT}\right) + n_i\exp\left(\frac{q(\phi_p - \psi)}{kT}\right) + N_D - N_A\right] \tag{55}$$

This non-linear equation can be solved using several approaches (for example Mayergoyz 1986).

One of the objectives of simulating devices is to determine the relationship between the terminal current and voltages. The current I associated with a contact may be obtained by integrating the total electric current density **J** across the a suitable surface surrounding the contact,

$$I = \int_{p} \mathbf{J} . ds \tag{56}$$

The total current **J** includes both the particle currents $\mathbf{J}_n$, $\mathbf{J}_p$ and displacement current.

Generation and Recombination

The significance of generation-recombination effects on device operation depends on the type of device being considered and the operating conditions. Bipolar devices, such as pn junction diodes and bipolar junction transistors are strongly influenced by generation-recombination mechanisms. In contrast unipolar devices such as Schottky barrier diodes and FET's can be modelled for most operating conditions without taking account of generation-recombination. For example, the majority of MOSFET models omit generation-recombination rates and obtain satisfactory results for most operating conditions. However, it is important to incorporate suitable generation-recombination models when investigating bipolar devices and devices operating with very high field regions where ionization and breakdown phenomena are present. Hence, in the case of sub-micron MOSFET's, where very high field regions may be encountered, it is important to use a full generation-recombination model to account for phenomena which can occur at relatively low voltages (and use the model to optimise the structure to minimise the onset of unwanted breakdown).

In thermal equilibrium, a dynamic balance exists between the generation and recombination of electrons and holes. Hence, for uniformly doped material, the electron and hole concentrations n_o and p_o respectively remain in equilibrium governed by the relationship,

$$n_i^2 = n_o p_o \tag{57}$$

The total generation-recombination process can be considered as result of several distinct mechanisms. The main generation-recombination processes are thermal (phonon transitions), impact ionization, Auger recombination, surface recombination and optical generation and recombination (photon transitions). The total generation-recombination rate R is the sum of the individual contributions from each of these processes,

$$R \approx R_{thermal} + R_{impact} + R_{Auger} + R_{surface} + R_{optical} \tag{58}$$

The thermal generation and recombination process is attributable to phonon transitions occuring as a result of traps. It is often represented using the well known Shockley-Read-Hall model. The thermal generation and recombination mechanism is an indirect process since it involves a trap centre in the energy band-gap with associated capture and emission mechanisms, Fig. 1. The Shockley-Read-Hall model is described by the equation,

$$R_{thermal} = \frac{pn - n_i^2}{\tau_n(p + p_t) + \tau_p(n + n_t)} \tag{59}$$

where τ_n and τ_p are the electron and hole lifetimes. The intrinsic carrier density n_i is commonly substituted for n_t and p_t. The lifetimes τ_n and τ_p typically lie in the range 100 ns to 10 μs.

The second contribution to generation-recombination processes is due to impact ionisation. This is a generation process, where electron-hole pairs are generated as the result of interaction with a third particle. Electron-hole pairs are generated by carriers moving directly across the band-gap. The impact ionisation generation rate is given by,

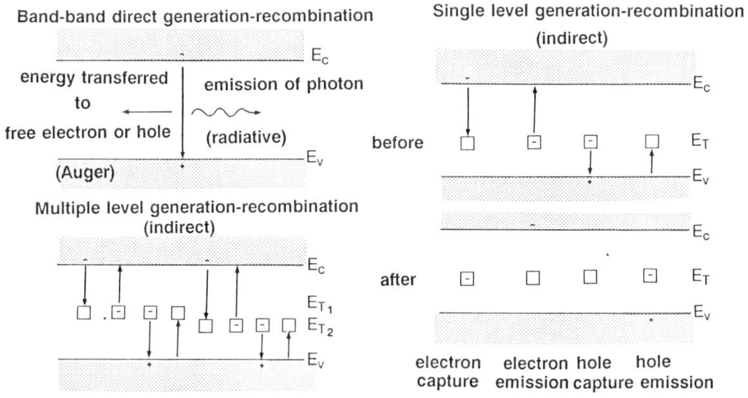

Figure 1 Generation-Recombination by direct and indirect processes

$$R_{impact} = -\alpha_n \frac{|\mathbf{J}_n|}{q} - \alpha_p \frac{|\mathbf{J}_p|}{q} = -\frac{1}{q}(\alpha_n |\mathbf{J}_n| + \alpha_p |\mathbf{J}_p|) \tag{60}$$

where α_n and α_p are the ionization coefficients and $\mathbf{J}_n$ and $\mathbf{J}_p$ are the electron and hole current densities. The ionization coefficients depend on the electric field component E parallel to the direction of current flow.

Auger recombination, which is the reverse of impact ionization, is again associated with three particles and involves the recombination of an electron-hole pair and the emission of energy to a third particle. Auger recombination is a result of direct band-gap recombination processes and indirect processes involving trap centres. The net Auger generation-recombination rate is given by,

$$R_{Auger} = (pn - n_i^2)(nC_n + pC_p) \tag{61}$$

where C_n and C_p are the Auger coefficients, which are of the order of $3\times10^{-43}\ m^6 s^{-1}$ and $10^{-43}\ m^6 s^{-1}$ respectively at room temperature.

Surface generation-recombination may be modelled using a modified form of equation (64), where the carrier lifetimes are replaced by reciprocals of surface recombination velocities.

The contribution to the generation-recombination process attributable to photon transitions, involves the direct transition of carriers between the valence and conduction bands. Electrons are excited to the conduction band from the valence band by gaining energy from incident photons. Alternatively, electrons lose energy, equivalent to the band-gap energy, which is emitted as a photon and the electron m the conduction band to the valence band. This optical process is significant in narrow band-gap and direct band-gap semiconductors such as GaAs, but has negligible effect in semiconductors such as silicon. The optical generation-recombination rate is,

$$R_{optical} = C_{opt}(pn - n_i^2) \tag{62}$$

where C_{opt} is the optical capture-emission rate.

Limitations of the drift-diffusion approximation

The drift-diffusion transport model derived above assumes that the carrier velocities are instantaneous functions of the electric field and that the mobility and diffusion coefficients are functions of the local electric field alone. In reality the carriers do not respond instantaneously to changes in the electric field and the mobility and diffusion coefficients are tensor quantities, dependent on several parameters in addition to the electric field. More detailed simulations often use mobility and diffusion coefficients which are functions of doping density and temperature (Freeman and Hobson 1972; Snowden et al 1983). A more detailed study of carrier transport requires a more rigorous treatment of energy and momentum relaxation effects. This is usually dealts with in the context of *hot electron* models.

ENERGY TRANSPORT, SEMICLASSICAL AND HOT ELECTRON MODELS

The classical carrier transport model considered so far is based on the fundamental assumption that the carrier energy distribution remains close to its equilibrium form. However, in circumstances where the electric fields, carrier gradients and current densities in the device become large, non-equilibrium transport conditions will prevail. The small geometries encountered in many modern semiconductor devices frequently have regions of very high electric fields and current densities associated with them which cause substantial electron 'heating'. This causes the electrons in these regions to attain very high energies relative to equilibrium conditions.

This non-equilibrium transport process, where the carrier velocity exceeds its steady-state value on a transient basis is also referred to as 'transient carrier transport'. It is particularly significant in devices with active channels of less than 1.0 μm and has a critical effect on the device characteristics when the channel is less than 0.1 μm in silicon and 0.5 μm in gallium arsenide. In these very small devices a significant portion of the current density can be attributed to the electron energy gradient term $\nabla\xi$.

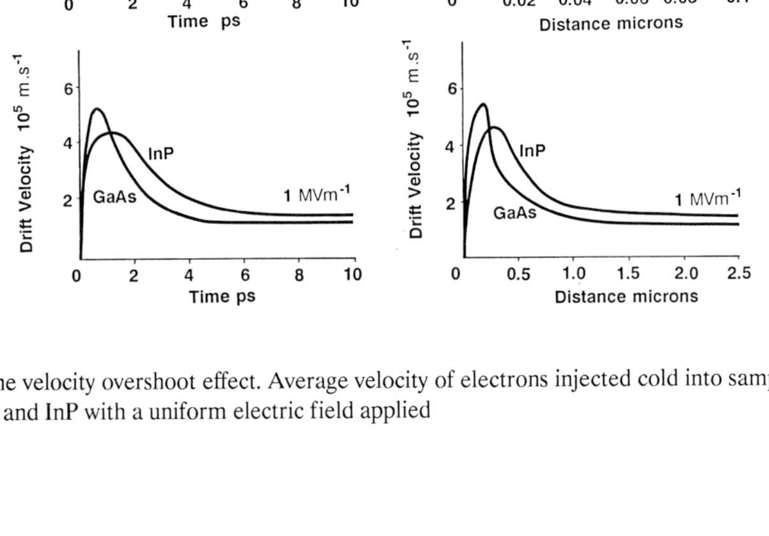

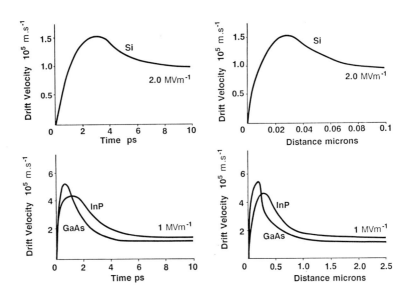

Figure 2 The velocity overshoot effect. Average velocity of electrons injected cold into samples of Si, GaAs and InP with a uniform electric field applied

The effects of non-equilibrium transport are shown in Fig. 2, illustrating the time dependence of the average velocity of electrons injected 'cold' into a semiconductor sample and drifting in a region of uniform electric field. The drift velocity initially overshoots to a value significantly higher than the equilibrium value for the applied electric field. This velocity overshoot effect can be attributed to the difference in the momentum and energy relaxation times. The momentum relaxation time is generally much smaller than the energy relaxation time, which causes the electron distribution to be perturbed first in momentum space. After a short period of time the energy relaxation becomes effective and the distribution function spreads out. This in turn causes the average drift velocity to decrease. Strong intervalley and acoustic phonon scattering in silicon causes the drift velocity to reach its equilibrium value in less than 0.1 ps. Transient carrier transport effects in gallium arsenide last substantially longer than in silicon because of the weaker influence of polar optical scattering which is the dominant scattering mechanism in GaAs.

Hydrodynamic and Semiclassical Semiconductor Equations

A more complete model for carrier transport is based on the full set of transport equations derived from Boltzmann Transport Equation, which take into account energy and momentum relaxation effects. Some models also attempt to incorporate effects omitted as a result of the approximations used during the derivation of the simplified Boltzmann model of equation (9). The semiclassical treatment of transport describes the motion of carriers in free-flight using classical mechanics with scattering events described by a quantum statistical treatment. The semiclassical model can be extended to bulk transport models of the type under consideration here by utilising data obtained from Monte Carlo simulations which consider scattering in the light of quantum-mechanical effects (which are approximated in varying degrees). The coefficients of the transport equations, which are obtained from the Monte Carlo simulations, are usually assumed to be a function of average carrier energy, although more detailed models have been proposed which account for the multi-valley nature of compound semiconductors (Blotekjaer 1970; Cook and Frey 1982). If it is assumed that the distribution function is symmetrical in momentum space, as it is for a displaced Maxwellian distribution, then higher order moments of the distribution function vanish. This approximation allows the equations obtained by taking the first three moments of the Boltzmann equation to adequately represent transient carrier transport effects. A set of equations for electrons are,

Particle conservation
$$\frac{\partial n}{\partial t} + \nabla.(n\mathbf{v}) = -R \tag{63}$$

Momentum conservation
$$\frac{\partial \mathbf{v}}{\partial t} + \mathbf{v}.\nabla\mathbf{v} + \frac{q\mathbf{E}}{m^*} + \frac{1}{m^*n}\nabla\left(nkT_e\right) = -\frac{v(\xi)}{\tau_p(\xi)} \tag{64}$$

Energy conservation
$$\frac{\partial \xi}{\partial t} + q\mathbf{v}E + \mathbf{v}\nabla\xi + \frac{1}{n}\nabla(n\mathbf{v}kT_e) = -\frac{\xi - \xi_o}{\tau_e(\xi)} \tag{65}$$

where k is Boltzmann's constant, ξ is the average electron energy, τ_p and τ_e are the momentum and energy relaxation times, which are themselves functions of the average electron energy. ξ_o is the equilibrium electron energy corresponding to the lattice temperature T_o. This set of transport equations are often referred to as the *hydrodynamic* equations. The total average electron energy ξ consists of kinetic and thermal energy components

$$\xi = \frac{1}{2}m^*v^2 + \frac{3}{2}kT_e \tag{66}$$

where m^* is the effective mass of the electron and T_e is the electron temperature. The kinetic energy term is often neglected as it is at least an order of magnitude smaller than the thermal energy in most circumstances (except in regions of low electric field and current density).

Since the energy conservation equation equation(assumes a displaced Maxwellian distribution. hence heat flow (energy flux) due to the electron gas is not included in equation (65). In practice the heat flow term may be significant for non-Maxwellian distributions, which are likely to occur in small geometry devices, and more rigourous derivations of the energy conservation equation include an energy flux term and incorporate the fourth moment equation (Blotekjaer 1970). An alternative energy transport equation incorporating the energy flux **s** is (Stratton 1962; Parker and Lowke 1969; Widiger et al 1985),

$$\frac{\partial n\xi}{\partial t} = \mathbf{J}.\mathbf{E} - nB - \nabla.\mathbf{s} \qquad (67)$$

where

$$\mathbf{s} + \frac{\partial(\tau_h \mathbf{s})}{\partial t} = \mu_\xi n\xi\nabla\psi - \nabla(D_\xi n\xi) \qquad (68)$$

B is the energy-dissipation factor, τ_h, μ_ξ and D_ξ are the high frequency factor, flux mobility and flux diffusivity respectively.

The extended set of semiconductor equations described above are difficult to solve in their full form. Numerical solutions for these equations require very small time and space steps to ensure numerical stability and convergence in the solutions (time steps of less than 5 fs are usually required for two-dimensional solutions) (Snowden and Loret 1987). The relaxation times associated with carrier transport lie in the range 0.01 ps to 1 ps ($\tau_p < 0.1\,ps$, $\tau_h = 0.1\,ps$ and $\tau_e \approx 1\,ps$). Most device simulators are interested in response times associated with circuit time constants, which are usually greater than 10 ps and often in the region of 10 ns or more. Simulations aimed at extracting information on these time scales (or DC) can take advantage of several simplifying approximations (termed *quasi-static* models). Many formulations of the transport equations neglect the products $\tau_p \mathbf{v}.\nabla\mathbf{v}$ and $\partial(\tau_h \mathbf{s})/\partial t$ because they are generally small compared with other terms.

In the case of energy transport equations incorporating the energy flux term, such as equation (68), the energy transport equation can be written in the form,

$$\frac{\partial n\xi}{\partial t} = \mathbf{J}.\mathbf{E} - nB = \nabla.\alpha[\mu n\xi\mathbf{E} + \nabla(Dn\xi)] \qquad (69)$$

where

$$\alpha = \frac{<\tau_p \varepsilon^2>}{<\tau_p \varepsilon><\varepsilon>} \qquad (70)$$

α is usually assumed to be constant, which is a reasonable approximation in the case of power-law scattering (Widiger et al 1985). This implies that the momentum relaxation time is isotropic (which strictly speaking is not a good approximation for polar-optical phonon scattering).

Quasi-static models based on the Maxwellian model can be obtained from equations (63) to (65) by neglecting the terms $\partial v/\partial t$, $\partial\xi/\partial t$ and $\nabla.n\mathbf{v}$ in the momentum and energy equations. If the kinetic energy term in equation (66) is neglected by assuming that $3/2kT_e \gg 1/2m^*v^2$, the equations can be further simplified. A quasi-static set of transport equations for electrons based on these simplifying assumptions is,

$$\frac{\partial n}{\partial t} = -\nabla.(n\mathbf{v}) - R \qquad (71)$$

$$\mathbf{v} = \frac{\tau_p}{m^*}\left[-q\mathbf{E} - \frac{2}{3}\nabla\xi - \frac{2\xi}{3n}\nabla n\right] \qquad (72)$$

$$\frac{5}{3}\mathbf{v}.\nabla\xi = q\mathbf{v}E - \frac{\xi - \xi_o}{\tau_e(\xi)} \qquad (73)$$

Equation (72) is equivalent to the more conventional form of the velocity equation,

$$\mathbf{v} = -\mu(\xi)\mathbf{E} - \frac{1}{n}\nabla(D(\xi)n) \tag{74}$$

where the mobility and diffusion coefficients are functions of average energy (rather than local electric field) and are given by,

$$\mu(\xi) = \frac{q\tau_p(\xi)}{m^*} \tag{75}$$

$$D(\xi) = \frac{kT_e(\xi)\mu(\xi)}{q} \tag{76}$$

The single-electron gas model described above is a satisfactory basis for deriving transport equations for silicon, but in the case of compound semiconductors, such as GaAs and InP which have multi-valley energy band structures, a more detailed treatment is required. Effective mass variations due to intervalley transfer have a strong influence on electron transport in compound semiconductors. A full multi-valley analysis would be formidable, and for the sake of simplicity the analysis is usually reduced to a two-valley model which assumes parabolic band structures. A full solution of this model would require the derivation of transport equations for carriers in each valley, incorporating a suitable treatment of intervalley scattering. A detailed analysis incorporating transport equations for multi-valley semiconductors was carried out by Blotekjaer (1970). This analysis showed that the mobility and diffusion coefficients could be treated as functions of local average velocity of electrons in the lower valley, rather than electric field. It is more convenient to simplify the solution of the transport equations by assuming an average electron temperature for the electron gases in the upper and lower valleys.

The velocity equation (72) can be re-expressed in terms of the average electron temperature as,

$$v = -\mu(\xi)\mathbf{E} - \left[\frac{k\mu(\xi)}{q}\nabla(T_e) + \frac{kT_e\mu(\xi)}{qn}\nabla n\right] \tag{77}$$

which leads to an electron current density equation of the form,

$$\mathbf{J} = qn\mu(\xi)\mathbf{E} + nk\mu(\xi)\nabla(T_e) + kT_e\mu(\xi)\nabla n \tag{78}$$

A comparison of this equation with the classical current density equation reveals that this expansion has an additional term for the spatial variation in electron temperature, which expresses the fact that the most energetic (hot) electrons will diffuse out of the collection. Curtice and Yun have shown that if the temperature gradients are negligible, then the term containing the electron temperature gradient in equation (78) may be omitted, yielding an expression identical to equation (79) (Curtice and Yun 1981). This approximation is only valid for devices where electron heating is not significant and the gradient term has been shown to be significant in sub-micron structures (Snowden and Loret 1987). The variation of average electron energy across the cross-section of sub-micron MESFET and HEMT devices are shown in Fig. 3.

The previous approximations applied to the momentum and energy conservation equations were based on a time-independent solution. An alternative model can be obtained by neglecting spatial gradient terms. This simplifies the equations to the following forms,

$$\frac{d(m^*(\xi)\mathbf{v})}{dt} = -q\mathbf{E} - \frac{m^*(\xi)\mathbf{v}}{\tau_p(\xi)} \tag{79}$$

$$\frac{d\xi}{dt} = -q\mathbf{E}.\mathbf{v} - \frac{\xi - \xi_o}{\tau_e(\xi)} \tag{80}$$

This approximation has been used in many models (Maloney and Frey 1977; Carnez et al 1980;

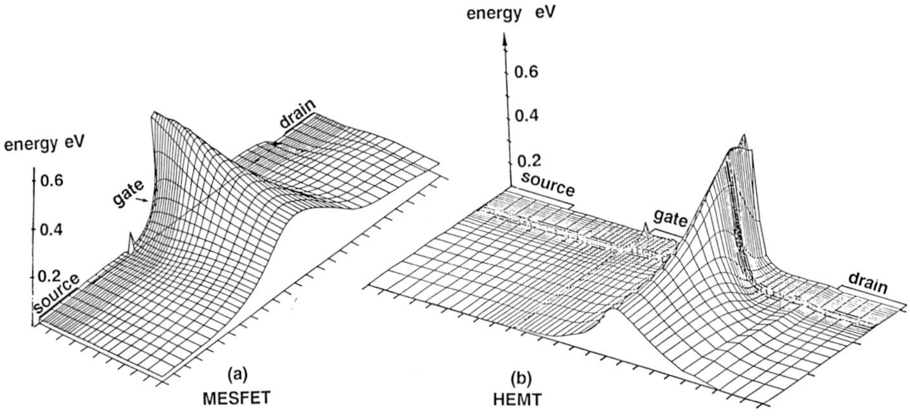

Figure 3 Two-dimensional distribution of average electron energy in (a) 0.3 μm gate length MESFET and (b) 0.5 μm gate length HEMT. ($V_{DS} = 3\,V$, $V_{GS} = 0\,V$). A full energy transport model was used (as in Snowden and Loret 1987).

Shur 1981; Snowden 1984) and is relatively easy to implement and solve. However, it is necessary to exercise caution in using this simple model as the energy gradient (electron temperature gradient) in some devices can account for a significant part of the particle current density (determined from the more complete velocity expression of equation 4.77) (Snowden and Loret 1987). Devices in which the electric field is relatively uniform and slowly varying may be modelled by neglecting both time and space variations, using a localised transport model,

$$\mathbf{v} = -\frac{q\tau_p(\xi)}{m^*}\mathbf{E} \tag{81}$$

$$\xi - \xi_o = -q\tau_e(\xi)\mathbf{E}.\mathbf{v} \tag{82}$$

The velocity equation (81) is directly equivalent to the steady-state drift velocity $\mathbf{v}_d = \mu\mathbf{E}$, and neglects diffusion as a consequence of the simplifying assumptions.

The electric and energy field-dependent transport characteristics of electrons and holes have been characterised using Monte Carlo particle simulation techniques. The extraction of the relaxation times for silicon is a relatively straightforward process. However, it is considerably more difficult to obtain these parameters for GaAs and InP because of the strong dependence of effective mass $m^*(\xi)$ on energy. Steady-state Monte Carlo results describing the electric field dependence of velocity, energy, effective mass and energy relaxation time for GaAs, are shown in Fig. 4 (Carnez et al 1980; Mains et al 1983; Snowden 1986)

BOUNDARY CONDITIONS

The semiconductor equations form a set of partial differential equations which require the domain, boundary and initial conditions to be specified before they can be solved. The domain of the model is defined by the geometry chosen to represent the actual device. This takes the form of a one-, two- or three-dimensional representation. The number of dimensions which are chosen for the model is largely determined by the device dimensions, geometry, contact and surface properties, and uniformity of field and carrier distributions. In the case of two- and three-dimensional models, the model may be planar or non-planar in form. Simple device structures, such as those found in vertical two-terminal devices (diodes), with large cross-sectional areas relative to there thickness, in which the current flow is predominantly one-dimensional through the

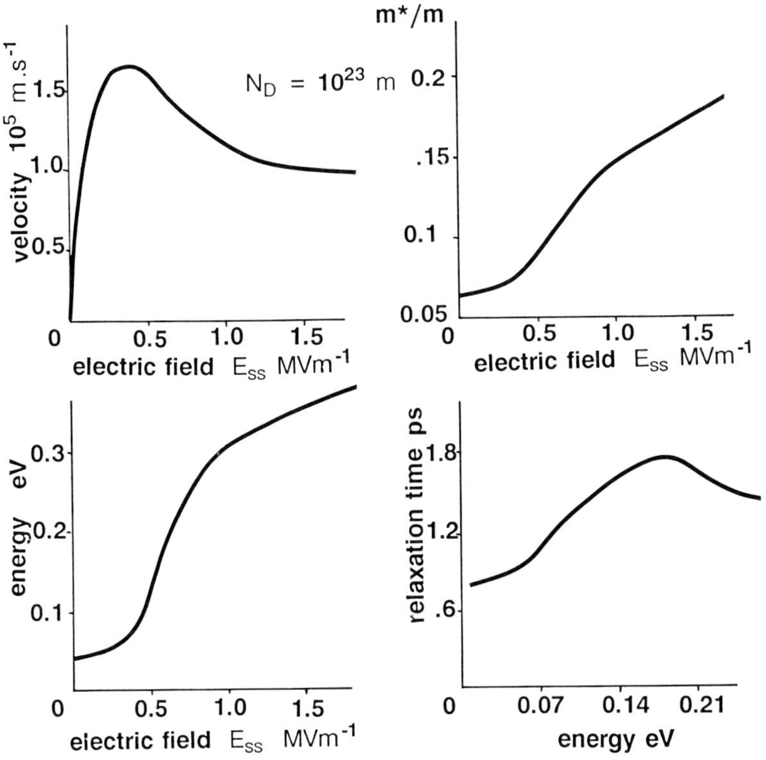

Figure 4 Velocity, energy, effective mass and energy relaxation time characteristics for GaAs as a function of static electric field.

structure, may be adequately represented using one-dimensional models, Fig. 5.

Many surface-orientated devices, including those found in modern integrated circuits, require at least two-dimensional models to account for the multi-dimensional nature of the electric field and carrier distributions Fig. 6. Early planar devices, such as long gate length FET's and large bipolar transistors were modelled using simplified one-dimensional models (such as the gradual channel FET model). However, this approach has been shown to have serious shortcomings for modelling sub-micron gate length FETs and high current density bipolar devices (where current crowding and base-spreading effects are important). In devices where the width of the active region is much greater than the length, which is usually the case for analogue devices, a two-dimensional model is adequate. However, in circumstances where the device width is of the same order as the length, such as in digital VLSI circuits, a three-dimensional model may be required to account for fringing fields and non-homogeneous current flow in the active channel, which can substantially modify the operation of these devices.

The boundary conditions are obtained from surface and contact properties. The presence of current-free boundary regions as well as conducting contacts on the surfaces leads to mixed boundary conditions. Ohmic contacts are usually described by Dirichlet boundary conditions where potential and carrier concentrations are pre-defined at the contacts. Schottky contacts may be modelled using Dirichlet boundary conditions which approximate the reverse bias condition or Neumann boundary conditions based on thermionic emission theory. The absence of current flow

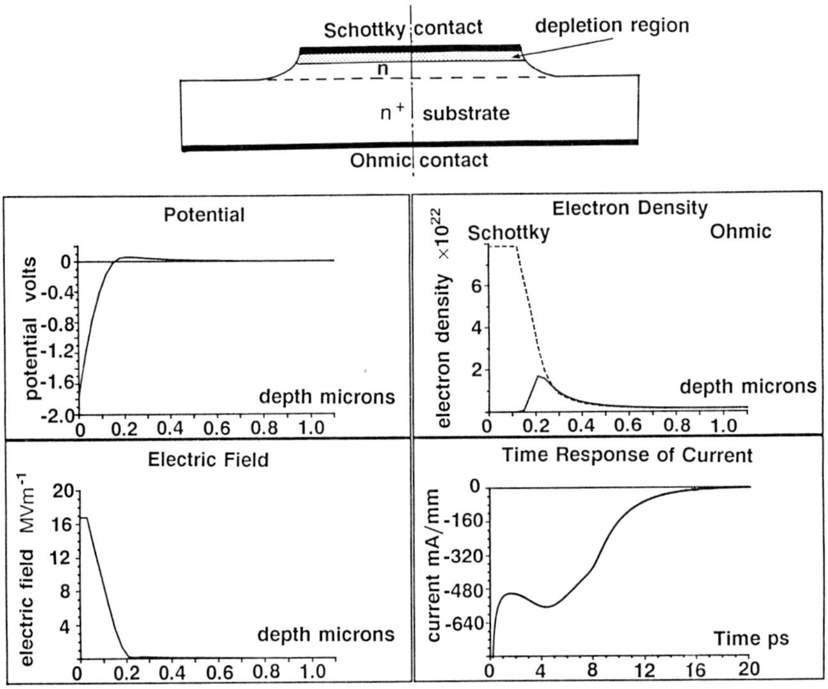

Figure 5 Vertical device structure. A Schottky varactor diode and one-dimensional simulation results for reverse-bias operation, showing the transient current response (drift-diffusion model).

through surfaces and internal boundaries may, in its simplest form, be modelled by assuming that the potential and carrier gradients normal to the surface are zero. The effects of surface potentials and surface recombination may also be included as boundary conditions. The influence of surface states on the operation of semiconductor devices can be significant in some devices.

The initial conditions required to solve the Poisson and current continuity equations may be satisfied by setting the potential ψ to zero (except at contacts) and the carrier density to the doping density throughout the device. Alternatively, steady-state potential and carrier density distributions obtained from previous simulations can be used as initial conditions, if they are available.

THERMAL CONDUCTIVITY AND HEAT FLOW

A complete model of a semiconductor device should account for thermal effects, which can play an important role in determining the device characteristics. The electrical characteristics of semiconductor devices are often a strong function of temperature. The mobility, diffusion and carrier velocity are a strong function of temperature, which has a direct influence on the current (for example the well known temperature dependence of bipolar devices). The effect of temperature distribution has an important influence on all semiconductor devices, but is particularly significant for power devices where increased power dissipation can cause thermal runaway and device failure.

A full solution for heat flow in a semiconductor device requires the solution of the heat flow equation,

$$\rho c \frac{\partial T}{\partial t} = \nabla . k(T) . \nabla T + H \tag{83}$$

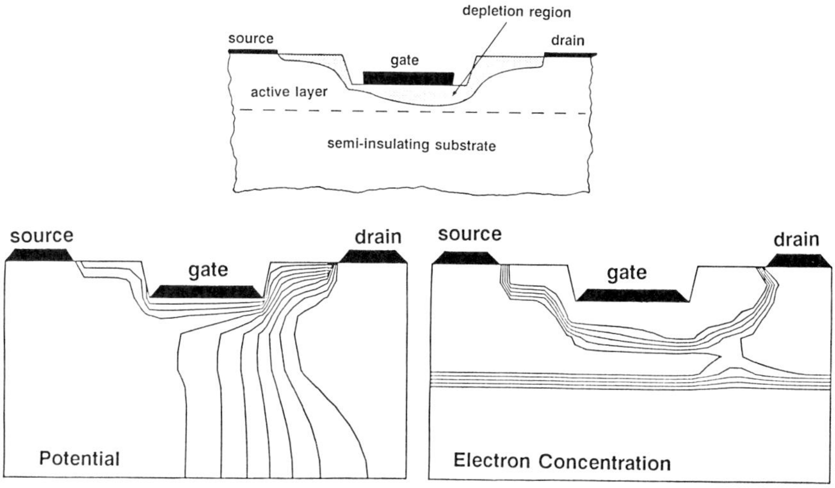

Figure 6 (a) Surface orientated device structure (GaAs MESFET) and (b) MESFET simulation results showing potential and carrier distributions and including the influence of surface depletion (drift-diffusion model).

where ρ is the density and c is the specific heat capacity of the material. ρ and c are usually assumed to be independent of temperature variations. $k(T)$ and H represent the thermal conductivity and the locally generated heat respectively.

Temperature gradients in the device contribute to the current flow. This requires extra terms in the current density equations which represent the temperature gradient ∇T,

$$\mathbf{J}_n = qn\mu_n\mathbf{E} + qD_n\nabla n + qnD_{Tn}\nabla T \tag{84}$$

$$\mathbf{J}_p = qp\mu_p\mathbf{E} - qD_p\nabla p - qpD_{Tp}\nabla T \tag{85}$$

D_{Tn} and D_{Tp} are the the thermal diffusion coefficients, which are functions of the more familiar diffusion coefficents D_n and D_p (Snowden 1984).

The thermal generation H may be modelled using the following expression which accounts for energy transfer to the lattice through generation and recombination (Adler 1978),

$$H = \nabla.\left(\frac{E_c}{q}\mathbf{J}_n + \frac{E_v}{q}\mathbf{J}_p\right) \tag{86}$$

where E_c and E_v are the conduction and valence band-edge energies

Models incorporating a rigourous treatment of thermal conductivity and heat flow are beginning to appear for a range of devices, where heat generation and conductance is important (for example Nathan et al 1988). More recently work has been published on including energy transfer between the carriers and the lattice in energy transport models (McAndrew et al 1988). This requires a comprehensive treatment of the energy and momentum conservation equations, accounting for variations in lattice temperature, heat generation and flow. This represents the most advanced bulk transport device model described to-date.

REFERENCES

Adler, M.S. (1978) Accurate calculations of the forward drop and power dissipation in thyristors, IEEE Trans. Electron Devices, ED-25, No.1, pp.16-22

Blotekjaer, K. (1970) Transport equations for electrons in two-valley semiconductors, IEEE Trans. Electron Devices, ED-17, pp.38-47

Carnez, B., Cappy, A., Kaszynski, A., Constant, E. and Salmer, G. (1980) Modeling of a submicrometer gate field-effect transistor including effects of nonstationary electron dynamics, J.Appl. Phys., 51, pp.784-790

Cook, R.K. and Frey, J. (1982) An efficient technique for two-dimensional simulation of velocity overshoot effects in Si and GaAs devices, Int. J. Comp. Math. Electron. Elec. Eng., Vol.1, pp.65-87

Curtice, W.R. and Yun, Y.H. (1981) A temperature model for the GaAs MESFET, IEEE Trans. Electron Devices, ED-28, pp.954-962

De Mari, A. (1968) An accurate numerical steady-state one-dimensional solution of the $p-n$ junction, Solid-State Electron., pp.33-58

Freeman, K.R. and Hobson, G.S. (1972) The V_{fT} relation of CW Gunn effect devices, IEEE Trans. Electron Devices, Vol.ED-19, No.1, pp.62-70

McAndrew, C.C., Heasell, E.L. and Singhal K. (1988) Modelling thermal effects in energy models of carrier transport in semiconductors, IoP, J. Semicond. Sci. Technol, 3, pp.758-765

Mains, R.K., Haddad, G.I and Blakey, P.A. (1983) Simulation of GaAs IMPATT Diodes including energy and velocity effects, IEEE Trans. Electron Devices, ED-30, pp.1327-1338

Maloney, T.J. and Frey, J. (1977) Transient and steady-state electron transport properties of GaAs and InP, J. Appl. Physics, 48, pp.781-787

Markowich, P.A. (1986) The stationary semiconductor equations, in Selberherr (Ed), Computational Microelectronics, (Springer Verlag, Wien, New York)

Mayergoyz, I.D. (1986) Solution of the non-linear Poisson equation of semiconductor device theory, J. Appl. Phys., 59, pp.195-199

Nathan, A., Allegretto, W., Chau, K. and Baltes, H.P., Electrical and Thermal Analysis of VLSI Contacts and Vias, Baccarani and Rudan (Eds.), Proc. Simulation of Semiconductor Devices and Processes, Vol. 3, (Bologna, Italy), pp.589-598

Parker, J.H. and Lowke, J.J (1969) Theory of electron diffusion parallel to electric fields. 1. Theory", Phys. Rev., Vol. 181, pp.290-301

Polak, S.J., Den Heijer, C. and Schilders, W.H.A. (1987) Semiconductor device modelling from the numerical view point, Int. J. Num. Meth. in Eng., 24, pp.763-838

Shur, M. (1981) Ballistic transport in a semiconductor with collisions, IEEE Trans. Electron Devices, ED-28, pp.1120-1130

Snowden, C.M., Howes, M.J. and Morgan, D.V. (1983) Large-signal modeling of GaAs MESFET operation, IEEE Trans. Electron Devices, ED-30, pp.1817-1824

Snowden, C.M. (1984) Numerical simulation of microwave GaAs MESFETs, Proc. Int. Conf. on Simulation of Semiconductor Devices and Processes, Swansea: Pineridge, pp.406-425

Snowden, C.M. (1986) Two-dimensional modelling of non-stationary effects in short gate length MESFETs, Proc. 2nd Int. Conf. on Simulation of Semiconductor Devices and Processes, Swansea: Pineridge, pp.544-559

Snowden, C.M. and Loret, D. (1987) Two-dimensional hot electron models for short gate length GaAs MESFETs, IEEE Trans. Electron Devices, ED-34, No.2

Snowden, C.M. (1988) Semiconductor Device Modelling, (Peter Peregrinus, London), pp.33-37

Stratton, R. (1962) Diffusion of hot and cold electrons in semiconductor barriers, Phys. Rev., Vol. 126, pp.2002-2014

Widiger, D.J., Kizilyalli, I.C., Hess, K. and Coleman, J.J. (1985) Two-dimensional transient simulation of an idealized high electron mobility transistor, IEEE Trans. Electron Devices, Vol. ED-32, pp.1092-1102

Numerical Techniques
Finite Difference and Boundary Element Methods

Derek B Ingham
University of Leeds

INTRODUCTION

It is impossible in one paper to fully describe the range of possible numerical techniques that are available in the mathematical solving of problems in semiconductor devices and in this paper we will restrict ourselves to the methods of finite difference and boundary element. However, invariably the problem of solving the Poisson equation, or some simple variation of it, is required as part of the full solution procedure. Usually the complete set of governing equations are non-linear and therefore accurate, fast solvers of the Poisson equation are required.

In this paper we will concentrate on three numerical aspects, namely:

(i) singularities,

(ii) the boundary element method,

(iii) the multigrid method,

and each topic will be described so as to be independent of the others. Further, we will simplify the mathematical detail involved in the problem as much as possible so that only the salient points of the method are presented. For example, although the solution of the Poisson equation is required we will discuss in more detail the solution of the Laplace equation since the solution techniques are invariably the same for the two equations, the only difference being the need to include a particular integral for the Poisson equation.

In solving the equation

$$\nabla^2 \psi(\underline{p}) = 0, \qquad \underline{p} \in \Omega \tag{1}$$

using the classical concept of finite differences in a rectangular shaped region we normally discretize on a rectangular grid throughout the device. The grid spacings depend on the estimate of the truncation error of the discretization scheme along the whole grid lines. The classical discretization of this two-dimensional Laplace operator, see Fig. 1, leads to

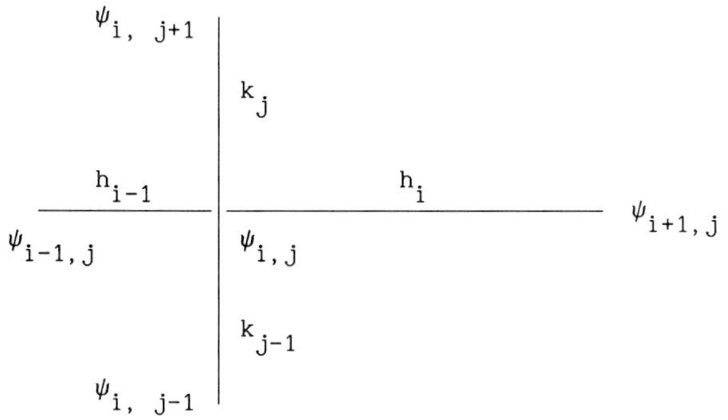

Fig 1 Classical 5 point discretization.

$$\nabla^2 \psi_{i,j} = \psi_{i+1,j} \frac{2}{h_i(h_i+h_{i-1})} + \psi_{i,j+1} \frac{2}{k_j(k_j+k_{j-1})}$$
$$+ \psi_{i-1,j} \frac{2}{h_{i-1}(h_i+h_{i-1})} + \psi_{i,j-1} \frac{2}{k_{j-1}(k_j+k_{j-1})}$$
$$- \psi_{i,j} \left[\frac{2}{h_i h_{i-1}} + \frac{2}{k_j k_{j-1}} \right] \qquad (2)$$

where the local truncation error is of at least O(h+k), with h=max h_i and k=max k_j . The details of the mathematics to derive expression (2) may be found in Smith (1985).

The finite difference formulation of the Laplace equation can be solved using either L-U factorisation or simple over-relaxation iterative schemes depending on whether static or ac solutions are required. The direct L-U factorisation techniques provide a faster solution for static solutions, where many iterations would otherwise be required during the first few time steps. In the case of ac simulations of several hundred picoseconds, where the device is driven from relatively slowing varying signals and there is only a small change in bias between time steps, the iterative scheme is more computationally efficient, requiring only a few iterations to reach convergence, see, for example, Snowden (1987).

SINGULARITIES

A common form of singularity arises in semiconductor devices when the boundary conditions change discontinuously or when the boundary suddenly changes in direction, in particular through an angle exceeding π as shown in Fig. 2.

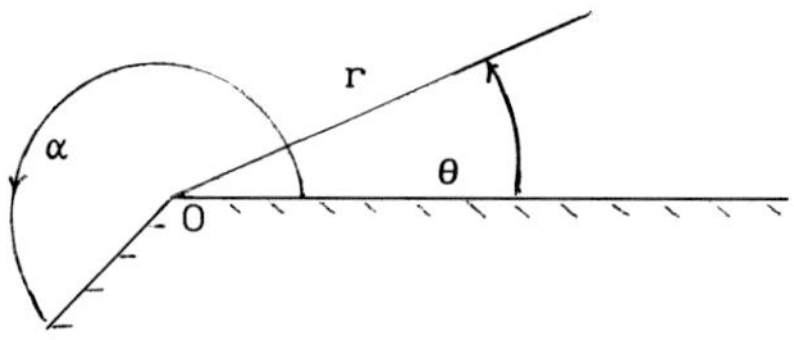

Fig 2 Geometry and co-ordinate system.

If we consider equation (1) in polar co-ordinates, *i.e.*

$$\frac{\partial^2\psi}{\partial r^2} + \frac{1}{r}\frac{\partial\psi}{\partial r} + \frac{1}{r^2}\frac{\partial^2\psi}{\partial\theta^2} = 0 \ , \tag{3}$$

we obtain, by the method of separation of variables, the solution in the form

$$\psi = \sum_{\text{all } s} r^s \left(A_s \cos s\theta + B_s \sin s\theta \right) . \tag{4}$$

Further if $\psi = 0$ or $\frac{\partial\psi}{\partial n} = 0$ on all the boundaries meeting at O then we have either

$$\psi = \sum_{m=1}^{\infty} b_m r^{m\pi/\alpha} \sin (m\pi\theta / \alpha) \ , \tag{5a}$$

or

$$\psi = \sum_{m=0}^{\infty} a_m r^{m\pi/\alpha} \cos (m\pi\theta / \alpha) \ . \tag{5b}$$

In either case it is clear that if $\alpha > \pi$ the exponent in r is less than unity for m=1 and the first derivative, $\frac{\partial\psi}{\partial n}$, does not exist at O.

In order to illustrate how this problem may be overcome we consider the problem described by Motz (1946). He concentrated on the situation $\alpha=2\pi$, see Fig. 3,

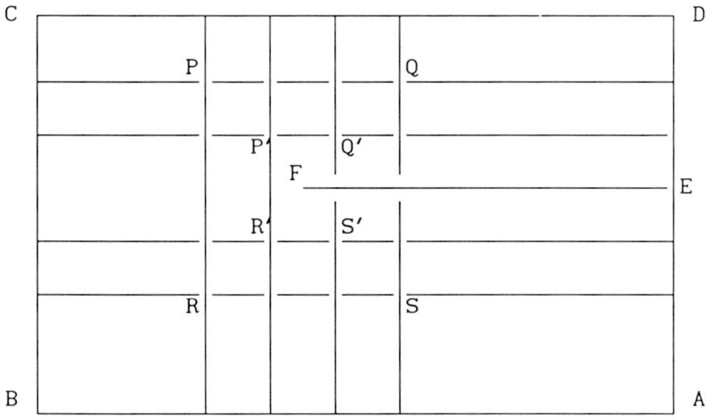

Fig 3 Notation for the Motz problem.

and assuming that ψ is given on ABCD and $\frac{\partial\psi}{\partial n} = 0$ on EF (both sides of the boundary). We can use, with confidence, result (2) everywhere except near the point F, where $\frac{\partial\psi}{\partial n}$ is infinite and ψ is given by

$$\psi = a_0 + a_1 r^{1/2}\cos\frac{\theta}{2} + a_2 r\cos\theta + a_3 r^{3/2}\cos\frac{3\theta}{2} + \ldots \tag{6}$$

Motz used the standard finite difference formula (2) at every mesh point except P′, Q′, R′ and S′. He then determined the coefficients a_0, a_1, a_2 and a_3 in equation (6) by using the values of ψ at four points near to F, say P, Q, R and S, *i.e.* solved four equations in four unknowns. Having determined a_0, a_1, a_2 and a_3 formula (6) can then be used to determine ψ at P′, Q′, R′ and S′. Thus we have obtained the same number of algebraic equations as unknowns but the special equations are used at the primed points and standard finite difference equations at all other points.

Although the Motz method enables one to obtain a more accurate solution to to the problem the final results are not as accurate as one may have wished. To illustrate this let us consider a simple one-dimensional equation in which the solution is of the same form as that given in equation (6), namely $\chi = Ar^{1/2}$ where A is an arbitrary constant. Then

$$\frac{d^2\chi}{dr^2} = -\frac{1}{4}\frac{A}{r^{3/2}}\,, \tag{7}$$

and assuming a constant mesh h, equation (7) can be written

$$\left.\frac{d^2\chi}{dr^2}\right|_r = \frac{\chi(r+h) - 2\chi(r) + \chi(r-h)}{h^2} . \tag{8}$$

Substituting into expression (8) the exact solution at r=nh gives

$$\left.\frac{d^2\chi}{dr^2}\right|_r = \frac{A}{h^{3/2}} \left[(n+1)^{1/2} - 2n^{1/2} + (n-1)^{1/2} \right] , \tag{9}$$

but from the result (7) the exact answer is

$$\left.\frac{d^2\chi}{dr^2}\right|_r = - \frac{1}{4} \frac{A}{h^{3/2}} \frac{1}{n^{3/2}} . \tag{10}$$

The percentage error in the representation of $\frac{d^2\chi}{dr^2}$ in finite difference form, even with the *exact* values of χ included is as follows:

value of r	h	2h	3h	4h	6h	8h
% error	134.3	9.0	3.7	2.0	0.9	0.5

Thus, although the method described by Motz will assist in obtaining a more accurate solution, at the points closest to r=0 there will always be inherently substantial errors (~ 10%) no matter how small the mesh size h is made. Probably the best method to deal with this problem is to use the method of singularity subtraction, see Ingham and Kelmanson (1984), where in the problem described by Motz we write

$$\bar{\psi} = \psi - (a_0 + a_1 r^{1/2} \cos\theta/2 + a_2 r \cos\theta + a_3 r^{3/2} \cos3\theta/2), \tag{11}$$

and then $\bar{\psi}$ is regular everywhere. In order to find the constants a_0, a_1, a_2 and a_3 we impose that $\bar{\psi} \equiv 0$ at the points P′, Q′, R′ and S′ .

It is probably also worth noting here that in some problems the method of collocation can give a very simple and rapid solution. In the problem considered by Motz for which the solution is given by equation (5b) we terminate the solution after M terms, *i.e.* we write

$$\psi = \sum_{m=0}^{M-1} a_m r^{m/2} \cos(m\theta/2) . \tag{12}$$

We then determine the constants a_m (m=0, 1, ..., M-1) by enforcing the boundary conditions at M points on the boundary ABCD and solving M equations for M unknowns. In the example problem given by Motz this method gives accurate results (to within 0.1% when $M = 8$) using a hand calculator whereas the finite difference method requires the solution of a large number, $O(10^2)$ or larger, of linear equations and gives much less accurate results.

THE BOUNDARY ELEMENT METHOD

The most advantageous uses of the boundary element method is when it is possible to transform the governing differential equation, which is valid in a region Ω, into a set of integral equations over the *boundary* of the solution domain $\partial\Omega$. A more complete review of the method may be found in Jaswon and Symm (1977), Banerjee and Butterfield (1981), Brebbia et. al. (1984) and Ingham and Kelmanson (1984). The main advantage of such a transformation is that the numerical solution of the boundary integral equation only requires subdividing the boundary surface of the solution domain. We will again illustrate this technique with reference to the two-dimensional Laplace equation (1).

We shall use the divergence theorem

$$\int_{\partial\Omega} \underline{R} \cdot \underline{n} \, dq = \int_{\Omega} \operatorname{div} \underline{R} \, dQ , \tag{13}$$

where $\underline{R}$ is a vector function of position, $\underline{n}$ is the unit outward normal vector to Ω, dq represents the length element at $\underline{q} \in \partial\Omega$ and dQ the surface element. Applying result (13) to the vectors $\psi\nabla G$ and $G\nabla\psi$, respectively gives

$$\int_{\partial\Omega} \psi \frac{\partial G}{\partial n} dq = \int_{\Omega} [\nabla\psi . \nabla G + \psi\nabla^2 G] dQ , \tag{14}$$

$$\int_{\partial\Omega} G \frac{\partial \psi}{\partial n} dq = \int_{\Omega} [\nabla\psi . \nabla G + G\nabla^2 \psi] dQ , \tag{15}$$

respectively, and the identity known as Green's Theorem can be obtained by subtracting equation (15) from (14), namely

$$\int_{\partial\Omega} \left[\psi \frac{\partial G}{\partial n} - G \frac{\partial \psi}{\partial n} \right] dq = \int_{\Omega} (\psi\nabla^2 G - G \nabla^2\psi) dQ. \tag{16}$$

Now if ψ satisfies the Laplace equation and we choose G such that

$$\nabla^2 G(\underline{p}_0 , \underline{p}) = \delta (\underline{p}_0 - \underline{p}) , \tag{17}$$

where δ is the Dirac delta function then identity (16) becomes

$$\int_{\partial\Omega}\left[\psi(\underline{q})\frac{\partial G}{\partial n}(\underline{p}_0,\underline{q})-G(\underline{p}_0,\underline{q})\frac{\partial\psi}{\partial n}(\underline{q})\right]dq=\eta(\underline{p}_0)\,\psi(\underline{p}_0) \qquad (18)$$

where $\underline{q}\in\partial\Omega$, $\underline{p}_0\in\Omega\cup\partial\Omega$ and $\eta\ (\underline{p}_0)$ is a constant which depends on the location of the point $\underline{p}_0$. Equation (18) is usually referred to as Green's Integral Formula.

The function $G(\underline{p}_0,\underline{p})$ is usually referred to as the *Fundamental Solution.* Taking $G(\underline{p}_0\ ,\ \underline{p})$ to be a function which depends simply on the distance from $\underline{p}_0$ to $\underline{p}$, *i.e.* $G = G(\ r\ )$ where $r = |\underline{p}_0 - \underline{p}|$ then

$$\nabla^2 G(r) = \delta(r)\ , \qquad (19)$$

and seeking a singular solution of $\nabla^2 G = 0$ gives $G(r) = A \ln r$ where A is a constant. Thus in order to satisfy (19) $G(r)=\ln r$ and hence

$$G(\underline{p}_0\ ,\ \underline{p}) = \ln|\underline{p} - \underline{p}_0|\ . \qquad (20)$$

It is now only necessary to determine $\eta(\underline{p}_0)$ and this depends on whether $\underline{p}\in\Omega$ or $\underline{p}\in\partial\Omega$. This can be determined, see references above, and the result is that

$$\eta\ (\underline{p}_0) = \begin{cases} 0 & \underline{p}\notin\Omega\cup\partial\Omega \\ 2\pi & \underline{p}\in\Omega \\ \theta & \underline{p}\in\partial\Omega \end{cases}\ , \qquad (21)$$

where θ is the angle included between the tangents to $\partial\Omega$ on either side of $\underline{p}_0$ — if $\partial\Omega$ is smooth $\theta = \pi$.

The problem now reduces to solving

$$\int_{\partial\Omega}\left[\psi(\underline{q})\,G'(\underline{p},\underline{q})-\psi'(\underline{q})\,G(\underline{p},\underline{q})\right]dq=\eta(\underline{p})\,\psi(\underline{p}), \qquad (22)$$

subject to the linear boundary conditions

$$\left.\begin{aligned} \psi &= \psi_0\ (\underline{q}) && \text{on } \partial\Omega_1 \\ \psi'_0 &= \psi'_0\ (\underline{q}) && \text{on } \partial\Omega_2 \\ \psi'_0 &= \alpha(\underline{q})\ \psi_0(\underline{q}) + \beta(\underline{q}) && \text{on } \partial\Omega_3 \end{aligned}\right\} \qquad (23)$$

where ψ_0 , ψ'_0 , α and β are prescribed functions and $\partial\Omega = \partial\Omega_1 + \partial\Omega_2 + \partial\Omega_3$. Non linear boundary conditions have been investigated by Ingham, Heggs and Manzoor (1981).

The simplest numerical approach to solving equation (22) subject to the boundary conditions (23) is to use the *classical* boundary element method and this involves:

(i) letting $\underline{p} \longrightarrow \bar{\underline{q}} \in \partial\Omega$ in equation (22) ,

(ii) discretising the boundary into N straight segments $\partial\Omega_j$ such that

$$\partial\Omega = \bigcup_{j=1}^{N} \partial\Omega_j \quad ,$$

(iii) using the constant element method where a node $\underline{q}_j$ is located at the mid point of each element $\partial\Omega_j$. It is now assumed that ψ and ψ' are constant over each element, *i.e.* piecewise constant over $\partial\Omega$, and take their values at the midpoints, *i.e.*

$$\left.\begin{aligned} \psi\,(\underline{q}) &= \psi\,(\underline{q}_j) = \psi_j \\ \psi'(\underline{q}) &= \psi'(\underline{q}_j) = \psi'_j \end{aligned}\right\} \quad \forall \;\; \underline{q} \in \partial\Omega_j \quad ,$$

(iv) take $\bar{\underline{q}} = \underline{q}_i$, where $\underline{q}_i$ is any of the N nodes $\underline{q}_j$.

The net effect is that equation (22) may be written

$$\sum_{j=1}^{N} \left[\psi_j \left(\int_{\partial\Omega_j} \ln' |\underline{q}_i - \underline{q}| dq \right) - \psi'_j \left(\int_{\partial\Omega_j} \ln \; |\underline{q}_i - \underline{q}| dq \right) \right] = \eta_i \; \psi_i \; , \qquad (24)$$

or alternatively,

$$\sum_{j=1}^{N} A_{i,j}\; \psi_j + \sum_{j=1}^{N} B_{i,j} \psi'_j = 0 \qquad i=1, \; \ldots, \; N \; , \qquad (25)$$

where

$$\left.\begin{aligned} A_{i,j} &= \int_{\partial\Omega_j} \ln' |\; \underline{q}_i - \underline{q} \;| dq - \eta_i \delta_{i,j} \\ B_{i,j} &= - \int_{\partial\Omega_j} \ln \; |\underline{q}_i - \underline{q} \;| dq \end{aligned}\right\} \; , \qquad (26)$$

and $A_{i,j}$ and $B_{i,j}$ are dependent solely on the geometry of $\partial\Omega$ and maybe determined analytically, see Ingham and Kelmanson (1984). Equation (25) is N equations invoving 2N unknowns, namely ψ_j and ψ'_j , j=1,...,N. However, at each node $\underline{q}_i$, ψ_i and ψ'_i or a linear combination of ψ_i and ψ'_i is known, see boundary condition (23), and therefore the problem reduces to solving N equations in N unknowns. The governing matrix tends to be dense in structure and hence direct, rather than indirect, methods of inversion are the most appropriate.

Once the values of ψ_j and ψ'_j are known then the potential $\psi(\underline{p})$ at any interior point can be computed quite simply from the discretised form of the Green's Integral formula, *i.e.*

$$\sum_{j=1}^{N} \left[\psi_j \int_{\partial\Omega_j} \ln' \, |\underline{p} - \underline{q}| dq - \psi'_j \int_{\partial\Omega_j} \ln \, |\underline{p}-\underline{q}| dq \right] = 2\pi \, \psi(\underline{p}) \; . \tag{27}$$

The accuracy of the method may be improved in several ways, see the review papers already quoted. The most common method is to approximate ψ and ψ' as linear, or quadratic functions over the elements $\partial\Omega_j$. On many occasions the increase in accuracy achieved by using the higher order methods does not justify the additional programming complexity. However the method is useful for problems where the boundary of the solution domain is not smooth. Further singularities frequently exist and the method of singularity subtraction, as described by Ingham and Kelmanson (1984), is the best way to deal with these difficulties.

In some problems the dielectric permittivity may depend on the potential distribution within the device and then the Laplace equation becomes

$$\nabla . (\; g(\psi) \; \nabla\psi \;) = 0 \; , \tag{28}$$

where $g(\psi)$ is some given function of ψ. If $g(\psi)$ is a constant then equation (28) reduces to the Laplace equation. The first step in solving equation (28) is to write

$$\nabla T = g(\psi) \; \nabla\psi \; , \tag{29}$$

so that

$$\nabla^2 T = 0 \; , \tag{30}$$

and Green's Integral Formula (18) for T becomes

$$\int_{\partial\Omega} \left\{ T(\underline{q}) \; \ln' | \; \underline{p} - \underline{q} \; | \; - T'(\underline{q}) \; \ln|\underline{p} - \underline{q}| \right\} dq = \eta(\underline{p}) \; T(\underline{p}) \; . \tag{31}$$

If either T or T′ are prescribed at each point $\underline{q} \in \partial\Omega$ then the solution technique is that previously described but the boundary conditions (23) are in terms of ψ and ψ' . Thus we define

$$h(\psi) = \frac{1}{\psi} \int^{\psi} g(\beta) \; d\beta \; , \tag{32}$$

and therefore

$$T = \psi \; h(\psi) \; , \qquad\qquad T' = \psi' \; g(\psi). \tag{33}$$

Combining equations (31) and (33) gives the nonlinear equation

$$\int_{\partial\Omega} \left[\psi(\underline{q})\ h(\psi(\underline{q}))\ \ln' \ |\underline{p} - \underline{q}| - \psi'(\underline{q})\ g(\psi(\underline{q}))\ \ln|\underline{p}-\underline{q}| \right] dq$$

$$= \eta(\underline{p})\ \psi(\underline{p})\ h(\psi(\underline{p}))\ , \qquad \underline{p} \in \Omega + \partial\Omega, \quad \underline{q} \in \partial\Omega\ . \tag{34}$$

The method of solution now proceeds in a similar manner to that previously described except that an iterative method of solution of the resulting nonlinear equations is necessary. The Newton method is the most appropriate to employ and an example of this problem is given in Ingham and Kelmanson (1984) which also includes boundary singularities.

THE MULTIGRID METHOD

The usual approach to solving partial differential equations is to discretize the problem in some preassigned manner (*e.g.* finite difference, finite element, boundary element) on a fixed grid and then submit the resulting algebraic equations to some numerical solver. In Multi-Level Adaptive Techniques (MLAT) the discretization and the solution process are intermixed.

With MLAT one has a very fast solver of the algebraic system of equations since relaxation on each grid is extremely efficient. Each grid is efficient at liquidating one particular wavelength component of the error and this does not depend on the problem being linear, the shape of the domain, the form of the boundary conditions or the smoothness of the solution. The solution procedure may be adaptive although we will only describe the general multigrid (MG) theory. In adaptive techniques fine grids are only introduced where needed. Grid adaption by certain criteria, derived from optimization conditions, automatically decide where and how to change the local discretization pattern. Coarse grid acceleration techniques have been known for a long time, Southwell (1946), but the current re-emergence of interest in this topic is due to Brandt (1973). Some of the ideas presented in this section are contained in the thesis by Keen (1988) and the paper by Falle and Wilson (1988).

In solving a set of equations similar to those given in equation (2) we will consider the problem in a more general form. Consider the system LU = F and $\Lambda U=\Phi$ on the domain Ω and its boundary $\partial\Omega$, respectively, where U is a variable on Ω and $\partial\Omega$, L and Λ are differential operators and F and Φ are source terms. These systems can be discretised on a grid G^k , approximating Ω, as follows

$$L^k U^k = F^k \ , \tag{35}$$

$$\Lambda^k U^k = \Phi^k \ , \tag{36}$$

where L^k and Λ^k are difference operators and U^k , F^k and Φ^k are the discrete approximations to u, F and Φ on G^k. Now if u^k is an approximation to U^k then

$$L^k u^k = F^k - f^k \ , \tag{37}$$

$$\Lambda^k u^k = \Phi^k - \phi^k \ , \tag{38}$$

where f^k and ϕ^k are the residuals, or the amounts by which u^k fails to satisfy the governing equation and boundary conditions, respectively. From now on a capital U is used to refer to a converged finite difference solution and a small u is used to refer to an unconverged finite difference solution. Now suppose there exist two grids G^h and G^H with mesh sizes h and H such that H=2h. One way to obtain a good approximation, u^h, on the finer grid, G^h , is first to obtain a solution, U^H, on the coarser grid, G^H ; $L^H U^H = F^H$, and then interpolate to the fine grid

$$u^h = I_H^h U^H \ , \tag{39}$$

where I_H^h represents a prolongation from G^H to G^h. This idea can be extended to obtain the G^H solution from even coarser grids, however, this theory does not exploit the proximity of G^H to G^h . Using multi-grids, not only can a first guess to u^h be generated using G^H, but improvements to u^h can also be achieved. In order to improve u^h the error function V^h must be used where

$$V^h = U^h - u^h \ . \tag{40}$$

If U^h is not smooth on scales of order h then, for arbitrary u^h, V^h will not be smooth. However, after a few relaxation sweeps on G^h the high frequency error components of V^h will be destroyed. At this point convergence slows down and coarser grids must be used. Assuming a coarser grid, G^H, has been used, V^h can be approximated by V^H as follows

$$V^h = I_H^h \ V^H \ , \tag{41}$$

and then u^h can be updated

$$(u^h)^{new} = (u^h)^{old} + I_H^h \ V^H \ . \tag{42}$$

The residual equations satisfied by V^h when L is nonlinear are as follows

$$\hat{L}^h V^h = r^h , \tag{43}$$

where $\hat{L}^h V^h = L^h (u^h + V^h) - L^h(u^h)$ and $r^h = F^h - L^h(u^h)$. Similar equations exist on the boundary, but for convenience, they will be ignored from now on, since they are of the same form as the interior equations.

The problem is solved using a scheme known as the full approximation storage (FAS) method. This is carried out as follows: approximating equation (43) on G^H gives

$$L^H (I_h^H u^h + V^H) - L^H(I_h^H u^h) = I_h^H r^h , \tag{44}$$

where I_h^H represents a contraction from G^h to G^H. Now introduce the new variables

$$u_h^H \equiv I_h^H u^h + V^H . \tag{45}$$

These new variables represent, on the coarse grid, the sum of the basic approximation, which can be thought of as being fixed, plus its correction V^H. Equation (44) now becomes

$$L^H u_h^H = F_h^H = L^H(I_h^H u^h) + I_h^H r^h \tag{46}$$

Equation (46) has the advantage that it resembles the original equation (43) except for a different right hand side, so the same relaxation routine can be used on all levels. The approximation u^h on G^h is corrected by solving (46) (or approximately solving it), and interpolating V^H on to G^H , updating u^h as indicated in equation (42).

When convergence is achieved $u^h = U^h$ and $V^h = 0$ so that

$$u_h^H = I_h^H U^h , \tag{47}$$

i.e. U_h^H is a coarse grid function which coincides with the projection of the fine grid solution. The solution process is started by producing an approximation on G^h by prediction on a coarser grid or grids (not MG but just straight interpolation of u from G^H to G^h). A few relaxation sweeps are then carried out on G^h to smoothen the approximation. The residuals r^h are then determined from (43) and equation (44) is completed by interpolating u^h on to G^H and operating upon it (*i.e.* $I_h^H u^h$) with L^H. This right hand side (*i.e.* $L^H(I_h^H u^h) + I_h^H r^h$) is now fixed for the first multigrid cycle (although it will be redefined in subsequent cycles). Smoothing now commences on G^H, starting with $I_h^H u^h$ as an initial value for $U_h^H = I_h^H u^h + V^H$ (*i.e.* V^H=0). When smoothing is complete (*i.e.* convergence slow) on G^H the correction

V^H (= $U_h^H - I_h^H u^h$) is interpolated onto G^h and used to correct u^h (*i.e.* $(u^h)^{new} = (u^h)^{old} + I_H^h V^H$). $(u^h)^{new}$ will be found to be a considerable improvement on $(u^h)^{old}$ if the multigrid cycle has worked. A few relaxation sweeps are now carried out on G^h and r^h is passed on to G^H again, etc.... The above theory can be extended to three or more grids, as explained in the following solution algorithm.

Suppose an unconverged solution u^M exists on G^M, the finest grid, with the coarser grids being represented by G^0, and the grid size ratio of G^{k+1} to G^k being equal to $\frac{1}{2}$. Then we proceed as follows:

(i) Interpolate u^M and r^M to each of the coarser grids. The correction equations are determined from equation (46) to be

$$L^k u^k = \bar{F}^k \,, \tag{48}$$

where $\bar{F}^k = L^k(I_{k+1}^k u^{k+1}) + I_{k+1}^k(\bar{F}^{k+1} - L^{k+1}u^{k+1})$, $k \neq M$,

$\bar{F}^M = F^M$ and $u^k = I_M^k u^M$,

plus the boundary conditions.

(ii) On each grid start with u^k as an initial state and perform one iteration.

(iii) Start iterating on the grid which is most efficient at converging the solution. This is the grid on which the quantity

$$\left\| \{u^k\}^{n+1} - \{u^k\}^n \right\| , \tag{49}$$

is largest. Here n+1 and n refer to two consecutive iterations and $\|\cdot\|$ is the 2-norm. The quantity (49) is generally, though not always, largest on the coarsest grid.

(iv) Assuming that (49) is largest on G^k , iterations are carried out using equation (48) on G^k until convergence slows down or the residual has decreased by a predetermined amount.

(v) The correction which is given by:

$$V^k = (u^k)^{new} - (u^k)^{old} , \tag{50}$$

is interpolated to the next finite difference grid, G^{k+1}, where u^{k+1} is updated according to equation (42).

(vi) Iterate on G^{k+1} and then correct to G^{k+2} , and so on to G^M.

(vii) Iterate on G^M until convergence slows down (usually only a few iterations).

Steps (i)-(vii) constitute one multi-grid cycle.

For the solution of the Laplace equation (1) in which $h_i=k_i=h$, a constant for all values of i, a typical fine to coarse interpolation function, I^k_{k+1}, may be taken to be the 9-point restriction used by Ghia *et. al.* (1982). That is, the value of ψ at the point (i+1, j+1) on the coarse mesh is given as follows

$$\left[I^k_{k+1}\psi^{k+1}\right]_{i+1,j+1} = \frac{1}{4}\psi^{k+1}_{2i+1,\,2j+1} + \frac{1}{8}\left[\psi^{k+1}_{2i+2,2j+1} + \psi^{k+1}_{2i+1,2j+2} + \psi^{k+1}_{2i,2j+1} + \psi^{k+1}_{2i+1,2j}\right] + \frac{1}{16}\left[\psi^{k+1}_{2i+2,2j+2} + \psi^{k+1}_{2i,2j+2} + \psi^{k+1}_{2i+2,2j} + \psi^{k+1}_{2i,2j}\right]. \tag{51}$$

The coarse to fine interpolation function, I^{k+1}_k, may also be taken to be the linear form used by Ghia *et al.*(1982), which is given as follows

$$\left.\begin{aligned}
\left[I^{k+1}_k\psi^k\right]_{2i+1,2j+1} &= \psi^k_{i+1,j+1}\,, \\
\left[I^{k+1}_k\psi^k\right]_{2i+2,2j+1} &= \frac{1}{2}\left(\psi^k_{i+1,j+1} + \psi^k_{i+2,j+1}\right), \\
\left[I^{k+1}_k\psi^k\right]_{2i+1,2j+2} &= \frac{1}{2}\left(\psi^k_{i+1,j+1} + \psi^k_{i+1,j+2}\right), \\
\left[I^{k+1}_k\psi^k\right]_{2i+2,2j+2} &= \frac{1}{4}\left(\psi^k_{i+1,j+1}+\psi^k_{i+2,j+1}+\psi^k_{i+1,j+2}+\psi^k_{i+2,j+2}\right).
\end{aligned}\right\} \tag{52}$$

Convergence to the exact solution is assumed to have occurred when the average change in the finite difference variable ψ between consecutive iterations has fallen below some preassigned small value, say, 10^{-6}. In all of the situations considered, the MG procedure is seen to considerably speed up the convergence of the solution of the finite difference equations. For example in solving the Laplace's equation with Dirichlet boundary conditions a converged solution was obtained on a single 64×64 grid and this took over 50 times longer to converge than a MG solution using five grids.

REFERENCES

Banerjee PK, Butterfield R (1981) Boundary Element Methods in Engineering Science. McGraw-Hill, New York.

Brandt A (1972) Lecture Notes in Physics 18, Spinger, Berlin Heidelberg New York.

Brebbia CA, Telles JCF, Wrobel LC (1984) Boundary Element Techniques: Theory and Applications in Engineering. Springer, Berlin New York.

Falle SAEF, Wilson M (1988) ICFD on Numerical Methods for Fluid Dynamics: IMA Conference Series. Clarendon Press, Oxford.

Ghia U, Ghia KN, Shin CT (1982) High-Re Solutions for Incompressible Flow using the Navier-Stokes Equations and a Multigrid Method. J. Comp. Phys. 48: 387-411.

Ingham DB, Heggs PJ, Manzoor M (1981) Boundary Integral Equation Solution of Nonlinear Plane Potential Problems. IMA J. Num. Anal. 1: 416-426.

Ingham DB, Kelmanson MA (1984) Lecture Notes in Engineering 7. Springer, Berlin Heidelberg New York Tokyo.

Jaswon MA, Symm GT (1977) Integral Equation Methods in Potential Theory and Electrostatics. Academic Press, London.

Keen DJ (1988) Combined Convection in Heat Exchanges. Ph.D Thesis, Leeds University.

Motz H (1946) Singularities in Elliptical Equations. Q. Appl. Math. 4: 371-378.

Smith GD (1985) Numerical Solution of Partial Differential Equations: Finite Difference Methods. Clarendon Press, Oxford.

Snowden GM (1987) Two-Dimensional Hot-Electron Models for Short-Gate-Length GaAs MESFET's IEEE Trans. Electron Devices. ED-34: 212-222.

Southwell RV (1946) Relaxation Methods in Theoretical Physics. Clarendon Press, Oxford.

Numerical Techniques - The Finite Element Method

Stephen D Mobbs
University of Leeds

1. INTRODUCTION

Numerical modelling of semiconductor devices requires the solution of a system of simultaneous, nonlinear partial differential equations. One method by which this may be achieved is the *Finite Element Method*. A brief description of the general principles behind the finite element method is presented here. Some simple but practical numerical schemes for semiconductor device modelling are presented. The advantages and disadvantages of using finite elements for numerical modelling of semiconductor devices are discussed.

2. THE BASIC EQUATIONS

The equations governing the voltage and the electron and hole concentrations in a semiconductor are:

$$\nabla^2 V = \frac{q}{\varepsilon}[n - p - N_d + N_a] \; ; \tag{1}$$

$$q\frac{\partial p}{\partial t} = -\nabla . \underline{J}_p - qG \; ; \tag{2}$$

$$q\frac{\partial n}{\partial t} = \nabla . \underline{J}_n - qG \; ; \tag{3}$$

$$\underline{J}_p = -q\mu_p p\nabla V - qD_p\nabla p \; ; \tag{4}$$

$$\underline{J}_n = -q\mu_n n\nabla V + qD_n\nabla n \; . \tag{5}$$

The variables in Eq. (1) - Eq. (5) are:

n = electron concentration,
p = hole concentration,
N_d = donor impurity density,
N_a = acceptor impurity density,
q = electron charge,
ε = dielectric permittivity,

$\underline{J}_p$ = hole current density,
$\underline{J}_n$ = electron current density,
G = hole-electron recombination rate,
μ_p = hole mobility,
μ_n = electron mobility,
D_p = hole diffusivity,
and D_n = electron diffusivity.

The mobilities and diffusivities are related by the Einstein relationships

$$\frac{\mu_p}{D_p} = \frac{kT}{q} \quad \text{and} \quad \frac{\mu_n}{D_n} = \frac{kT}{q} \tag{6}$$

where k is Boltzmann's constant and T is the absolute temperature. Substituting the expressions for $\underline{J}_p$ and $\underline{J}_n$ from Eq. (4) and Eq. (5) into Eq. (2) and Eq. (3) we get the continuity equations for holes and electrons:

$$\frac{\partial p}{\partial t} = \nabla.[\mu_p p\nabla V + D_p\nabla p] - G \; ; \tag{7}$$

$$\frac{\partial n}{\partial t} = \nabla.[-\mu_n n\nabla V + D_n\nabla n] - G \; . \tag{8}$$

We shall be concerned with the numerical solution of the Poisson equation, Eq. (1), and the continuity equations, Eq. (7) and (8). This is to be performed in a region of semiconductor which in general will have two types of boundary, namely:

1. Electrode (conducting) boundaries. On these surfaces, the boundary conditions

$$V = V_0 \; ; \quad p = p_0 \; ; \quad n = n_0 \tag{9}$$

must be applied. V_0, p_0 and n_0 are known, constant values.

2. Insulating boundaries. On these surfaces, the boundary conditions

$$\nabla V.\hat{\underline{n}} = \underline{J}_p.\hat{\underline{n}} = \underline{J}_n.\hat{\underline{n}} = 0 \tag{10}$$

must be applied, where $\hat{n}$ is the outward unit vector.

Steady State Solutions

Under steady state conditions the equations to be solved are Eq. (1) along with:

$$\nabla.[\mu_p p\nabla V + D_p\nabla p] = G \; ; \tag{11}$$

$$\nabla.[-\mu_n n\nabla V + D_n \nabla n] = G \ . \qquad (12)$$

There are two important features of Eq. (1), Eq. (11) and Eq. (12) which affect the method of solution. These are:

1. The equations are coupled and so all three equations must be solved simultaneously.
2. The equations are nonlinear, for two reasons. Firstly, the current densities involve the products $n\nabla V$ and $p\nabla V$. Secondly, the mobilities, diffusivities and electron-hole recombination rate will in general depend on the electric field strength $|\nabla V|$.

Both of these features will be discussed in subsequent sections. Firstly, however, the essential features of the finite element method will be illustrated for the Poisson equation by writing Eq. (1) in the simplified form

$$\nabla^2 V = -\frac{\rho}{\varepsilon} \qquad (13)$$

where ρ is the charge density. We will assume initially (and unrealistically) that the spatial distribution of ρ is constant and known.

3. FINITE ELEMENT SOLUTION OF THE POISSON EQUATION

Basic descriptions of the finite element method are given by Zienkiewicz (1977), Owen and Hinton (1980), Chung and Yeo (1979) and many others. An introduction to the application of the finite element method to semiconductor device modelling is given by Snowden (1988). The essential idea is that the solution to a differential equation is approximated by simple functions over small sub-domains of the total simulation domain. These sub-domains are usually called *elements*. By matching together the solutions in individual elements, the total solution is built up. For example, in two dimensions the domain may be sub-divided into a mesh of triangles:

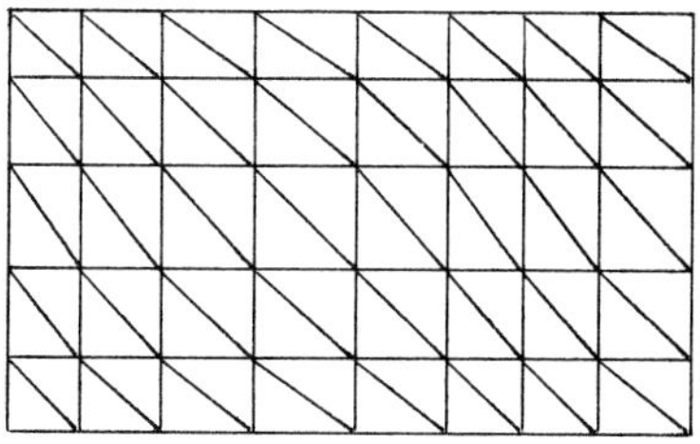

Figure 1. Finite element subdivision of the domain into triangular elements.

The solution within an element is often approximated by a low order polynomial. For example, in the Poisson equation, Eq. (13), the voltage may be approximated by a linear function:

$$V_{element} = a_0 + a_1x + a_2y \ . \tag{14}$$

An essential feature of the finite element method is that special significance is attached to the solution at certain points within elements, called the *nodes*. For triangular elements and linear approximations to the solution, it is conventional to take the nodes to be at the vertices of the triangle (Fig. 2).

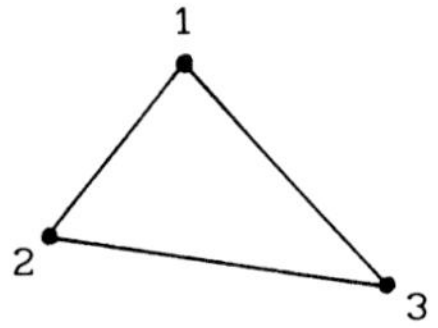

Figure 2. A triangular element with three nodes.

Suppose the values of V at the three nodes in Fig. 2 are V_1, V_2 and V_3. In the finite element method, Eq. (14) is rewritten in the form

$$V_{element} = \sum_{i=1}^{3} N_i(x,y)V_i \ . \tag{15}$$

The functions $N_i(x,y)$ are called the *shape functions* and must, if Eq. (14) and Eq. (15) are to be equivalent, be linear in x and y. They must also have the property that N_i is 1 at the ith node and zero at the others. The whole expression for V in (15) is called a *trial function*.

We now substitute the approximate solution (15) into the Poisson equation (13) giving

$$\sum_{i=1}^{3} V_i\nabla^2N_i + \frac{\rho}{\varepsilon} = R \ . \tag{16}$$

The right hand side of Eq. (16) would be zero if the approximation (15) were an exact solution but in general it will not be, giving rise to an error, or *residual*, R. The aim is now to make the solution be as accurate as possible by minimising R. There are several techniques by which this may be achieved (see, for example, Zienkiewicz 1977). Here we will illustrate the *Galerkin Method*.

Galerkin Method

This approach involves multiplying Eq. (16) by each of the three shape functions N_j (j=1,2,3) in turn and integrating over the element:

$$\int_\Omega \sum_{i=1}^{3} N_j\left[V_i\nabla^2N_i + \frac{\rho}{\varepsilon}\right]ds = \int_\Omega N_jR\,ds \tag{17}$$

(Ω denotes the element).

The next step is to set the right hand side of Eq. (17) to zero and to apply the divergence theorem to the first term on the left hand side, giving:

$$\int_{\Omega} \sum_{i=1}^{3} \left[-V_i \nabla N_j . \nabla N_i + N_j \frac{\rho}{\varepsilon} \right] ds + \int_{\partial\Omega} V_i N_j \nabla N_i . \hat{\underline{n}} d\ell = 0 \qquad (18)$$

where $\partial\Omega$ denotes the boundary of the element. It is evident that Eq. (18) is actually three simultaneous equations for the three unknown nodal voltages V_i. In principle, given suitable boundary conditions, these could be solved by simple algebraic means. However, we have still to match together the solutions in differrent elements. This is achieved as follows.

Combining the Element Equations

It will be noted, that each of the nodal values V_i in Eq. (18) will in fact appear in several elements. The equations for each element are coupled by the following method. For the kth node in the whole domain, all the elements sharing that node are identified. Within each of these elements, there will be one equation of the form (18) for which the local node numbered j is actually the node having global node number k. These equations are added togethe to give a global equation at the kth node. If the procedure is repeated for all the nodes (M say), then there will result a system of M simultaneous equations for the M unknown nodal voltages. The operation of combining equations from elements surrounding each node is known as *element assembly*.

It is necessary to consider the boundary integral in Eq. (18). Firstly, on inter-element boundaries, it turns out that these boundary integral contributions cancel exactly when the summation over the elements is performed. On the outer boundaries of the total domain, the boundary conditions (9) or (10) must be used. For insulating boundaries, the condition $\nabla V . \hat{\underline{n}} = 0$ means that the boundary integral is zero. In other words, setting the integral to zero is all that is necessary in order to satisfy the boundary condition. On electrode boundaries, the approach is rather different. At all nodes on such boundaries, the condition $V = V_0$ is applied explicitly, replacing the finite element equations at those nodes. Thus there is no need to evaluate the boundary integrals along electrode boundaries, since Eq. (18) is not used at the nodes on these boundaries. It follows that no boundary integrals need ever be evaluated.

Numerical Integration

The integration is Eq. (18) is normally carried out by a process of *Gauss quadrature* (see, for example, Smith 1982 for an elementary description). For a given order of shape function, this can normally be made exact.

Order of Shape Functions

For a particular differential equation, the shape functions must satisfy certain conditions in order that the finite element solution should converge towards the true solution as the number of elements is increased and the mesh refined. The two conditions are:

1. Completeness. The shape functions and their derivatives up to the order of the highest derivative appearing in Eq. (18) must be continuous and non-zero within the element. First derivatives appear in (18) and so at least first order (linear) shape functions are needed.
2. Compatibility. The shape functions and their derivatives up to one order less than the highest derivative appearing in Eq. (18) must be continuous across the element boundaries. Thus for the Poisson equation, the trial functions must be continuous across element boundaries. This is guaranteed by having V_i as nodal variables.

Solution of the Simultaneous Equations

The global system of equations can be solved by several methods. Usually, direct methods such as Gaussian elimination (see any elementary text on numerical analysis) are used although iterative techniques are possible.

4. FINITE ELEMENT SOLUTION OF THE STEADY CONTINUITY EQUATIONS

The finite element solution of the two continuity equations (11) and (12) proceeds in a similar manner to the Poisson equation. It is helpful initially to write them as

$$\nabla . \underline{J}_p + G = 0 \; ; \tag{19}$$

$$\nabla . \underline{J}_n + G = 0 \; . \tag{20}$$

Each of the current densities may be approximated by a trial function in a similar manner to the voltage in Eq. (15):

$$\underline{J}_p = \sum_{i=1}^{3} \underline{J}_{pi} N_i(x,y) \quad \text{and} \quad \underline{J}_n = \sum_{i=1}^{3} \underline{J}_{ni} N_i(x,y) \; . \tag{21}$$

The Galerkin method may then be applied by multiplying each of Eq. (19) and Eq. (20) by the shape functions and integrating over the element. Thus

$$\int_\Omega N_j\left[\nabla.\underline{J}_p + G\right]ds = 0\ ; \tag{22}$$

$$\int_\Omega N_j\left[\nabla.\underline{J} + G\right]ds = 0\ . \tag{23}$$

Applying the divergence theorem, we get

$$\int_\Omega\left[-\nabla N_j.\underline{J}_p + N_j G\right]ds + \int_{\partial\Omega} N_j\underline{J}_p.\hat{\underline{n}}d\ell = 0\ ; \tag{24}$$

$$\int_\Omega\left[-\nabla N_j.\underline{J}_n + N_j G\right]ds + \int_{\partial\Omega} N_j\underline{J}_n.\hat{\underline{n}}d\ell = 0\ . \tag{25}$$

On insulating boundaries, the boundary integrals vanish by virtue of the boundary conditions (10). On electrode boundaries they do not vanish. However the equations for the nodes on electrode boundaries are replaced by the boundary conditions (9) and so no boundary integrals need be calculated in order to solve the finite element problem. However, if the contact currents are required, the boundary integrals must be calculated *after* the finite element equations have been solved. These are given by

$$I_p = \int\sum_{i=1}^{3} N_i\underline{J}_{pi}.\hat{\underline{n}}d\ell\ ; \tag{26}$$

$$I_n = \int\sum_{i=1}^{3} N_i\underline{J}_{ni}.\hat{\underline{n}}d\ell\ . \tag{27}$$

Further details concerning contact currents are given by Barnes and Lomax (1977).

Using the definitions (4) and (5) for the current densities and the trial functions

$$p = \sum_{i=1}^{3} N_i(x,y)p_i\ ; \tag{28}$$

$$n = \sum_{i=1}^{3} N_i(x,y)n_i\ , \tag{29}$$

Eq. (24) and (25) become

$$\int_{\Omega}\left[-\nabla N_j\cdot\left\{-q\mu_p\sum_{i=1}^{3}p_iN_i\ \nabla V - D_p\sum_{i=1}^{3}p_i\nabla N_i\right\} + N_jG\right]ds = 0\ ; \tag{30}$$

$$\int_{\Omega}\left[-\nabla N_j\cdot\left\{-q\mu_n\sum_{i=1}^{3}n_iN_i\ \nabla V + D_n\sum_{i=1}^{3}n_i\nabla N_i\right\} + N_jG\right]ds = 0\ . \tag{31}$$

The same element assembly proceedure as was used for the Poisson equation can be applied to the element equations (30) and (31). If the voltage V is also expressed in terms of a trial function (using Eq. (15)), then Eq. (30) and (31) give rise to 2M simultaneous equations in the 3M unknown nodal values of V, p and n (where M is the number of nodes). A further M equations are provided by the Poisson equation as described in the previous section. In principle, the 3M algebraic equations could be solved to provide the finite element solution. However, in practice this is virtually impossible since Eq. (30) and Eq. (31) are nonlinear, involving products of p and n with ∇V.

Coping with Nonlinearity

The nonlinear equations (30) and (31) can be linearised if it is assumed that V is already known. Let the known distribution of V be V^*. Since the mobilities, diffusivities and G may also depend on V, we will write them as μ_p^*, μ_n^*, D_p^*, D_n^* and G^* when evaluated at $V = V^*$. Then (30) and (31) become

$$\int_{\Omega}\left[-\nabla N_j\cdot\left\{-q\mu_p^*\left(\sum_{i=1}^{3}p_iN_i\right)\nabla V^* - D_p^*\sum_{i=1}^{3}p_i\nabla N_i\right\} + N_jG^*\right]ds = 0\ ; \tag{32}$$

$$\int_{\Omega}\left[-\nabla N_j\cdot\left\{-q\mu_n^*\left(\sum_{i=1}^{3}n_iN_i\right)\nabla V^* + D_n^*\sum_{i=1}^{3}n_i\nabla N_i\right\} + N_jG^*\right]ds = 0\ . \tag{33}$$

However, in Eq. (18), the substitution $V = V^*$ is not made; V is regarded as unknown in that equation. The procedure for solving the system of 3M equations arising from the assembly of Eq. (18), (32) and (33) is:

1. Make an initial guess for V^*.
2. Using V^*, solve the 3M linear equations arising from element assembly.
3. Put V^* equal to the newly calculated V.
4. Repeat steps 3 and 4 until there is insignificant change.

5. TIME DEPENDENT SOLUTIONS

For modelling unsteady problems, the fully continuity equations (7) and (8) must be used. In this case, the nodal values p_i and n_i in Eq. (28) and (29) must be time dependent, so that

$$\frac{\partial p}{\partial t} = \sum_{i=1}^{3} N_i(x,y)\frac{dp_i}{dt} ; \qquad (34)$$

$$\frac{\partial n}{\partial t} = \sum_{i=1}^{3} N_i(x,y)\frac{dn_i}{dt} . \qquad (35)$$

Using these expressions, the continuity equations (7) and (8) become, after applying the Galerkin method as in section 4

$$\int_{\Omega}\sum_{i=1}^{3} N_j N_i \frac{dp_i}{dt} ds = \int_{\Omega}\left[-\nabla N_j \cdot \left\{q\mu_p\left(\sum_{i=1}^{3} p_i N_i\right)\nabla V + D_p \sum_{i=1}^{3} p_i \nabla N_i\right\} - N_j G\right] ds; \qquad (36)$$

$$\int_{\Omega}\sum_{i=1}^{3} N_j N_i \frac{dn_i}{dt} ds = \int_{\Omega}\left[-\nabla N_j \cdot \left\{-q\mu_n\left(\sum_{i=1}^{3} n_i N_i\right)\nabla V + D_n \sum_{i=1}^{3} n_i \nabla N_i\right\} - N_j G\right] ds. \qquad (37)$$

For numerical solution, the time derivatives in Eq. (36) and (37) are discretised in finite difference form. Finite elements are never used in time, since this leads to undesirable dependence of the behaviour on later behaviour. Details concerning the application of finite difference methods can be found in the chapter by D.B. Ingham. Applications of the finite element method to unsteady semiconductor device modelling are given by Barnes and Lomax (1977) and Buturla *et al.* (1981).

6. ADVANTAGES AND DISADVANTAGES OF THE FINITE ELEMENT METHOD

It is useful to compare the finite element solution of semiconductor device modelling with finite difference methods. The advantages of the finite element method are

1. The application to problems with irregular geometry and different sizes of element in different regions of the domain is no more difficult than for regular geometry and meshes. Such problems are more complex when tackled using finite difference methods.
2. Once a computer code has been written for a particular order of trial function, only very minor changes to the code are needed in order to change the order of trial function. Conversely, however, in order to change the accuracy of a finite difference code, major modifications are usually required.
3. Procedures involving successive mesh refinements are more easily incorporated into finite element schemes.

Other considerations, however, may favour finite difference methods. For example:

1. It is generally much easier to write a finite difference rather than

a finite element computer code.

2. Error analysis is much easier to perform for finite difference methods.
3. Finite difference methods generally require smaller computer memory than finite element methods.

7. THREE DIMENSIONAL PROBLEMS

Although we have been concerned principally with two dimensional problems, the description of the finite element method given in sections 3, 4 and 5 is equally valid for three dimensional problems. However, in three dimensions the elements will usually take the form of tetrahedra or cuboids. The boundary integrals discussed above will of course become surface integrals and the integrals over the elements will be volume integrals.

8. PROGRAMMING FINITE ELEMENTS

This article does not attempt to give advice on approaches to writing computer codes for finite element methods. It is well known, however, that there is an enormous gulf between developing a suitable finite element scheme on paper and having a working code. Anyone not familiar with the practical aspects of implementing finite element schemes is advised to consult a text concerned principally with programming techniques for finite elements (for example, Smith 1982 or Hinton and Owen 1977).

9. REFERENCES

Barnes JJ, Lomax, RJ (1977) Finite-element methods in semiconductor device simulation. IEEE Trans., ED-24: 1082-1089

Buturla EM, Cottrell PE, Grossman BM, Salsburg KA (1981) Finite-element analysis of semiconductor devices: the FIELDAY program. IBM J. Res. Develop 25: 218-231

Chung TJ, Yeo MF (1979) A practical introduction to finite element analysis. Pitman, London

Hinton E, Owen DRJ (1977) Finite element programming. Academic Press, London, New York, San Francisco

Owen DRJ, Hinton E (1980) A simple guide to finite elements. Pineridge Press, Swansea.

Smith IM (1982) Programming the finite element method. Wiley, Chichester, New York, Brisbane, Toronto, Singapore

Snowden CM (1988) IEE materials and devices series 5: Semiconductor device modelling. Peter Peregrinus Ltd, London

Zienkiewicz OC (1977) The finite element method. McGraw-Hill, London

Gallium Arsenide versus Silicon - Applications and Modelling

Michael Shur

University of Minnesota

Silicon is the most important material for semiconductor electronics. Silicon devices and integrated circuits are dominant in most applications -- from consumer electronics to computers, from automotive electronics to sensors and imagers. However, in certain areas, silicon is being challenged by other semiconductor materials, such as gallium arsenide and related compounds. Gallium arsenide devices and monolithic microwave integrated circuits excel in ultra-high frequency applications. Gallium arsenide digital integrated circuits have emerged as leading contenders for ultrahigh speed applications in the next generation of supercomputers and communication equipment. These new technologies are located much earlier on the learning curve and their relative importance is expected to increase sharply in the near future.

In a silicon crystal, each atom has four valence electrons forming four bonds with the nearest neighbors. This corresponds to so-called tetrahedral configuration in which each silicon atom is located at the center of the tetrahedron formed by four nearest atoms (see Fig. 1a). The tetrahedral bond configuration is repeated, forming the same crystal structure as in a diamond crystal. This structure (called the diamond structure) is formed by two interpenetrating face-centered cubic sublattices of atoms, shifted with respect to each other by one fourth of the body diagonal. Perhaps, we should call this structure the silicon structure instead, because we can argue that silicon is much more important for our civilization than diamond. (Another important semiconductor — germanium (Ge) — has exactly the same crystal configuration as well.)

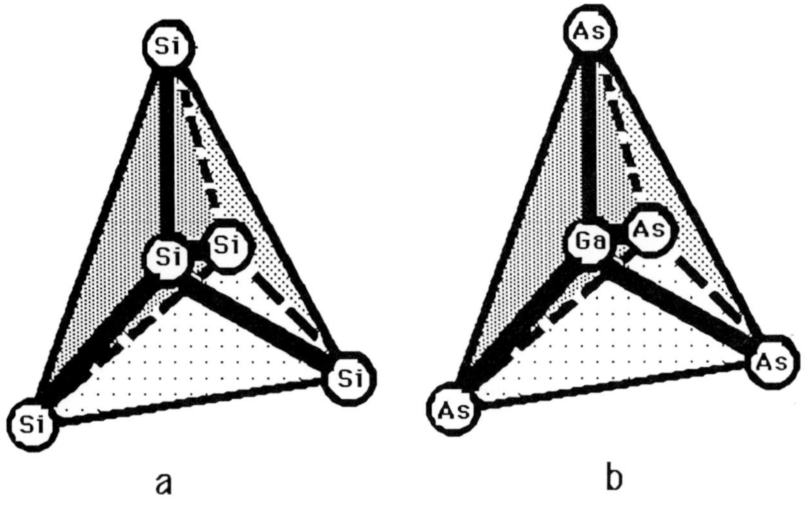

Fig. 1. Tetrahedral bonds in silicon and GaAs

Compound semiconductors (such as GaAs, InP, ZnSe, AlGaAs, InGaAs, and many others) "emulate" silicon by having the same **average** number of valence electrons per atom. Many of compound semiconductors have a chemical structure $A_{III}B_V$ where A_{III} represents an element from the third column of the periodic table (3 valence electrons) and B_V represents an element from the fifth column of the periodic table (5 valence electrons). Examples are GaAs, InP, InSb, GaSb, GaP, AlAs, etc. Another chemical formula for compound semiconductors is $A_{II}B_{VI}$. Again, the total number of valence electrons per atom is four. Examples are ZnSe, ZnTe, ZnS, etc.

In compound semiconductors the chemical bond between the nearest neighbors is partially heteropolar. Nevertheless, in many such compounds, this bond is still more or less covalent leading to the tetrahedral bond configuration similar to that in a silicon crystal (see Fig. 1b). As a consequence, most III-V compounds crystallize in the so-called zinc blende crystal structure that is very similar to the diamond structure. The primitive cell of the zinc blende structure contains two atoms A and B which are repeated in space, with each species forming a face-centered cubic lattice. It can be described as mutually penetrating face-centered cubic (fcc) lattices of element A and element B (gallium (Ga) and arsenic (As) in case of GaAs) shifted relative to each other by a quarter of the body diagonal of the unit cell cube. Numerous compound semiconductors give a material scientist and a device designer a rich variety of choices. However, gallium arsenide technology is the most developed compound semiconductor technology today.

The energy bands of silicon and GaAs and related compounds is also similar (see Fig. 2). In all important cubic semiconductors, the top of the highest filled (valence) band corresponds to the Γ point of the first Brillouin zone, i.e. to point where where wave vector k = (0,0,0). The tops of the valence bands are located at point Γ; two of these bands (light and heavy holes) have the same energy at this point (i.e., are degenerate), while the third one is split (it is called a split-off band). The lowest minimum of the conduction band in different semiconductors corresponds to different points of the first Brillouin zone. For example, in germanium, the L minimum (corresponding to the point (1,1,1) of the first Brillouin zone) is the lowest. The lowest minimum of the conduction band in silicon is located along the Δ axis (corresponding to the direction (k,0,0) of the first Brillouin zone), close to the X-point (corresponding to the point (1,0,0) of the first Brillouin zone). In GaAs, the lowest minimum of the conduction band is at point Γ(0,0,0), i.e. at the same value of the wave vector, k, as the top of the valence band. Such a semiconductor is called a direct gap semiconductor. Silicon and germanium are indirect gap semiconductors.

Gallium arsenide and related compounds have several other advantages compared to silicon. The electron effective mass in the Γ-minimum of the conduction bands is much smaller than in X or L minima. This leads to higher electron velocity (see Fig. 3) which is extremely important for ultra-high frequency and ultra-high-speed applications. The lighter effective mass is especially important in short semiconductor devices where so-called overshoot and ballistic effects lead to an additional velocity enhancement (see Fig. 4) and in a new generation of quantum devices (such as resonant tunneling devices, superlattice structures, heterostructure transistors utilizing two-dimensional electron gas, etc.). Also, as can be seen from Fig.2, in many of these compounds the electron velocity decreases with electric field in a certain range of applied electric fields. This negative differential mobility (related to the electron transfer from the Γ-minimum of the conduction band to the X and L minima) has been utilized in so-called Transferred Electron devices.

GaAs is a "direct gap" semiconductor which means that it can be used to generate and detect both incoherent and coherent (laser) light and, hence, is well suited for applications in optoelectronics. These applications are further enhanced by an ability to grow layered structures of gallium arsenide and aluminum gallium arsenide and use such structures for the light confinement. Because of the direct band-gap, recombination of electrons and holes in gallium arsenide is more efficient. As a consequence, when gallium arsenide is subjected to radiation, electron-hole pairs produced by radiation recombine much faster than in silicon. Hence, gallium arsenide is "radiation-hard" and devices and circuits can withstand much more radiation than silicon.

Gallium arsenide can be grown as a highly resistive material (semi-insulating GaAs) which can be used as a substrate. A whole integrated circuit can be made using a direct ion implantation into such a substrate - a relatively simple technology with few fabrication steps. Alternatively, using Molecular Beam Epitaxy (MBE) or Metal Organic Chemical Vapor Deposition (MO-CVD) one can deposit gallium arsenide and related compounds on a gallium arsenide highly resistive substrate controlling the composition and impurity profiles literally within one atomic distance. This opens up many exiting opportunities for making new kind of ultra-fast heterostructure devices using what has been called "energy band engineering". These devices include Heterostructure Field Effect Transistors, Heterojunction Bipolar Transistors, Tunneling Emitter Bipolar Transistors, Induced Base Transistors, Hot Electron Transistors, Planar Doped Barrier Transistors, and Resonant Tunneling devices. All these devices will compete for applications in ultra-high speed digital and ultra-high frequency microwave devices and integrated circuits.

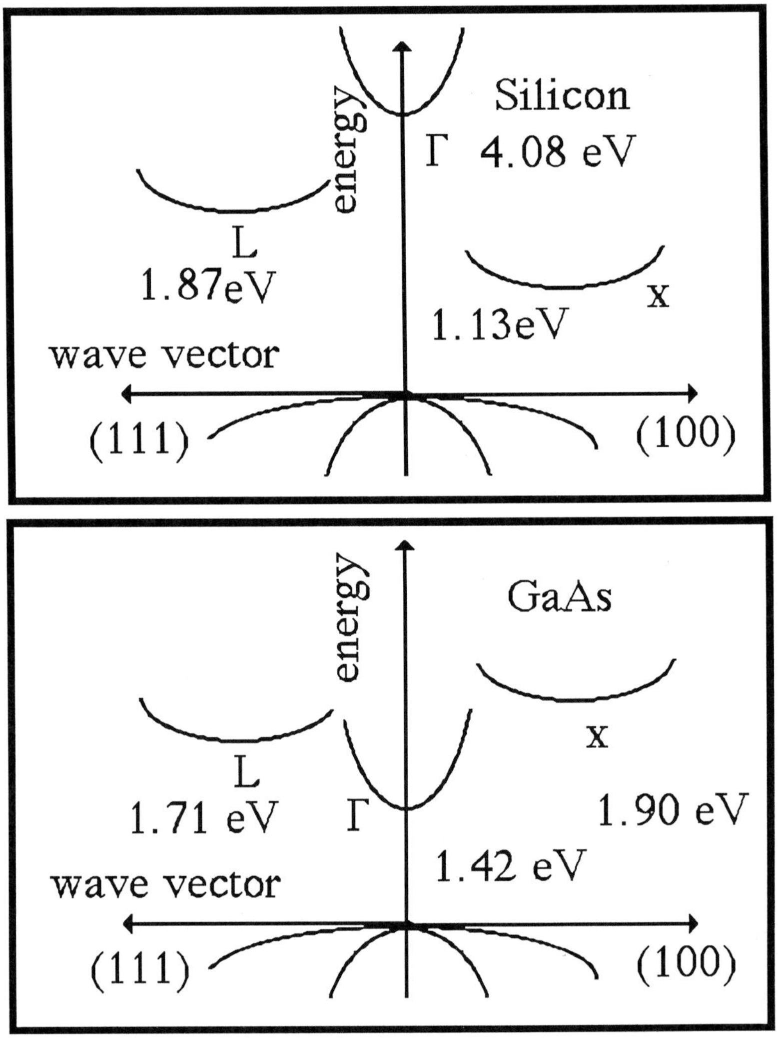

Fig. 2. Minima of conduction band and maxima of valence band in Si and GaAs. Numbers show the energy separations from the bottom of the minima of the conduction band from the top of the valence band.

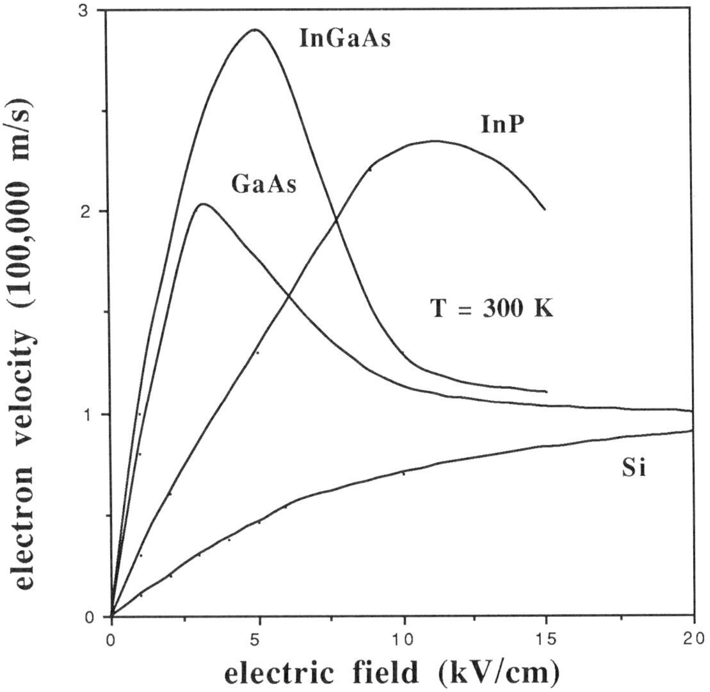

Fig. 3. Electron drift velocity versus electric field.

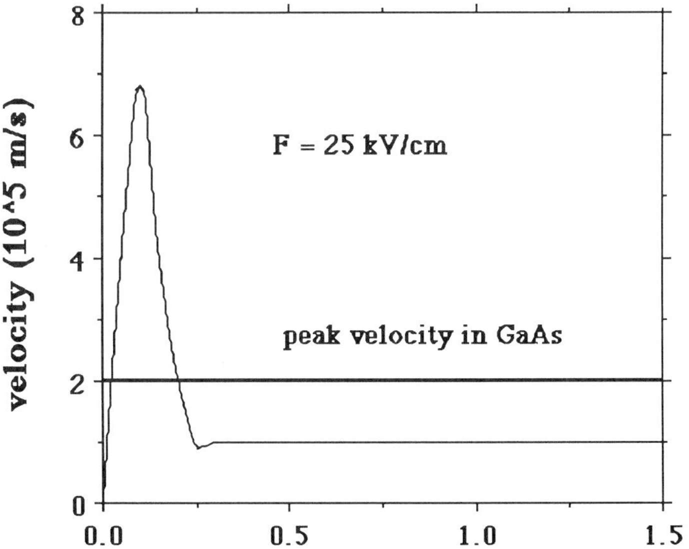

Fig. 4. Electron velocity vs. distance for electrons injected into region of constant electric field (from Shur (1976).

There are also disadvantages of gallium arsenide technology compared to silicon. Here we can only repeat Aleksandr Pushkin (1822)

Oh? What fault was there to find?
None to speak of, never mind.
Or at most the merest tittle.
Still a flaw (though very little).

Indeed, in an elemental semiconductor, such as silicon, one has to be concerned with a purity of a single element. In a compound semiconductor, such as gallium arsenide, the material composition is also extremely important. There are more parameters to control, and it is more difficult and expensive to obtain pure and defect-free material. Silicon is plentiful and cheap. Gallium and arsenic are relatively scarce and expensive. Finally, silicon is blessed with a stable high quality native oxide which has good dielectric quality. This property is utilized to make silicon Metal-Oxide-Semiconductor Field Effect Transistors (MOSFETs) - a workhorse of silicon electronics. Gallium arsenide native oxide is poor, and gallium arsenide MOSFETs are practically non-existent. Only very recently, the new approach of oxidizing a thin silicon layer grown by Molecular Beam Epitaxy on the GaAs surface offered hope for the development of a viable GaAs MOSFET technology (see Tiwari et al. (1988)). Schottky barrier metal semiconductor field effect transistors (MESFETs), junction field effect transistors (JFETs), and heterostructure AlGaAs/GaAs transistors are the most commonly used GaAs devices.

Hence, it is unlikely that gallium arsenide will ever replace silicon in electronics. Rather, it will be dominant in microwave and optoelectronic applications and in very high speed and/or radiation hard circuits interfaced with silicon. Relatively small gallium arsenide integrated circuits (ICs) will be combined with silicon ICs to achieve a much higher level of an overall system performance. Professor Kroemer of University of California in Santa Barbara, once compared a future role of gallium arsenide in electronics to a role played by specialty steels in steel industry.

The progress in gallium arsenide technology has been quite remarkable (see, for example, Shur (1987)). Gallium arsenide MEtal- Semiconductor Field Effect Transistors (MESFETs) made by a direct ion implantation into a highly resistive (semi-insulating) gallium arsenide substrate are used today to produce Medium Scale Integrated (MSI) Circuits for the next generation of supercomputers (see Welch et al. (1988)). Yields of 60 to 80 percent or higher may be achieved for gallium arsenide circuits of hundred to two hundred gate complexity. Much more complicated gallium arsenide Large Scale Integrated (LSI) circuits, such as 16x16 multiplier (containing about 10,000 transistors), a 16 kilobit Static Random Access Memory (SRAM) have been demonstrated in laboratory environment. Sub-0.1 micron-gate-length Heterostructure Field Effect Transistors have exhibited low-noise high-gain operation at frequencies as high s 92 GHz (see Chao et al. (1989)). After many years of research and development the prospects of gallium arsenide technology today seem to be better than ever.

Computer-Aided Design (CAD) tools have long become an integral part of both silicon and gallium arsenide technology. Fig. 5 (from Shur (1989)) illustrates a typical development sequence of an integrated circuit technology. Nearly all these development steps rely on simulation and modeling and utilize Computer-Aided Design (CAD) tools. These tools include device models, both analytic and numerical, device simulators, circuit simulators, logic simulators, layout and routing programs, process simulation programs, etc. Below we briefly discuss several device simulation and process simulation programs that include SEDAN , BIPOLE, PISCES, BAMBI, SUPREME, FISHID, PUPHS1D, PUPHS2D, DEMON, and SEQUAL.

SEDAN (SEmiconductor Device ANalysis) program has been developed and distributed by the Integrated Circuit Laboratory at Stanford University (see Yu and Dutton (1985)). It computes potential, electron and hole concentration profiles as functions of time for one-dimensional device structures. The SEDAN III version simulates both silicon (including transistors with polysilicon emitters) and GaAs (or AlGaAs/GaAs) devices.

BIPOLE is a program that computes the characteristics of bipolar junction transistors from the information about the fabrication process and mask dimensions. It was developed at the Department of Electrical Engineering, University of Waterloo, Canada and is distributed by Waterloo Engineering Software, 180 Columbia Street, West, Waterloo, Ontario, Canada N2L 3L3. The program takes into

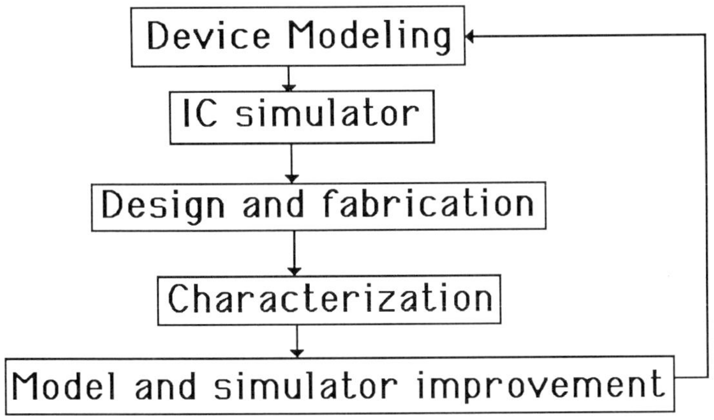

Fig. 5 . Typical development sequence of integrated circuit technology (from Shur (1989)) .

account band-gap narrowing in heavily doped emitter regions, the dependence of the lifetime on doping, temperature, electric field and doping dependences of electron and hole mobilities. It is applicable for simulation of integrated structures and GaAs and AlGaAs/GaAs Heterostructure Bipolar Transistors.

PISCES is a two-dimensional device simulator developed by Stanford Electronics Laboratories, Department of Electrical Engineering, Stanford University, Stanford, CA 94305 (see Pinto et al. (1985)). The program solves basic semiconductor equations based on field-dependent electron and hole velocities. BAMBI (Basic Analyzer of Mos and BIpolar DEvices) is a two-dimensional device simulator developed by Franz and Franz (1985) and distributed by Dr. Siegfrid Selbergeher, Institute fur Allgemeine Electrotechnik und Elektronik, Technische Universitatet Wien, Gusshausstraße 27-29, A-1040 Wien, Austria. It handles both DC and transient problems.

In silicon technology computer programs for modeling device and circuit fabrication processes play a crucial role. A very powerful program for simulating fabrication processes is called SUPREME. It has been developed and distributed by Stanford University (see Ho et al. (1983), Plummer (1986)). It includes models for simulating diffusion, ion implantation, and oxidation in silicon and, in later versions, gallium arsenide. In particular, the program generates the doping profiles resulting from given processing steps. These dopant profiles may be directly used as an input to device simulation programs SEDAN and PIESCES. SUPREME is based on one-dimensional models. However, SUPREME-4 will include two-dimensional models as well. This will be especially useful for modeling the fabrication processes for short-channel devices.

SIMPL (SIMulated Profiles from the Layout) is a computer program that generates device cross-sections from the circuit layout (see Grimm et al. (1983) and Neureuther (1986)). The input data to the program include process and mask information. SAMPLE (Simulation and Modeling of Profiles in Lithography and Etching) (see Neureuther (1986)) simulates lithography, etching, and deposition processes. Both SIMPL and SAMPLE were developed at Berkeley and more information may be obtained from EECS Industrial Liaison Program, 457 Cory Hall, University of California, Berkeley, CA 94720.

Programs FISHID, PUPHS1D, PUPHS2D, DEMON, and SEQUAL have been developed by Professor Mark Lundstrom of Purdue University. FISHID solves Poisson's equation in compositionally non-uniform semiconductors (such as GaAs/AlGaAs device structures) both under equilibrium conditions and under bias (assuming zero current densities). The program allows you to plot energy band diagrams, potential, field and carrier concentration profiles, and compute C-V characteristics (see Gray and Lundstrom (1988)).

PUPHS1D, PUPHS2D are one-dimensional and two-dimensional versions of a device simulator, resperctively. These programs solve the the electron and hole continuity equations for compositionally non-uniform semiconductors in equilibrium and under bias. These programs have been used to simulate solar cells, diodes, field-effect transistors, bipolar junction transistors, and heterostructure bipolar junction transistors (see Lundstrom and Schulke (1983)).

Demon simulates electron transport in compositionally non-uniform semiconductors using one-dimensional Monte-Carlo simulation (see Maziar et al. (1986) and Klausmeier-Brown (1986)).

SEQUAL solves a one-dimensional Schrödinger equation for a given one-dimensional band diagram (obtained, for example, using FISH1D or PUPHS1D. This program was described by Cahay et al. (1987) and by McLennan (1987).

UM-SPICE is a GaAs circuit simulator. It has models for GaAs MEtal-Semiconductor Field Effect Transistors, Heterostructure Field Effect Transistors (HFETs or MODFETs), Superlattice Heterostructure Field Effect Transistors, n-channel and p-channel Heterostructure Insulated Gate Field Effect transistors, and other field-effect transistors (see, for example Hyun et al. (1986) and Hyun (1985)). As an example, we show in Fig.6 inverter transfer curves simulated by UM-SPICE. In Fig. 7. we compare experimental and simulated propagation delays versus power dissipation for 25-stage HFET ring oscillators.

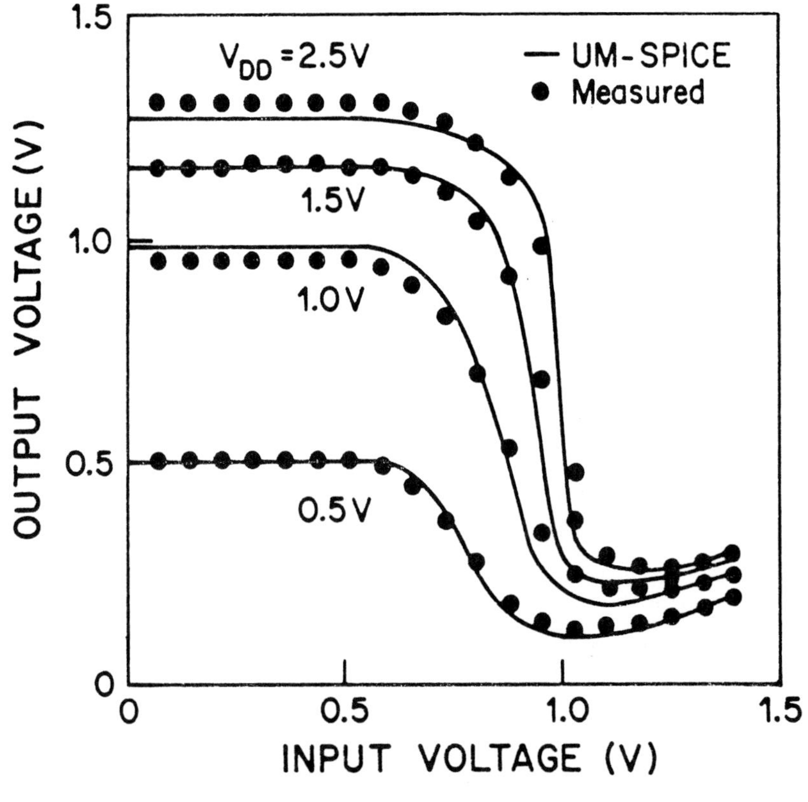

Fig.6. Inverter transfer curves simulated by UM-SPICE (from Hyun et al. (1986).

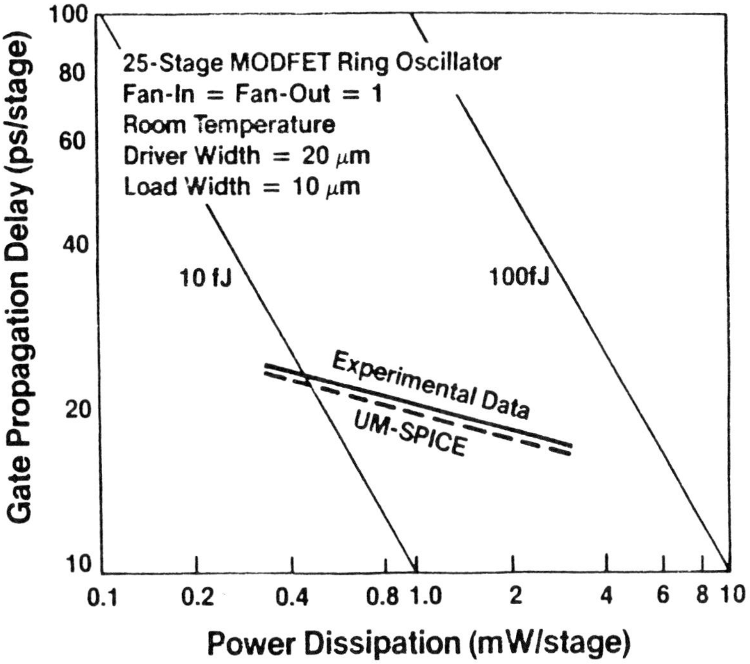

Fig. 7. Experimental and simulated propagation delays versus power dissipation for 25-stage HFET ring oscillators.

The differences between silicon and gallium arsenide technologies lead to somewhat different objectives and priorities in simulating and modeling silicon and gallium arsenide devices and circuits. The physics of ultra-short channel gallium arsenide devices is more complicated. Overshoot and ballistic effects make conventional diffusion-drift equations invalid. In devices based on the heterostructures, utilizing quantum wells, superlattices and such, quantum effects become dominant or very important. Hence, complicated computations based on the self-consistent solution of the Boltzmann equation and Poisson's equation are often needed in order to elucidate complicated device physics. On the other hand, technology of gallium arsenide device and circuits is less developed than silicon technology. Numerous "non-ideal" effects such as effects of traps, space-charge injection, non-uniform distribution of dopants, surface states and surface leakage, passivation, etc. play an important role making characteristics of nominally identical devices quite different, especially from a wafer to a wafer. Therefore, simple semi-empirical models based on charge-control considerations are quite useful for determining general trends, scaling considerations, and device characterization. The goal here is to establish the feedback loop shown schematically in Fig. 5. Using such an approach we can develop useful gallium arsenide CAD capitalizing on our successes and learning from our mistakes.

Acknowledgement. I would like to thank Professor Mark Lundstrom of Purdue University for providing information about device simulation programs developed by him and his students.

References.

Cahay M., M. McLennan, S. Datta, and M. Lundstrom, Importance of Space-Charge Effects in Resonant Tunneling Devices, Appl. Phys. Lett., vol. 50, pp. 612-614 (1987)

P. C. Chao, M. Shur, , R. C. Tiberio, K.H. G. Duh, P. M. Smith, J. M. Ballingall, P. Ho, and A. A. Jabra, DC and Microwave Characteristics of Sub-0.1μm gate-length planar-doped pseudomorphic HEMTS, IEEE Trans. Electron Devices, to be published (1989)

Franz A. F. and G. A. Franz, in Simulation of Semiconductor Devices and Processes, Swansea, Pineridge , pp. 204-218 (1984)

Gray J. L. and M. S. Lundstrom, Solution to Poisson's Equation with Application to C-V Analysis of Heterostructure Capacitors, IEEE Trans. Electron Devices, vol. ED-32, pp. 2102-2109 (1985)

Grimm M. A., K. Lee, A. R. Neureuther, IEDM Technical Digest, pp.255-258 (1983)

Ho C. P. , J. D. Plummer, S. E. Hansen, R. W. Dutton, IEEE Trans. Electron Devices, vol. ED-30, No. 11, pp. 1438-1453, Nov. (1983)

Hyun C. H., Analysis and Design of Novel Semiconductor Devices and Integrated Circuits, Ph. D. Thesis, University of Minnesota, Sep. (1985)

Hyun C. H., M. Shur, and A. Peczalski, Analysis of Noise Margin and Speed of GaAs MESFET DCFL using UM-SPICE", IEEE Trans. CAD ICAS, vol.ED-33, No.10, pp.1421-1426, April 1986

Klausmeier-Brown M. E., Monte Carlo Studies of Electron Transport in III-V Heterostructures, MSEE Thesis, Purdue University, May (1986)

Lundstrom M.S. and R. J. Schulke, Numerical Simulation of Heterostructure Devices, IEEE Trans. Electron Devices, ED-30, No. 11, pp. 1151-1159 (1983)

Maziar C., M.E. Klausmeier-Brown, and M. S. Lundstrom, Proposed Structure for Transit Time Reduction in AlGaAs/GaAs Bipolar Transistors, IEEE Electron Dev. Lett., EDL-8, pp. 486-486, (1986)

McLennan M., Quantum Ballistic Transport in Semiconductor Heterostructures, MSEE Thesis, Purdue University, May (1987).

Neureuther A. R. , Solid State Technology, 29, No. 3, pp. 71-75 (1986)

Pinto M. R., C. S. Rafferty, H. R. Yeager, R. W. Dutton, PISCES-IIB, Supplementary report, Stanford Electronics Laboratories, Department of Electrical Engineering, Stanford University, Stanford, CA 94305 (1985)

Plummer J. D. , Solid State Technology, 29, No. 3, pp. 61-66 (1986)

Pushkin Aleksandr , Tsar Nikita (1822), quoted from Aleksandr Pushkin, Collected Narrative and Lyrical Poetry, Edited and Translated by Walter Arndt, Ardis, Ann Arbor, 1984

Shur M. ,GaAs Devices and Circuits, Plenum, New York, 1987

Shur M., Modeling of GaAs and AlGaAs/GaAs Field Effect Transistors, in Introduction to GaAs Technology, J. Wiley, ed. C. T. Wang, to be published (1989)

Tiwari S., S. L. Wright, and J. Batey, Unpinned GaAs MOS capacitors and transistors, IEEE Electron Device Lett., vol. 9, No.9, pp. 488-489 (1988)

Welch B.M, R. C. Eden, F. Lee, GaAs IC Technology Circuits and Systems, in Proceedings of 1988 IEEE International Symposium on Circuits and Systems, pp. 1709-1713, Helsinki (1988)

Yu Z. and R. W. Dutton, Sedan III - A general electronic material device analysis program, Program manual, Stanford University, July (1985)

Physical Models for Silicon VLSI

Siegfried Selberherr
Technical University Vienna

1. INTRODUCTION

Device Modeling based on the self-consistent solution of fundamental semiconductor equations dates back to the famous work of Gummel in 1964 [31]. Since then numerical device modeling has been applied to nearly all important devices. For some citations regarding the history of modeling the interested reader is refered to [55].

2. BASIC TRANSPORT EQUATIONS

The model for hot carrier transport used in any numerical device simulation is based on the well known fundamental semiconductor equations (1)-(5). There are ongoing arguments in the scientific community whether these equations are adequate to describe transport in submicron devices. Particularly the current relations (4) and (5) which are the most complex equations out of the set of the basic semiconductor device equations undergo strong criticism in view of, for instance, ballistic transport [35], [49]. Their derivation from more fundamental physical principles is indeed not at all straightforward. They appear therefore with all sorts of slight variations in the specialized literature and a vast number of papers has been published where some of their subtleties are dealt with. The interested reader is referred to, e.g., [9], [13], [17], [26], [56]. Anyway, recent investigations on ultra short MOSFET's [50] do not give evidence that it is necessary to waive these well established basic equations for silicon devices down to feature sizes in the order of $0,1$ microns [58].

$$\mathrm{div}(\varepsilon \cdot \mathrm{grad}\,\psi) = -\rho \tag{1}$$

$$\mathrm{div}\,\vec{J}_n - q \cdot \frac{\partial n}{\partial t} = q \cdot R \tag{2}$$

$$\mathrm{div}\,\vec{J}_p + q \cdot \frac{\partial p}{\partial t} = -q \cdot R \tag{3}$$

$$\vec{J}_n = q \cdot \mu_n \cdot n \cdot \left(\vec{E} + \frac{1}{n} \cdot \mathrm{grad}\left(n \cdot \frac{k \cdot T_p}{q}\right)\right) \tag{4}$$

$$\vec{J}_p = q \cdot \mu_p \cdot p \cdot \left(\vec{E} - \frac{1}{p} \cdot \mathrm{grad}\left(p \cdot \frac{k \cdot T_n}{q}\right)\right) \tag{5}$$

These equations include a set of parameters which have to be appropriately modeled in order to describe the various transport phenomena qualitatively and quantitatively correctly.

3. SPACE CHARGE

Poisson's equation (1) requires a model for the space charge ρ which makes use of only the dependent variables ψ, n, p and material properties. The well established approach for this model is to sum up the concentrations with the adequate charge sign multiplied with the elementary charge (6).

$$\rho = q \cdot (p - n + N_D^+ - N_A^-) \tag{6}$$

This approach holds well for room temperature. At low ambient temperature, for instance, at liquid nitrogen temperature, some enhancements have to be made to this model. The doping concentration $N_D^+ - N_A^-$ is usually assumed to be fully ionized at room temperature which intuitively does not hold for low temperature analysis. The classical way to describe partial ionization is based on the formulae (7).

$$N_D^+ = \frac{N_D}{1 + 2 \cdot \exp\left(\frac{E_{fn} - E_D}{k \cdot T}\right)}, \qquad N_A^- = \frac{N_A}{1 + 4 \cdot \exp\left(\frac{E_A - E_{fp}}{k \cdot T}\right)} \tag{7}$$

E_D and E_A are the ionization energies of the respective donor and acceptor dopant. Typical values for $E_c - E_D$ and $E_A - E_v$ for the most common dopants in silicon are: 0.054eV for arsenic, 0.045eV for phosphorus, 0.039eV for antimony and 0.045eV for boron. A quite complete list can be found in [63]. These ionization energies are recommended to be modeled doping dependent in [18], however it seems not to be important for MOSFET's regarding my experience. Note, that only energy differences can be given (E_c and E_v are the conduction band and the valence band energy, respectively). Next the Fermi levels E_{fn} and E_{fp} have to be appropriately related to the dependent variables by making use of Fermi statistics.

$$n = N_c \cdot \frac{2}{\sqrt{\pi}} \cdot F_{1/2}\left(\frac{E_{fn} - E_c}{k \cdot T}\right), \qquad p = N_v \cdot \frac{2}{\sqrt{\pi}} \cdot F_{1/2}\left(\frac{E_v - E_{fp}}{k \cdot T}\right) \tag{8}$$

N_c and N_v are the density of states in the conduction band and the valence band, respectively. The classical formulae for the density of states are given by (9).

$$N_c = 2 \cdot \left(\frac{2 \cdot \pi \cdot k \cdot T \cdot m_n^*}{h^2}\right)^{3/2}, \qquad N_v = 2 \cdot \left(\frac{2 \cdot \pi \cdot k \cdot T \cdot m_p^*}{h^2}\right)^{3/2} \tag{9}$$

$F_{1/2}(x)$ is the Fermi function of order 1/2 which is defined by (10).

$$F_{1/2}(x) = \int_0^\infty \frac{\sqrt{y}}{1 + e^{y-x}} \cdot dy \tag{10}$$

The parameters m_n^* and m_p^* which are the effective masses for electrons and holes have now to be modeled to be able to evaluate the formulae for the density of states (9). The probably most elaborate models which are fits to experimental values date back to Gaensslen et al. [27], [29].

$$m_n^* = m_o \cdot \left(1,045 + 4,5 \cdot 10^{-4} \cdot \left(\frac{T}{\mathrm{K}}\right)\right) \tag{11}$$

$$m_p^* = m_o \cdot \left(0,523 + 1,4 \cdot 10^{-3} \cdot \left(\frac{T}{\mathrm{K}}\right) - 1,48 \cdot 10^{-6} \cdot \left(\frac{T}{\mathrm{K}}\right)^2\right) \tag{12}$$

These fitting expressions are claimed to be valid over the range 50-350K.

It is worthwhile to note that the ratio of the density of states depends only on the ratio of the effective masses.

$$\frac{N_c}{N_v} = \left(\frac{m_n^*}{m_p^*}\right)^{3/2} \tag{13}$$

By means of some simple algebraic manipulation with the expressions for the carrier concentrations (8) we obtain:

$$\frac{E_{fn} - E_D}{k \cdot T} = G_{1/2}\left(\frac{n}{N_c}\right) + \frac{E_c - E_D}{k \cdot T} \tag{14}$$

$$\frac{E_A - E_{fp}}{k \cdot T} = G_{1/2}\left(\frac{p}{N_v}\right) + \frac{E_A - E_v}{k \cdot T} \tag{15}$$

$G_{1/2}(x)$ is the inverse Fermi function of order 1/2 defined with (16).

$$G_{1/2}\left(\frac{2}{\sqrt{\pi}} \cdot F_{1/2}(x)\right) = x \tag{16}$$

A convenient fit to (16) is given by (19).

$$G_{1/2}(x) = \frac{\ln(x)}{1 - x^2} + \frac{\left(\frac{3 \cdot \sqrt{\pi} \cdot x}{4}\right)^{2/3}}{1 + \frac{1}{\left(0,24 + 1,08 \cdot \left(\frac{3 \cdot \sqrt{\pi} \cdot x}{4}\right)^{2/3}\right)^2}} \tag{19}$$

The first term in (19) has to be replaced by a truncated series expansion if the argument x is in the vicinity of 1.

$$\frac{\ln(x)}{1 - x^2} = \frac{x - 2}{2} + O((x-1)^2) \tag{20}$$

A review about approximations to Fermi functions and their inverse functions can be found in [12], [19], [55].

It is now possible by evaluating the expressions for the density of states (9) with the fits to the effective masses (11) and (12) to compute numerical values for the ionized impurity concentrations (7) using only the carrier concentrations which are the dependent variables in the basic equations. However, comparisons to experiment indicate that it is better to compute the density of states from relation (13) and a fit to the intrinsic carrier concentration (19).

$$n_i = \sqrt{N_c \cdot N_v} \cdot \exp\left(-\frac{E_g}{2 \cdot k \cdot T}\right) \tag{19}$$

E_g is the band gap $E_c - E_v$ which can be modeled temperature dependent with the fit provided by Gaensslen et al. [27], [29]. Note that most publications which present equation (20) contain a typographical error. The linear temperature coefficient for E_g below 170K is $1,059 \cdot 10^{-5}$eV and not as usually found $1,059 \cdot 10^{-6}$eV [39].

$$E_g = \begin{cases} 1,17\text{eV} + 1,059 \cdot 10^{-5}\text{eV} \cdot \left(\frac{T}{\text{K}}\right) - 6,05 \cdot 10^{-7}\text{eV} \cdot \left(\frac{T}{\text{K}}\right)^2 & T \leq 170\text{K} \\ 1,1785\text{eV} - 9,025 \cdot 10^{-5}\text{eV} \cdot \left(\frac{T}{\text{K}}\right) - 3,05 \cdot 10^{-7}\text{eV} \cdot \left(\frac{T}{\text{K}}\right)^2 & T > 170\text{K} \end{cases} \tag{20}$$

The prefactor in (19) can be fitted to experiments by (21).

$$\sqrt{N_c \cdot N_v} = \exp\left(45,13 + 0,75 \cdot \ln\left(\frac{m_n^*}{m_o} \cdot \frac{m_p^*}{m_o} \cdot \left(\frac{T}{300\text{K}}\right)^2\right)\right) \text{cm}^{-3} \tag{21}$$

With (13) and (21) it is now straightforward to compute the numerical values for the density of states. At room temperature we have $N_c = 5,1 \cdot 10^{19}\text{cm}^{-3}$ and $N_v = 2,9 \cdot 10^{19}\text{cm}^{-3}$; at liquid nitrogen temperature we obtain $N_c = 5,8 \cdot 10^{18}\text{cm}^{-3}$ and $N_v = 2,5 \cdot 10^{18}\text{cm}^{-3}$.

It should be noted that (7) are only valid for moderate impurity concentrations. For heavy doping the assumption of a localized ionization energy does definitely not hold. Instead an impurity band is formed which may merge with the respective band edge, e.g. [36], [53]. Just modeling a temperature dependence of the ionization energies will not account adequately for the underlying physics in this case. It appears to be appropriate to assume total ionization for concentrations above some threshold value and to account for a suitable functional transition between the classical formulae (7) and total ionization. All concepts to tackle this problem which have come to my attention so far, however, make use of a very simplistic, not to say alchemical, approach. Anyway, it should be noted that freeze-out is of major importance only for depletion mode devices and devices with a partially compensated channel doping [30].

In view of this dilemma with heavy doping one may for many applications well use asymptotic approximations for the Fermi function (10) and its inverse (16) (see [55]). This is not in contradiction to the partial ionization model given with (6)-(21).

4. CARRIER MOBILITIES

The next set of physical parameters to be considered carefully for silicon device simulation consists of the carrier mobilities μ_n and μ_p in (4) and (5). The models for the carrier mobilities have to take into account a great variety of scattering mechanisms the most basic one of which is lattice scattering. The lattice mobility in pure silicon can be fitted with simple power laws.

$$\mu_n^L = 1430 \frac{\text{cm}^2}{\text{Vs}} \cdot \left(\frac{T}{300\text{K}}\right)^{-2}, \qquad \mu_p^L = 460 \frac{\text{cm}^2}{\text{Vs}} \cdot \left(\frac{T}{300\text{K}}\right)^{-2,18} \tag{22}$$

The expressions (22) fit well experimental data of [2], [15] and [46].

The next effect to be considered is ionized impurity scattering. The best established procedure for this task is to take the functional form (23) of the fit provided by Caughey and Thomas

[16] and use temperature dependent coefficients.

$$\mu_{n,p}^{LI} = \mu_{n,p}^{min} + \frac{\mu_{n,p}^{L} - \mu_{n,p}^{min}}{1 + \left(\dfrac{CI}{C_{n,p}^{ref}}\right)^{\alpha_{n,p}}} \tag{23}$$

$$\mu_n^{min} = \begin{cases} 80\dfrac{\mathrm{cm}^2}{\mathrm{Vs}} \cdot \left(\dfrac{T}{300\mathrm{K}}\right)^{-0,45} & T \geq 200\mathrm{K} \\ 80\dfrac{\mathrm{cm}^2}{\mathrm{Vs}} \left(\dfrac{200\mathrm{K}}{300\mathrm{K}}\right)^{-0,45} \cdot \left(\dfrac{T}{200\mathrm{K}}\right)^{-0,15} & T < 200\mathrm{K} \end{cases} \tag{24}$$

$$\mu_p^{min} = \begin{cases} 45\dfrac{\mathrm{cm}^2}{\mathrm{Vs}} \cdot \left(\dfrac{T}{300\mathrm{K}}\right)^{-0,45} & T \geq 200\mathrm{K} \\ 45\dfrac{\mathrm{cm}^2}{\mathrm{Vs}} \left(\dfrac{200\mathrm{K}}{300\mathrm{K}}\right)^{-0,45} \cdot \left(\dfrac{T}{200\mathrm{K}}\right)^{-0,15} & T < 200\mathrm{K} \end{cases} \tag{25}$$

$$C_n^{ref} = 1,12 \cdot 10^{17}\mathrm{cm}^{-3} \cdot \left(\frac{T}{300\mathrm{K}}\right)^{3,2}, \qquad C_p^{ref} = 2,23 \cdot 10^{17}\mathrm{cm}^{-3} \cdot \left(\frac{T}{300\mathrm{K}}\right)^{3,2} \tag{26}$$

$$\alpha_{n,p} = 0,72 \cdot \left(\frac{T}{300\mathrm{K}}\right)^{0,065} \tag{27}$$

The fits (24)-(27) are from [34]. Similar data have been provided in [6] and [23].

In view of partial ionization, i.e., for low temperature simulation, one should consider neutral impurity scattering [55]. However, considering the uncertainty of the quantitative values for ionized impurity scattering it seems not to be worthwhile to introduce another scattering mechanism with additional fitting parameters. Furthermore, partial ionization appears to be a second order effect even at liquid nitrogen temperature. It seems therefore justified to include partial ionization only in the space charge model and not in the carrier mobilities.

I prefer to model surface scattering with an expression suggested by Seavey [52].

$$\mu_{n,p}^{LIS} = \frac{\mu_{n,p}^{ref} + (\mu_{n,p}^{LI} - \mu_{n,p}^{ref}) \cdot (1 - F(y))}{1 + F(y) \cdot \left(\dfrac{S_{n,p}}{S_{n,p}^{ref}}\right)^{\alpha_{n,p}}} \tag{28}$$

$$\mu_n^{ref} = 638\frac{\mathrm{cm}^2}{\mathrm{Vs}} \cdot \left(\frac{T}{300\mathrm{K}}\right)^{-1,19}, \qquad \mu_p^{ref} = 160\frac{\mathrm{cm}^2}{\mathrm{Vs}} \cdot \left(\frac{T}{300\mathrm{K}}\right)^{-1,09} \tag{29}$$

with:

$$F(y)=\frac{2\cdot\exp\left(-\left(\frac{y}{y^{ref}}\right)^2\right)}{1+\exp\left(-2\cdot\left(\frac{y}{y^{ref}}\right)^2\right)} \tag{30}$$

$$S_n=\max\,(0,\frac{\partial\psi}{\partial y}),\qquad S_p=\max\,(0,-\frac{\partial\psi}{\partial y}) \tag{31}$$

S_n^{ref} is assumed to be $7\cdot 10^5\frac{\mathrm{V}}{\mathrm{cm}}$; S_p^{ref} is $2,7\cdot 10^5\frac{\mathrm{V}}{\mathrm{cm}}$ and y^{ref} is 10nm.

The formulae for surface scattering are definitely not the ultimate expressions. They just fit quite reasonably experimental observations. Other approaches with the same claim can be found in, e.g., [7], [37], [38], [47]. A u-shaped mobility behavior for surface scattering at low ambient temperature as found in [8], [10] seems not to be worthwhile to synthesize because I believe in a different origin than surface scattering for this experimental observation. It should however be noted that soft turn-on at liquid nitrogen temperature has been successfully simulated with a u-shaped mobility expression [25].

Velocity saturation is modeled with formulae (32). These are again fits to experimental data with, however, a theoretical background considering their functional form [2], [41], [42].

$$\mu_n^{LISE}=\frac{2\cdot\mu_n^{LIS}}{1+\sqrt{1+\left(\frac{2\cdot\mu_n^{LIS}\cdot E_n}{v_n^{sat}}\right)^2}},\qquad \mu_p^{LISE}=\frac{\mu_p^{LIS}}{1+\frac{\mu_p^{LIS}\cdot E_p}{v_p^{sat}}} \tag{32}$$

E_n and E_p are the effective driving forces given by (33). Their derivation can be found in [32].

$$E_n=|\mathrm{grad}\ \psi-\frac{1}{n}\cdot\mathrm{grad}\ (Ut_n\cdot n)|,\qquad E_p=|\mathrm{grad}\ \psi+\frac{1}{p}\cdot\mathrm{grad}\ (Ut_p\cdot p)| \tag{33}$$

The temperature dependent saturation velocities are given in the following.

$$v_n^{sat}=1,45\cdot 10^7\frac{\mathrm{cm}}{\mathrm{s}}\cdot\sqrt{\tanh\left(\frac{155\mathrm{K}}{T}\right)},\qquad v_p^{sat}=9,05\cdot 10^6\frac{\mathrm{cm}}{\mathrm{s}}\cdot\sqrt{\tanh\left(\frac{312\mathrm{K}}{T}\right)} \tag{34}$$

The functional form of these fits is after [2]; the experimental data matched are from [2], [14], [15], [21]. An eventual dependence on the crystallographic orientation which one would deduce from [3], [5], [44] is presently not taken into account.

5. CARRIER TEMPERATURES

To describe carrier heating properly one has to account for local carrier temperatures $T_{n,p}$ in the current relations (4) and (5). This can be achieved by either solving energy conservation equations self consistently with the basic transport equations, or by using a model obtained by series expansions of the solution to the energy conservation equations [32]. I believe that the latter is sufficient for silicon devices. For the electronic voltages we have (35) as an approximation. Confirming theoretical investigations can be found in [1].

$$Ut_{n,p} = \frac{k \cdot T_{n,p}}{q} = Ut_o + \frac{2}{3} \cdot \tau_{n,p}^{\epsilon} \cdot \left(v_{n,p}^{sat}\right)^2 \cdot \left(\frac{1}{\mu_{n,p}^{LISE}} - \frac{1}{\mu_{n,p}^{LIS}}\right) \tag{35}$$

The energy relaxation times $\tau_{n,p}^{\epsilon}$ are in the order of 0,5picoseconds and just weakly temperature dependent [11]. They should however be modeled as functions of the local doping concentration as motivated by the following reasoning. The product of carrier mobility times electronic voltage which symbolizes a diffusion coefficient must be a decreasing function with increasing carrier voltage (see also [11]). Its maximum is attained at thermal equillibrium. Relation (36) must therefore hold.

$$\mu_{n,p}^{LISE} \cdot Ut_{n,p} \leq \mu_{n,p}^{LIS} \cdot Ut_o \tag{36}$$

Note that models for carrier diffusion coefficients are not required in the basic current relations (4), (5).

Substituting (35) into (36) and rearranging terms one obtains relation (37) for the energy relaxation times.

$$\tau_{n,p}^{\epsilon} \leq \frac{3}{2} \cdot Ut_o \cdot \frac{\mu_{n,p}^{LIS}}{\left(v_{n,p}^{sat}\right)^2} \tag{37}$$

In, e.g. MINIMOS 4, the energy relaxation times are modeled on the basis of (37) with a fudge factor γ in the range $[0,1]$ and a default value of 0,8.

$$\tau_{n,p}^{\epsilon} = \gamma \cdot \frac{3}{2} \cdot Ut_o \cdot \frac{\mu_{n,p}^{LIS}}{\left(v_{n,p}^{sat}\right)^2} \tag{38}$$

For vanishing doping one obtains the maximal energy relaxation times which are at 300K $\tau_n^{\epsilon} = 4,44 \cdot 10^{-13}$s, $\tau_p^{\epsilon} = 2,24 \cdot 10^{-13}$s and at liquid nitrogen temperature $\tau_n^{\epsilon} = 8,82 \cdot 10^{-13}$s, $\tau_p^{\epsilon} = 8,68 \cdot 10^{-13}$s.

6. CARRIER GENERATION/RECOMBINATION

Carrier generation/recombination in silicon devices is generally modeled as the simple sum of several partial generation/recombination rates (39).

$$R = R^{SRH} + R^{OPT} + R^{AU} + R^{II} \tag{39}$$

The partial rates are thermal generation/recombination R^{SRH} usually named after the scientists Shockley, Read and Hall. Optical generation/recombination R^{OPT} which is definitely not of relevance for silicon devices, except for very special devices like light-triggered thyristors, Auger generation/recombination R^{AU} and impact ionization R^{II}.

(40) is the established formula for thermal generation/recombination which, however, should only be appropriate for moderate recombination/generation rates [55].

$$R^{SRH} = \frac{n \cdot p - n_i^2}{\tau_p \cdot (n + n_1) + \tau_n \cdot (p + p_1)} \tag{40}$$

The coefficients for the established model of Auger recombination (41) can be made weakly temperature dependent with (42). The fit has been made to the data of [24].

$$R^{AU} = (C_{cn} \cdot n + C_{cp} \cdot p) \cdot (n \cdot p - n_i^2) \tag{41}$$

$$C_{cn} = 2,8 \cdot 10^{-31} \frac{\mathrm{cm}^6}{\mathrm{s}} \cdot \left(\frac{T}{300\mathrm{K}}\right)^{0,14}, \qquad C_{cp} = 9,9 \cdot 10^{-32} \frac{\mathrm{cm}^6}{\mathrm{s}} \cdot \left(\frac{T}{300\mathrm{K}}\right)^{0,2} \tag{42}$$

A particular comment should be made on the model for the impact ionization rate R^{II}. It still seems, though under heavy dispute of the scientific community, that the old Chynoweth formulation (43) of impact ionization can be used quite satisfactorily for device simulation.

$$R^{II} = -\alpha_n \cdot \frac{|\vec{J}_n|}{q} - \alpha_p \cdot \frac{|\vec{J}_p|}{q} \tag{43}$$

with:

$$\alpha_{n,p} = \alpha_{n,p}^{\infty} \cdot \exp\left(-\frac{\beta_{n,p}}{E}\right) \tag{44}$$

The coefficients of (44) can be modeled temperature dependent by (45) and (46) to fit experimental data [20], [22], [48]. It should be noted that there is some lack of data for liquid nitrogen temperature, cf. [62]. However it seems that this impact ionization model is probably somewhat too pessimistic for a proper quantitative prediction of substrate currents as already stated in [45], [57].

$$\alpha_n^{\infty} = 7 \cdot 10^5 \text{cm}^{-1} \cdot \left(0,57 + 0,43 \cdot \left(\frac{T}{300\text{K}}\right)^2\right)$$
$$\alpha_p^{\infty} = 1,58 \cdot 10^6 \text{cm}^{-1} \cdot \left(0,58 + 0,42 \cdot \left(\frac{T}{300\text{K}}\right)^2\right) \tag{45}$$

$$\beta_n = 1,23 \cdot 10^6 \frac{\text{V}}{\text{cm}} \cdot \left(0,625 + 0,375 \cdot \left(\frac{T}{300\text{K}}\right)\right)$$
$$\beta_p = 2,04 \cdot 10^6 \frac{\text{V}}{\text{cm}} \cdot \left(0,67 + 0,33 \cdot \left(\frac{T}{300\text{K}}\right)\right) \tag{46}$$

7. SOME RESULTS

Results of investigations about submicron n-channel enhancement mode MOSFET's at room and liquid nitrogen ambient temperature are presented. A single drain technology designed with a 3/4 micron coded channel length (0, 39 micron metallurgical channel length) has been analyzed where the geometric channel length has been shrunk to 0, 51 micron (0, 15 micron metallurgical channel length). The gate oxide thickness is 9nm and the work function of the donor doped gate polysilicon is -570mV. A window of the critical drain profile corner is depicted in Fig.1.

The actual analysis has been carried out with MINIMOS 4 which is a further development of the MINIMOS program [32], [51], [54] to include also quantitative capabilities for low temperature simulation.

Figure 2 shows the simulated subthreshold characteristics for two different drain biases (UD=0.1 UD=2V, UB=0V) at room and liquid nitrogen temperature. The subthreshold slope is obviously much steeper at liquid nitrogen temperature with about 25mV/decade compared to 95mV/decade at room temperature. It is interesting that the improvement of the slope is almost as good as 3,9 the ratio of 300K/77K [40]. Furthermore, the shift of the subthreshold characteristics between low and high drain bias which should be primarily due to drain induced surface barrier lowering is about 50mV larger at 77K temperature compared to the room temperature shift. To have a larger influence of drain induced barrier lowering at lower temperature is in contradiction with the sound results of [65]. The observed phenomenon must therefore be of different origin. Detailed investigations have brought up several interacting causes. One is partial freeze-out of acceptors in the bulk below the channel which leads to an increase of built-in potential and thus to increasing depletion widths with decreasing temperature [43], [64]. This reasoning is partially confirmed in [28]. The second cause is the formation of a sort of parasitic channel by impact ionization which has also been reported in [51].

Figure 3 shows the simulated output characteristics for four different gate biases. If we take current output for the same gate drive as measure of device quality, the low temperature operation resulted in approximately 50% improvement compared to room temperature operation. Similar results have been experimentally obtained (cf.[60]). This improvement decreases with shrinking channel length as observed in [50].

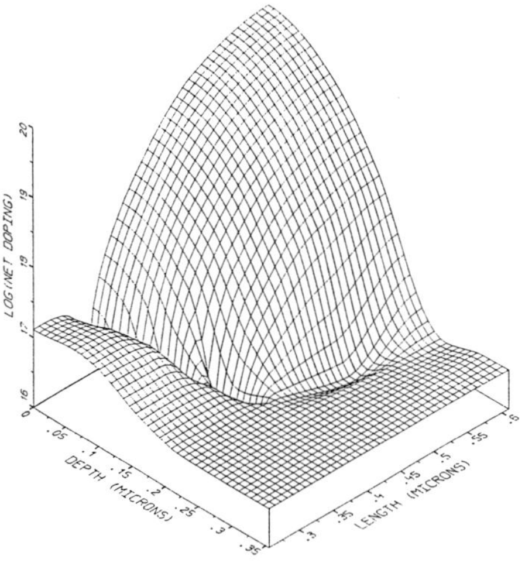

Fig.1: Detail of Doping Profile

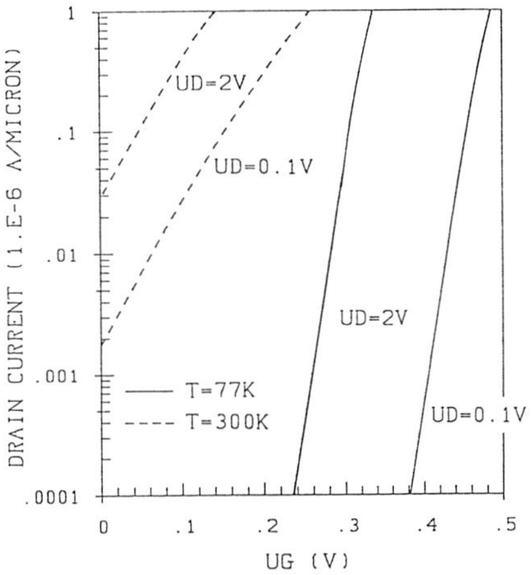

Fig.2: Subthreshold Characteristics

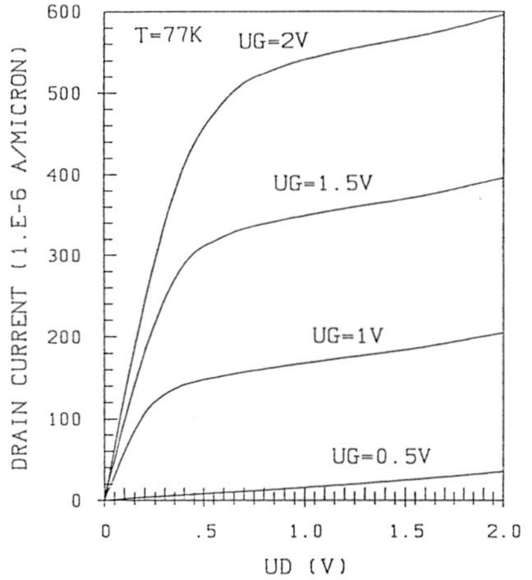

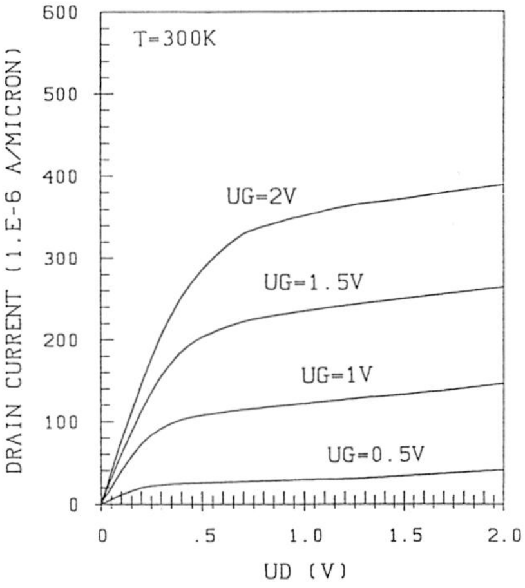

77K Fig.3: Simulated Output Characteristics 300K

In the following a few results about the distribution of the various physical quantities in the interior of the device will be presented. The off-state at UG=0V, UD=2V is despicted with the electron concentration given by Fig.4 at liquid nitrogen temperature and room temperature. It is easily visible that the device is not satisfactorily off at room temperature. The channel is perfectly depleted at 77K.

The on-state is documented with a bias of UG=2V, UD=2V. Fig.5 shows again the electron concentration at liquid nitrogen temperature and room temperature. One can nicely observe the inversion layer which is much steeper for the low temperature simulation. Furthermore, one can see that there are considerably more electrons generated by impact ionization close to the drain.

The impact ionization rates are shown in Fig.6. The peak concentration which occurs in both cases at the surface is almost two orders of magnitude higher for low temperature operation. The substrate current to drain current ratio is increased by a factor of 5.2 which is fairly high for n-channel devices [4].

Figure 7 shows the distribution of electron temperature at 77K and 300K ambient temperature. The maximum temperature is 213K at 77K and 2220K at room temperature This maximum is in both cases located in the reverse biased drain substrate diode with a smooth transition into the channel. The position of the maximum is deeper in the substrate and closer to the drain area at 77K compared to the room temperature result. The phenomenon of smaller carrier heating at liquid nitrogen temperature is a result of a smoother distribution and a smaller maximum of the driving force. Smaller carrier heating at 77K has been confirmed by many simulations; it is not really expected, particularly in view of larger energy relaxation times at liquid nitrogen temperature.

Similar investigations for a lightly doped drain technology can be found in [57].

The question remains how good these simulation results agree with measurements. The device presented has not been fabricated with 0,51 micron channel length. Satisfactory agreement has been achieved for devices down to 3/4 micron channel length. To be able to judge rigorously the agreement between measurement and simulation at low temperatures one should also look at results obtained with different programs. These can be found, e.g., in [33] for a modified version of CADDET, in [50], [61] for a modified version of FIELDAY and in [65] for a modified version of GEMINI.

ACKNOWLEDGMENT

This work is considerably supported by the research laboratories of SIEMENS AG at Munich, FRG, the research laboratories of DIGITAL EQUIPMENT CORPORATION at Hudson, USA, and the "Fond zur Förderung der wissenschaftlichen Forschung" under contract S43/10. I am indebted to Prof.H.Pötzl for many critical and stimulating discussions.

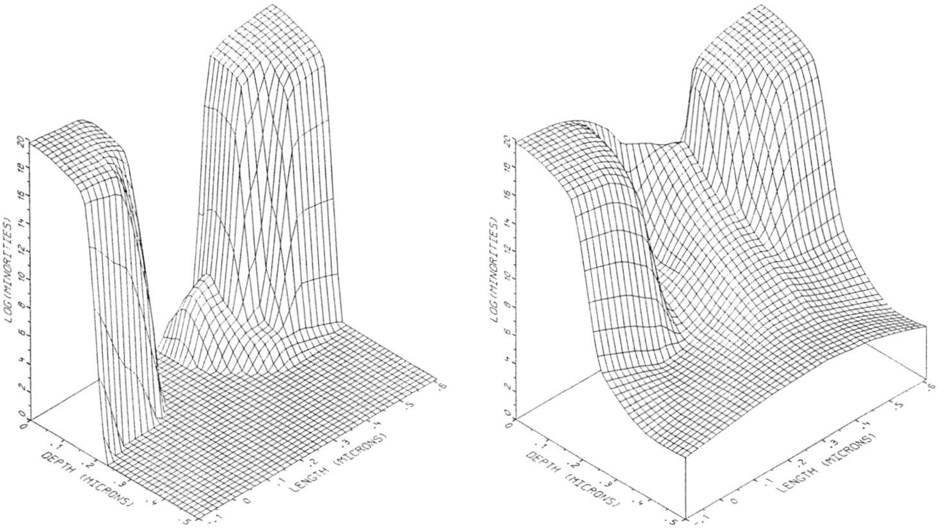

77K Fig.4: Electron Concentration (UG=0V, UD=2V) 300K

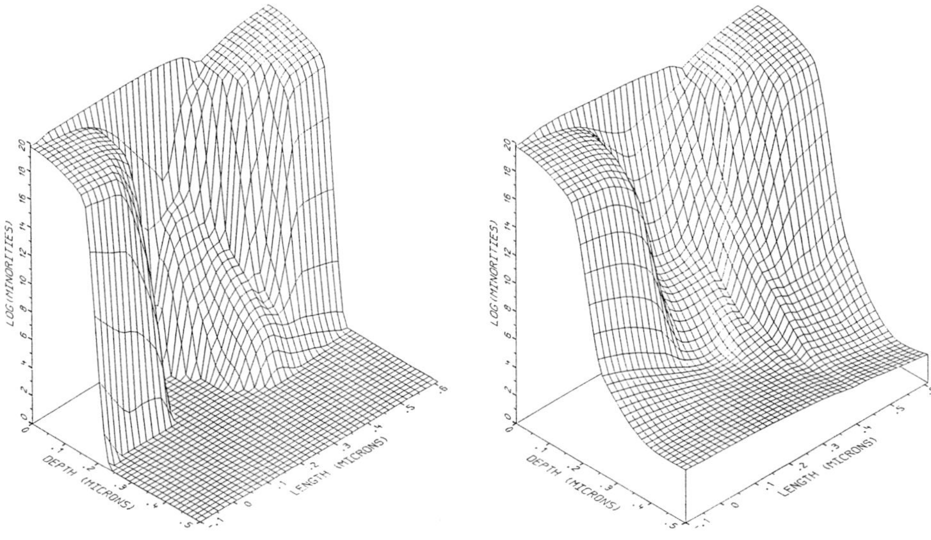

77K Fig.5: Electron Concentration (UG=2V, UD=2V) 300K

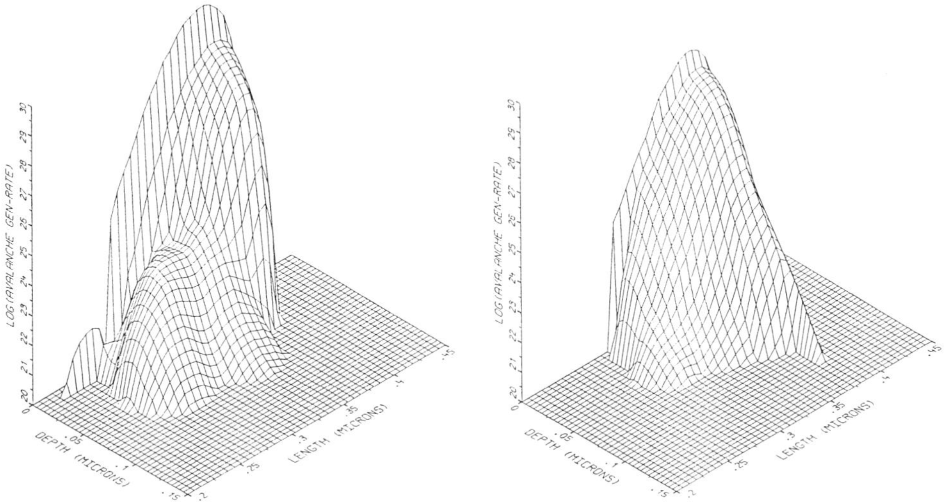

77K Fig.6: Impact Ionization Rate (UG=2V, UD=2V) 300K

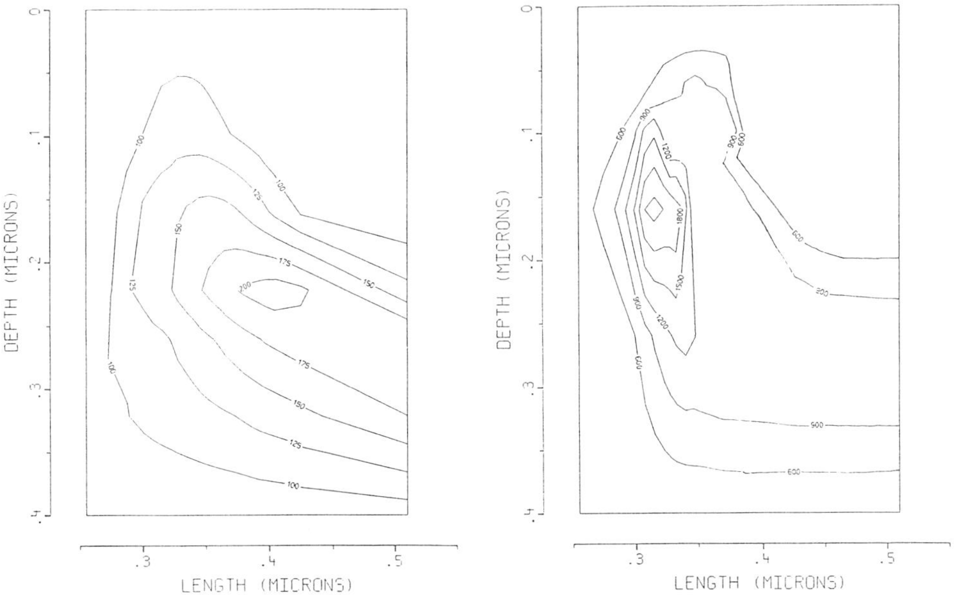

77K Fig.7: Electron Temperature (UG=2V, UD=2V) 300K

REFERENCES

[1] Ahmad N., Arora V.K.,
Velocity-Field Profile of n-Silicon: A Theoretical Analysis,
IEEE Trans.Electron Devices, Vol.ED-33, pp.1075-1077, 1986.

[2] Ali-Omar M., Reggiani L.,
Drift and Diffusion of Charge Carriers in Silicon and Their Empirical Relation to the Electric Field,
Solid-State Electron., Vol.30, pp.693-697, 1987.

[3] Aoki M., Yano K., Masuhara T., Ikeda S., Meguro S.,
Optimum Crystallographic Orientation of Submicron CMOS Devices,
Proc.IEDM, pp.577-580, 1985.

[4] Aoki M., Hanamura S., Masuhara T., Yano K.,
Performance and Hot-Carrier Effects of Small CRYO-CMOS Devices,
IEEE Trans.Electron Devices, Vol.ED-34, pp.8-18, 1987.

[5] Aoki M., Yano K., Masuhara T., Ikeda S., Meguro S.,
Optimum Crystallographic Orientation of Submicrometer CMOS Devices Operated at Low Temperatures,
IEEE Trans.Electron Devices, Vol.ED-34, pp.52-57, 1987.

[6] Arora N.D., Hauser J.R., Roulston D.J.,
Electron and Hole Mobilities in Silicon as a Function of Concentration and Temperature,
IEEE Trans.Electron Devices, Vol.ED-29, pp.292-295, 1982.

[7] Arora N.D., Gildenblat G.S.,
A Semi-Emperical Model of the MOSFET Inversion Layer Mobility for Low-Temperature Operation,
IEEE Trans.Electron Devices, Vol.ED-34, pp.89-93, 1987.

[8] Baccarani G., Wordeman M.R.,
Transconductance Degradation in Thin-Oxide MOSFET's,
Proc.IEDM, pp.278-281, 1982.

[9] Baccarani G.,
Physics of Submicron Devices,
Proc.VLSI Process and Device Modeling, pp.1-23, Katholieke Universiteit Leuven, 1983.

[10] Baccarani G., Wordeman M.R.,
Transconductance Degradation in Thin-Oxide MOSFET's,
IEEE Trans.Electron Devices, Vol.ED-30, pp.1295-1304, 1983.

[11] Baccarani G., Wordeman M.R.,
An Investigation of Steady-State Velocity Overshoot in Silicon,
Solid-State Electron., Vol.28, pp.407-416, 1985.

[12] Blakemore J.S.,
Approximations for Fermi-Dirac Integrals, especially the Function F1/2(x) used to Describe Electron Density in a Semiconductor,
Solid-State Electron., Vol.25, pp.1067-1076, 1982.

[13] Blotekjaer K.,
Transport Equations for Electrons in Two-Valley Semiconductors,
IEEE Trans.Electron Devices, Vol.ED-17, pp.38-47, 1970.

[14] Canali C., Ottaviani G.,
Saturation Values of the Electron Drift Velocity in Silicon between 300K and 4.2K,
Physics Lett., Vol.32A, pp.147-148, 1970.

[15] Canali C., Majni G., Minder R., Ottaviani G.,
Electron and Hole Drift Velocity Measurements in Silicon and Their Empirical Relation to Electric Field and Temperature,
IEEE Trans.Electron Devices, Vol.ED-22, pp.1045-1047, 1975.

[16] Caughey D.M., Thomas R.E.,
Carrier Mobilities in Silicon Empirically Related to Doping and Field,
Proc.IEEE, Vol.52, pp.2192-2193, 1967.

[17] Cheng D.Y., Hwang C.G., Dutton R.W.,
PISCES-MC: A Multiwindow, Multimethod 2-D Device Simulator,
IEEE Trans.CAD of Integrated Circuits and Systems, Vol.CAD-7, pp.1017-1026, 1988.

[18] Chrzanowska-Jeske M., Jaeger R.C.,
Modeling of Temperature Dependent Transport Parameters for Low Temperature Bipolar Transistor Simulation,
Proc.Symposium on Low Temperature Electronics and High Temperature Superconductors, The Electrochemical Society, Vol.88-9, pp.30-38, 1988.

[19] Cody W.J., Thacher H.C.,
Rational Chebyshev Approximations for Fermi-Dirac Integrals of Orders -1/2, 1/2 and 3/2,
Math.Comp., Vol.21, pp.30-40, 1967.

[20] Crowell C.R., Sze S.M.,
Temperature Dependence of Avalanche Multiplication in Semiconductors,
Appl.Phys.Lett., Vol.9, pp.242-244, 1966.

[21] Debye P.P., Conwell E.M.,
Electrical Properties of N-Type Germanium,
Physical Review, Vol.93, pp.693-706, 1954.

[22] Decker D.R., Dunn C.N.,
Temperature Dependence of Carrier Ionization Rates and Saturated Velocities in Silicon,
J.Electronic Mat., Vol.4, pp.527-547, 1975.

[23] Dorkel J.M., Leturcq Ph.,
Carrier Mobilities in Silicon Semi-Empirically Related to Temperature, Doping and Injection Level,
Solid-State Electron., Vol.24, pp.821-825, 1981.

[24] Dziewior J., Schmid W.,
Auger Coefficients for Highly Doped and Highly Excited Silicon,
Appl.Phys.Lett., Vol.31, pp.346-348, 1977.

[25] Faricelli J.,
Private Communication,
1987.

[26] Frey J.,
Transport Physics for VLSI,
in: Introduction to the Numerical Analysis of Semiconductor Devices and Integrated Circuits, pp.51-57, Boole Press, Dublin 1981.

[27] Gaensslen F.H., Jaeger R.C., Walker J.J.,
Low-Temperature Threshold Behavior of Depletion Mode Devices,
Proc.IEDM, pp.520-524, 1976.

[28] Gaensslen F.H., Rideout V.L., Walker E.J., Walker J.J.,
Very Small MOSFET's for Low Temperature Operation,
IEEE Trans.Electron Devices, Vol.ED-24, pp.218-229, 1977.

[29] Gaensslen F.H., Jaeger R.C.,
Temperature Dependent Threshold Behaviour of Depletion Mode MOSFET's,
Solid-State Electron., Vol.22, pp.423-430, 1979.

[30] Gaensslen F.H., Jaeger R.C.,
Behavior of Electrically Small Depletion Mode MOSFET's at Low Temperature,
Solid-State Electron., Vol.24, pp.215-220, 1981.

[31] Gummel H.K.,
A Self-Consistent Iterative Scheme for One-Dimensional Steady State Transistor Calculations,
IEEE Trans.Electron Devices, Vol.ED-11, pp.455-465, 1964.

[32] Hänsch W., Selberherr S.,
MINIMOS 3: A MOSFET Simulator that Includes Energy Balance,
IEEE Trans.Electron Devices, Vol.ED-34, pp.1074-1078, 1987.

[33] Henning A.K., Chan N., Plummer J.D.,
Substrate Current in n-Channel and p-Channel MOSFET's between 77K and 300K: Characterization and Simulation,
Proc.IEDM, pp.573-576, 1985.

[34] Henning A.K., Chan N.N., Watt J.T., Plummer J.D.,
Substrate Current at Cryogenic Temperatures: Measurements and a Two-Dimensional Model for CMOS Technology,
IEEE Trans.Electron Devices, Vol.ED-34, pp.64-74, 1987.

[35] Hess K., Iafrate G.J.,
Theory and Applications of Near Ballistic Transport in Semiconductors,
Proc.IEEE, Vol.76, pp.519-532, 1988.

[36] Heywang W., Pötzl H.,
Bandstruktur und Stromtransport,
Springer, Berlin, 1976.

[37] Hiroki A., Odanaka S., Ohe K., Esaki H.,
A Mobility Model for Submicrometer MOSFET Device Simulations,
IEEE Electron Device Lett., Vol.EDL-8, pp.231-233, 1987.

[38] Hiroki A., Odanaka S., Ohe K., Esaki H.,
A Mobility Model for Submicrometer MOSFET Simulations Including Hot-Carrier-Induced Device Degradation,
IEEE Trans.Electron Devices, Vol.ED-35, pp.1487-1493, 1988.

[39] Jaeger R.C.,
Private Communication,
October 1987.

[40] Jaeger R.C., Gaensslen F.H.,
Low Temperature MOS Microelectronics,
Proc.Symposium on Low Temperature Electronics and High Temperature Superconductors, The Electrochemical Society, Vol.88-9, pp.43-54, 1988.

[41] Jaggi R., Weibel H.,
High-Field Electron Drift Velocities and Current Densities in Silicon,
Helv.Phys.Acta, Vol.42, pp.631-632, 1969.

[42] Jaggi R.,
High-Field Drift Velocities in Silicon and Germanium,
Helv.Phys.Acta, Vol.42, pp.941-943, 1969.

[43] Kamgar A.,
Miniaturization of Si MOSFET's at 77K,
IEEE Trans.Electron Devices, Vol.ED-29, pp.1226-1228, 1982.

[44] Kinugawa M., Kakumu M., Usami T., Matsunaga J.,
Effects of Silicon Surface Orientation on Submicron CMOS Devices,
Proc.IEDM, pp.581-584, 1985.

[45] Lau D., Gildenblat G., Sodini G.G., Nelson D.E.,
Low-Temperature Substrate Current Characterization of n-Channel MOSFET's,
Proc.IEDM, pp.565-568, 1985.

[46] Li S.S., Thurber W.R.,
The Dopant Density and Temperature Dependence of Electron Mobility and Resistivity in n-Type Silicon,
Solid-State Electron., Vol.20, pp.609-616, 1977.

[47] Nishida T., Sah C.T.,
A Physically Based Mobility Model for MOSFET Numerical Simulation,
IEEE Trans.Electron Devices, Vol.ED-34, pp.310-320, 1987.

[48] Okuto Y., Crowell C.R.,
Threshold Energy Effect on Avalanche Breakdown Voltage in Semiconductor Junctions,
Solid-State Electron., Vol.18, pp.161-168, 1975.

[49] Robertson P.J., Dumin D.J.,
Ballistic Transport and Properties of Submicrometer Silicon MOSFET's from 300 to 4.2K,
IEEE Trans.Electron Devices, Vol.ED-33, pp.494-498, 1986.

[50] Sai-Halasz G.A.,
Processing and Characterization of Ultra Small Silicon Devices,
Proc.ESSDERC Conf., pp.71-80, 1987.

[51] Schütz A., Selberherr S., Pötzl H.,
Analysis of Breakdown Phenomena in MOSFET's,
IEEE Trans.CAD of Integrated Circuits and Systems, Vol.CAD-1, pp.77-85, 1982.

[52] Seavey M.,
Private Communication,
1987.

[53] Seeger K.,
Semiconductor Physics,
Springer, Wien, 1973.

[54] Selberherr S., Schütz A., Pötzl H.,
MINIMOS - A Two-Dimensional MOS Transistor Analyzer,
IEEE Trans.Electron Devices, Vol.ED-27, pp.1540-1550, 1980.

[55] Selberherr S.,
Analysis and Simulation of Semiconductor Devices,
Springer, Wien New-York, 1984.

[56] Selberherr S., Griebel W., Pötzl H.,
Transport Physics for Modeling Semiconductor Devices,
in: Simulation of Semiconductor Devices and Processes, pp.133-152, Pineridge Press, Swansea, 1984.

[57] Selberherr S.,
Low Temperature Mos Device Modeling,
Proc.Symposium on Low Temperature Electronics and High Temperature Superconductors, The Electrochemical Society, Vol.88-9, pp.43-86, 1988.

[58] Shahidi G.G., Antoniadis D.A., Smith H.I.,
Electron Velocity Overshoot at 300K and 77K in Silicon MOSFET's with Submicron Channel Length,
Proc.IEDM, pp.824-825, 1986.

[59] Solomon P.M.,
Options for High Speed Logic at 77K,
Proc.Symposium on Low Temperature Electronics and High Temperature Superconductors, The Electrochemical Society, Vol.88-9, pp.3-17, 1988.

[60] Sugano T.,
Low Temperature Electronics Research in Japan,
Proc.Symposium on Low Temperature Electronics and High Temperature Superconductors, The Electrochemical Society, Vol.88-9, pp.18-29, 1988.

[61] Sun Y-C.J., Taur Y., Dennard R.H., Klepner S.P.,
Submicrometer-Channel CMOS for Low-Temperature Operation,
IEEE Trans.Electron Devices, Vol.ED-34, pp.19-27, 1987.

[62] Sutherland A.D.,
An Improved Empirical Fit to Baraff's Universal Curves for the Ionization Coefficients of Electron and Hole Multiplication in Semiconductors,
IEEE Trans.Electron Devices, Vol.ED-27, pp.1299-1300, 1980.

[63] Sze S.M.,
Physics of Semiconductor Devices,
Wiley, New York, 1969.

[64] Watt J.T., Fishbein B.J., Plummer J.D.,
A Low-Temperature NMOS Technology with Cesium-Implanted Load Devices,
IEEE Trans.Electron Devices, Vol.ED-34, pp.28-38, 1987.

[65] Woo J.C.S., Plummer J D.,
Short Channel Effects in MOSFET's at Liquid-Nitrogen Temperature,
IEEE Trans.Electron Devices, Vol.ED-33, pp.1012-1019, 1986.

Physical Models for Compound Semiconductor Devices

Michael Shur
University of Minnesota

Introduction.

GaAs technology has a significant advantage over silicon in terms of speed and radiation hardness. However, compound semiconductor technology is much less developed. Device and integrated circuit fabrication is expensive and the fabrication cycle usually takes several weeks. All this makes accurate simulation of device fabrication and realistic device and circuit modeling especially important for compound semiconductor technology, even more so than for its silicon counterpart.

In this chapter we will describe analytic models for GaAs MEtal Semiconductor Field Effect Transistors (MESFETs) and heterostructure FETs (HFETs) that have been incorporated into circuit simulators and used for a Computer-Aided Design (CAD) and optimization of GaAs devices and circuits. These models are based on the concept of charge control. According to a charge control model, the current in the channel of a field-effect transistor is controlled by the charge induced into the channel by the gate voltage. The device capacitances are found as derivatives of this charge with respect to the terminal voltages. Such a model can be modified to account for important non-ideal effects, such as space charge injection into a substrate, channel length modulation, effect of traps, effect of the piezoelectric stress, etc.

Charge-control model for GaAs MESFETs.

Silicon has a very stable natural oxide that can be grown with a very low density of traps, and the problems related to the gate insulators, such as hot electron trapping, threshold voltage shift due to charge trapped in the gate insulator, etc. have been minimized for silicon Metal Oxide Semiconductor Field Effect Transistors (MOSFETs) . However, compound semiconductors, such as GaAs, do not have a stable oxide, so most of compound semiconductor field effect transistors use a Schottky gate. In GaAs MESFETs the Schottky barrier gate contact to the GaAs channel is used in order to modulate channel conductivity (see Fig. 1). The drawback of MESFET technology is a limitation related to the gate voltage swing, limited by the turn-on voltage of the Schottky gate. However, this limitation is less important in low power circuits operating with a low supply voltage. Low-noise microwave GaAs transistors operate at negative gate voltages so that the inability to apply large positive gate voltages does not present a problem.

The depletion region under the gate of a MESFET for a finite drain-to-source voltage is schematically shown in Fig. 1. The depletion region is wider closer to the drain because the positive drain voltage provides an additional reverse bias across channel-to-gate junction. The analytical expressions describing the shape of the depletion region and the device current-voltage characteristics may be found using the so-called gradual channel approximation. According to the gradual channel approximation, the thickness of the depletion region at every point under the gate may be found using conventional one dimensional equations for the metal-semiconductor junction (in case of MESFETs) or a p-n junction (in case of JFETs) by assuming the potential drop $V_G - V(x)$ across the junction where V_G is the gate potential, $V(x)$ is the channel potential, and x is the longitudinal coordinate (i.e. coordinate along the channel). This approximation is valid when the derivative of the electric field with respect to the coordinate along the channel is much smaller than the derivative of the electric field in the depletion region with respect to the coordinate perpendicular to the channel. We also assume that the conducting channel is neutral and the space charge region between the gate and the channel is totally depleted. The boundary between the neutral conducting channel and the depleted region is assumed to

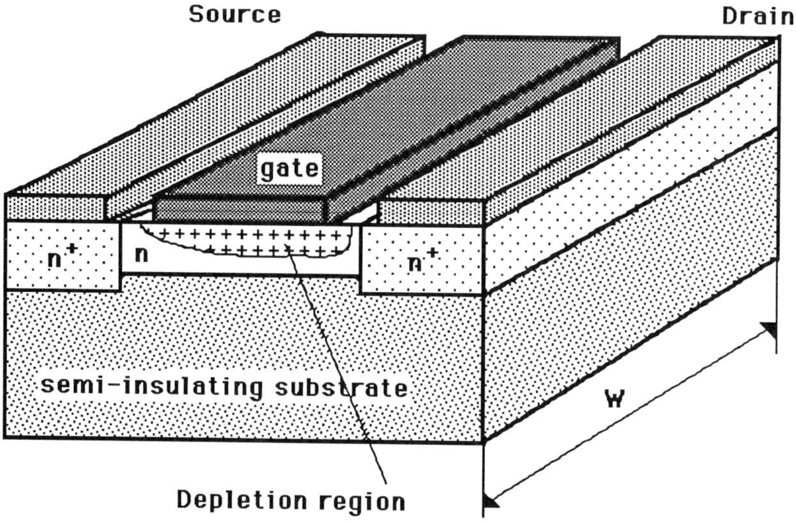

Fig. 1. Schematic diagram of GaAs MESFET

be sharp. (In fact, the transition region between the conducting channel and the space charge region is on the order of three Debye radii, $L_D = [\varepsilon k_B T/(q^2 N_d)]^{1/2}$. Here ε is the dielectric permittivity, k_B is the Boltzmann constant, T is the device temperature (in degrees K), q is the electronic charge, N_d is the concentration of ionized shallow donors in the channel that is equal to the electron density, n.)

According to the gradual channel approximation the potential across the channel varies slowly with distance, x, so at each point the width of the depleted region can be found from the solution of Poisson's equation valid for a one-dimensional junction. We first consider a constant mobility model (also called the Shockley model), i. e. we assume that the electron velocity, v, in the channel is proportional to the electric field, F ($v = \mu F$ where μ is the electron mobility). Using these assumptions we find the incremental change of channel potential

$$dV = I_{ds} dR = \frac{I_{ds} dx}{q \mu_n N_d W [A - A_d(x)]} \qquad (1)$$

where I_{ds} is the channel current, dR is the incremental channel resistance, A is the thickness of the active layer, x is the coordinate (x = 0 corresponds to the source side of the channel and x=L corresponds to the drain side of the channel), $A_d(x) = \{2\varepsilon[V(x) + V_{bi} - V_g]/(qN_d)\}^{1/2}$ is the thickness of the depletion layer, V_{bi} is the built-in voltage of the Schottky barrier, V_g is the gate potential, V(x) is the potential of the neutral channel (here we neglect the variation of the potential in the conducting channel in the direction y perpendicular to x), and W is the gate width. Substituting the expression for $A_d(x)$ into eq. (1) and integrating with respect to x from x = 0 to x = L we derive the equation called the fundamental equation of MESFETs:

$$I_{ds} = g_o\{V_D - \frac{2\,[(V_D + V_{bi} - V_g)^{3/2} - (V_{bi} - V_g)^{3/2}]}{3\,V_{po}^{1/2}}\} \qquad (2)$$

where $g_o = q\mu_n N_D WA/L$ is the conductance of the undepleted doped channel, L is the gate length, $V_{po} = qN_D A^2/2\varepsilon$ is the channel pinch-off voltage, and V_D is the voltage drop in the channel under the gate.

Eq. (2) is applicable only for such values of the gate voltage, V_G, and drain-to-source voltage, V_D, that the neutral channel still exists even in the most narrow spot at the drain side ,i.e., $A_d(L) < A$. If we neglect the effects of velocity saturation in the channel, then we may assume that when the pinch-off condition, $A_d(L) = A$, is reached, the drain-to-source current saturates. Hence, the saturation voltage, V_{Dsat}, predicted by the constant mobility model is given by $V_{Dsat} = V_{po} - V_{bi} + V_g$. At this voltage, the conducting channel is pinched-off (i.e. has zero cross-section) at the drain side. Thus, the electron velocity has to be infinitely high in order to maintain a finite drain-to-source current.

In reality, the electron velocity saturates in a high electric field. The velocity saturation leads to the saturation of the current at smaller values of the drain voltage when the channel cross-section calculated using the gradual channel approximation is still finite. The importance of the field dependence of the electron mobility for the understanding of the current saturation in field effect transistors was first mentioned by Dasey and Ross (1955). This concept was later developed in many theoretical models used to describe FET characteristics and interpret experimental results. Here we consider a very simple two piece-wise linear approximation for the field dependence of electron velocity ($v = \mu F$ for $F<F_s$ and $v = v_s$ for $F \geq F_s$ where F_s is the velocity saturation field).

Velocity saturation is first reached at the drain side of the gate (i.e at the point x =L) where the electric field is the highest. It occurs when $F(L) = F_s$. Using this condition one can calculate the drain-to-source saturation voltage and the drain saturation current. Though the results can be only obtained in a numerical form, the dependence of the drain saturation current on the gate voltage can be fairly close approximated by the following interpolation formula (see, for example, Shur (1987))

$$(I_{ds})_{sat} = \beta(V_g - V_T)^2 \qquad (3)$$

where the transconductance parameter

$$\beta = \frac{2\varepsilon\mu v_s W}{A(\mu V_{po} + 3 v_s L)} \qquad (4)$$

and $V_T = V_{bi} - V_{po}$. Eq. (4) can be used to determine the dependence of β and device transconductance on channel doping, gate length, electron mobility, and saturation velocity. As can be seen from eq. (4) the values of β (and, hence, the values of the transconductance for a given voltage swing) increase with the decrease of active layer thickness and with the increase of doping. This increase of β is accompanied by a similar increase in gate capacitance, $C_g = \varepsilon LW/A$. However, parasitic capacitances do not increase with an increase in doping or with a decrease in device thickness, hence thin and highly doped active layers should lead to a higher speed of operation. In highly doped devices the active layer thickness (for a given value of pinch-off voltage) is reduced. This allows one to minimize short channel effects, which become quite noticeable when $L/A < 3$ (see Dambkes et al.(1983)).

The "intrinsic" device transconductance $g_{mo} = d(I_{ds})_{sat}/dV_g = 2\beta(V_g - V_T)$. (Eqs. (3) and (4) do not account for parasitic series resistances.) We can roughly estimate the maximum transconductance, g_m, as βV_{po}. Using eq. (4), we find

$$g_m = g_{mo}\frac{1}{1 + 3 v_s L/\mu V_{po})} \qquad (5)$$

where $g_{mo} = (4/3)\varepsilon v_s W/A$. The dependence of g_m/g_{mo} on L for different values of V_{po} is shown in Fig. 2.

As can be seen from the figure, the transconductance increases with an increase in the gate length and with the increase in the pinch-off voltage (for a fixed channel depth).

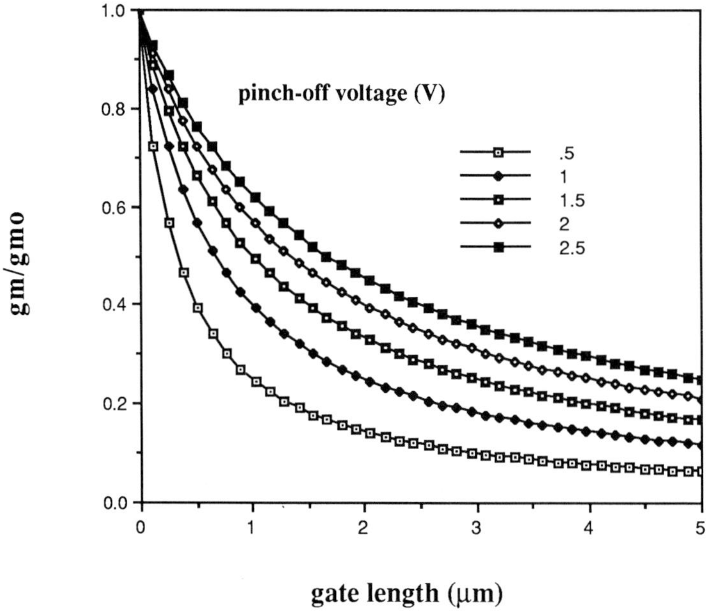

Fig. 2. Dependence of dimensionless transconductance on gate length for different pinch-off voltages. ($v_s = 2x10^5$ m/s.)

Eq. (4) shows that low-field mobility becomes increasingly important in low pinch-off voltage devices (se Fig. 3). For enhancement mode devices with low pinch-off voltages, the high values of low field mobility in GaAs (up to 4500 cm^2/Vs in highly doped active layers compared to silicon where $\mu \approx$ 1000 - 1200 cm^2/Vs) lead to a substantial improvement in performance even for short channel devices where velocity saturation effects are very important.

In practice the "square law" (i.e. eq. (2-37)) is fairly accurate for devices with relatively low pinch-off voltages ($V_{po} \leq 1.5 \sim 2$ V) . For devices with higher pinch-off voltages an empirical expression proposed by Statz et al. (1987)

$$(I_{ds})_{sat} = \frac{\beta(V_G - V_T)^2}{1 + b(V_G - V_T)} \qquad (6)$$

provides an excellent fit to the experimental data.

The analysis given above applies only to the values of drain-to-source current and transconductance at the onset of the current saturation when the electron velocity becomes equal to the electron saturation velocity at the drain side of the gate. At higher drain-to-source voltages the electric field in the conducting channel increases, leading to velocity saturation in a larger fraction of the channel (so-called gate length modulation effect). The implications of gate length modulation were considered, for example, by Grebene and Ghandi (1969) and Pucel et al. (1975).

Source and drain series resistances, R_s and R_d, may play an important role in determining the current-voltage characteristics of GaAs MESFETs. These resistances can be taken into account by substituting $V_g = V_{GS} - (I_{ds})_{sat}R_s$ (where V_{GS} is the gate-to-source voltage) into eq.(3) and solving for $(I_{ds})_{sat}$

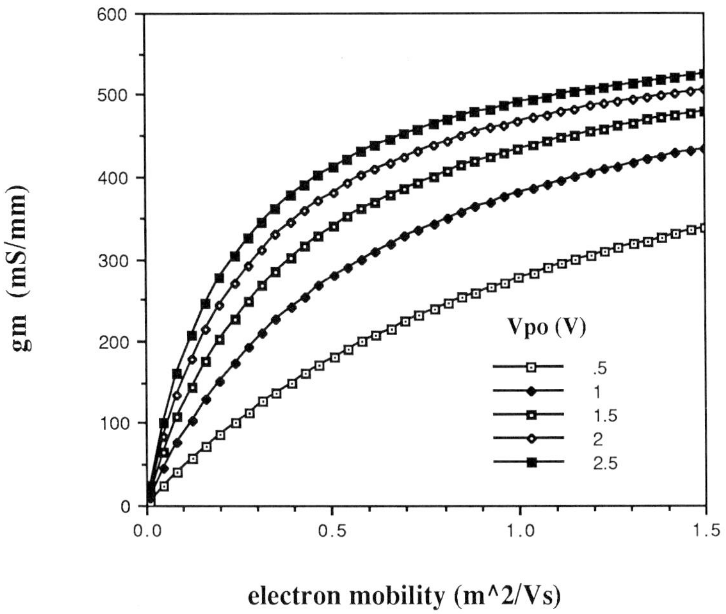

Fig. 3. Maximum transconductance ($g_m = 2\beta V_{po}$) versus electron mobility for different values of pinch-off voltage.
Parameters used in the calculation: $v_s = 2x10^5$ m/s, A=0.05 μm, $\varepsilon = 1.14x10^{-10}$ F/m

$$(I_{ds})_{sat} = \frac{1 + 2\beta R_s(V_{GS} - V_T) - [1 + 4\beta R_s(V_{GS} - V_T)]^{1/2}}{2\beta R_s^2} \tag{7}$$

Curtice (1980) proposed the use a hyperbolic tangent function for the interpolation of MESFET current-voltage characteristics. Based on this idea, Shur (1985) used the following expression:

$$I_{ds} = \beta(V_g - V_T)^2 (1 + \lambda V_{DS}) \tanh[g_{ch} V_{DS}/(I_{ds})_{sat}] \tag{8}$$

where $g_{ch} = g_{chi}/[1 + g_{chi}(R_s + R_d)]$ is the channel conductance at low drain-to-source voltages, and $g_{chi} = g_o\{1 - [(V_{bi} - V_G)/V_{po}]^{1/2}\}$ is the intrinsic channel conductance at low drain-to-source voltages predicted by the Shockley model. The drain-to-source voltage is given by $V_{DS} = V_D + I_{ds}(R_s + R_d)$ where V_D is the "intrinsic" drain voltage.

As previously mentioned, gate current also plays an important role in GaAs MESFET circuits. It becomes important when a positive gate voltage forward biases the Schottky gate junction (see Fig.4). An accurate analytical model for the gate current has not yet been developed. A practical

approach used in order to fit the experimental data and to account for gate current in circuit simulations is to introduce two equivalent diodes connecting the gate contact with the source and drain contacts respectively and use the empirical diode equations.

Fig. 5 compares the I_{DS} vs. V_{DS} characteristics calculated using this model with the experimental data. MESFET device parameters used in these calculations were determined from

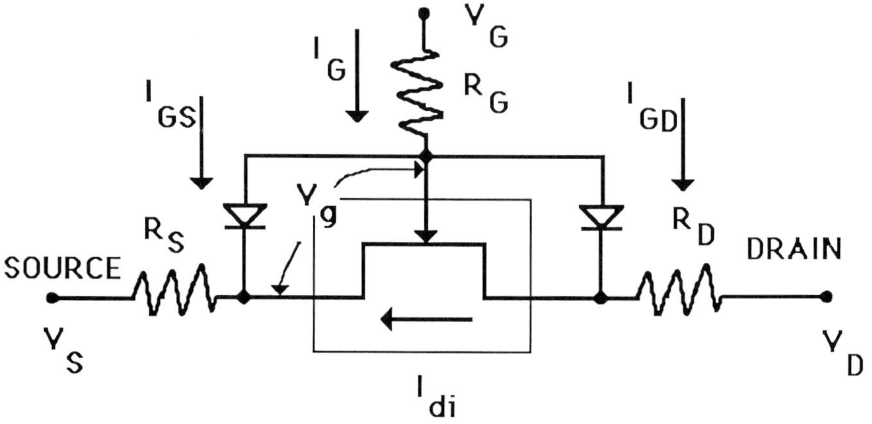

Fig. 4. Equivalent circuit of GaAs MESFET.

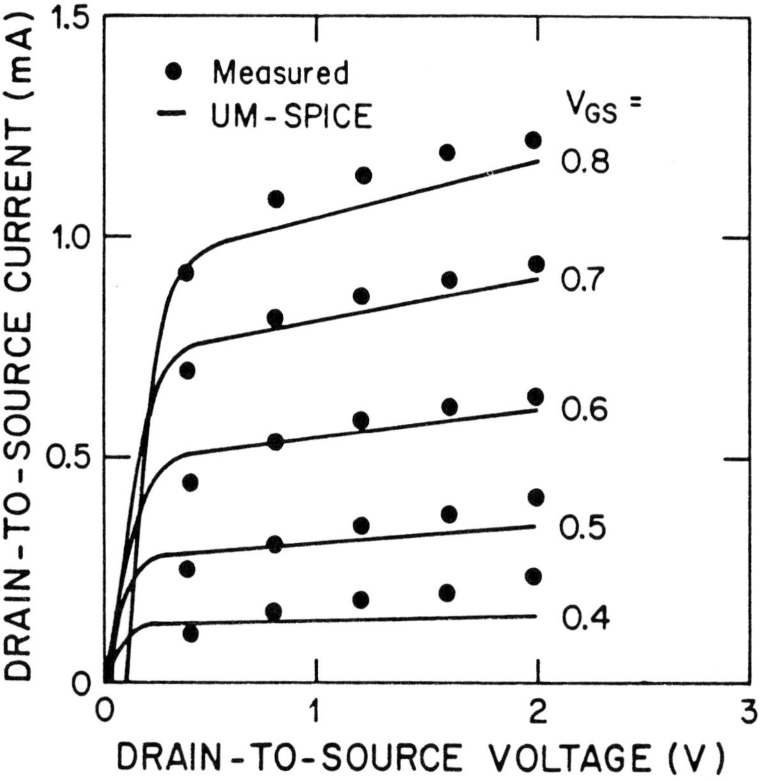

Fig. 5. Current-voltage characteristics of GaAs MESFET (from Hyun et al. (1986)).

the measured I-V characteristics using the following procedure. First, an appropriate value of λ was chosen from the I_{DS} vs. V_{DS} curve for one value of V_{GS}. Then saturation currents were extrapolated to $V_{DS}=0$ for all values of V_{GS}, keeping λ constant. To obtain the source resistance R_s, we plotted $\sqrt{I_{DS}}$ vs. $V_{GS} - I_{DS}R_s$ for different values of R_s, until a best least square fit is obtained. The slope and intercept of this line gave us β and V_T respectively. The value of built-in voltage, V_{bi}, was determined from the gate I-V characteristic. Channel thickness and doping were determined using the value of $V_{po} = V_{bi} - V_T = qN_dA^2/(2\varepsilon)$ and the implant dose (i.e. N_dA) data. Assuming $R_s=R_d$, we obtained the intrinsic channel resistance, R_i, from the slope of the I_{DS} vs. V_{DS} characteristic in the linear region at large V_{GS}: $R_i = R_{DS} - R_s - R_d$ where R_i is related to the gate voltage as follows

$$R_i = \frac{L}{qA\mu N_d W\{1 - [(V_{bi} - V_{GS})/V_{po}]^{1/2}\}} \tag{9}$$

This equation was used to determine μ. Once μ was known, the saturation velocity was calculated using eq. (3).

For more accurate device characterization, more sophisticated characterization techniques may be required such as "end" resistance measurements (see Lee et al. (1985) and Yang and Long (1986)), channel resistance measurements (see Hower and Bechtel (1973)), Gated Transmission Line Model (GTLM) measurements (see Baier et al. (1985)), and geometric magnetoresistance measurements (see Jay and Wallis (1981)).

The ability to approximate experimental data and the simple procedure for parameter determination make the simple "square law" model attractive for use in computer simulations of GaAs MESFET circuits with relatively low pinch-off voltage FETs. However, this model does not work very well for some ion-implanted devices where the non-uniformity of the doping profile may be important. The transfer characteristics (i.e $(I_{ds})_{sat}$ vs. V_{GS} curves) for transistors with low pinch-off voltages are relatively insensitive to the doping profiles. However, the channel conductance at low drain-to-source voltages is very dependent on the doping profile. Peczalski et al. (1987) proposed an extension of the analytical model described in this section that takes into account the effects of the non-uniform doping profile on the channel conductance. This improved model is in good agreement with experimental data for a large variety of different devices.

The analytical model discussed above is suitable for Computer-Aided Design of GaAs MESFETs and GaAs MESFET circuits. However, this model does not explicitly take into account many important and complicated effects, such as deviations from the gradual channel approximation (which may be especially important at the drain side of the channel), see Pucel (1975), possible formation of a high field region (i.e. a dipole layer) at the drain side of the channel (see Engelman and Liehti (1977), Shur and Eastman (1978), and Fjeldly (1986)), inclusion of diffusion and incomplete depletion at the boundary between the depletion region and conducting channel (see Yamaguchi and Kodera (1976)), ballistic or overshoot effects (Ruch (1972), Maloney and Frey (1975), Shur (1976), Warriner (1977), Cappy (1980)), effects of donor diffusion from the n^+ contact regions into the channel (Chen et al. (1987)), effects of the passivating silicon nitride layers (Asbeck et al. (1984) and Chen et al. (1987)), and effects related to traps (Chen et al. (1986)). These effects may be still included indirectly by adjusting the model parameters such as μ, v_s, N_d, etc. In a rigorous way, these effects can be only treated using numerical solutions. Such solutions provide an insight into the device physics. However, for a practical circuit simulator used in the the circuit design analytical or very simple numerical models remain a must. Besides considerations related to computer time involved in the simulation of hundreds transistors in a circuit, simple models make device characterization easier because of a relatively small number of parameters. Hence, these parameters can be readily measured and compared for different wafers, fabrication processes, etc.

In a short channel GaAs MESFET, effects related to the overshoot transport become especially important. These effects occur because the electron transit time under the gate becomes comparable to the energy relaxation time. Their importance was recently demonstrated by Snowden and Loret (1987) who compared the current-voltage characteristics of a GaAs MESFET, with 0.3 micron gate length, with and without taking into account the overshoot effects. They showed that considerably larger currents were predicted by a more accurate model.

Internal device capacitances play a role in determining the speed of GaAs MESFETs and GaAs MESFET circuits. The simplest approach is to model gate-to-source capacitance, C_{gs}, and gate-to-drain capacitance, C_{ds}, as capacitances of equivalent Schottky barrier diodes connected between the gate and source and drain respectively. However, this model is inadequate for gate voltages smaller than the threshold voltage. Indeed, below the threshold the channel under the gate is totally depleted and the variation of the charge under the gate is only related to the fringing (sidewall) capacitance (see Fig. 6).

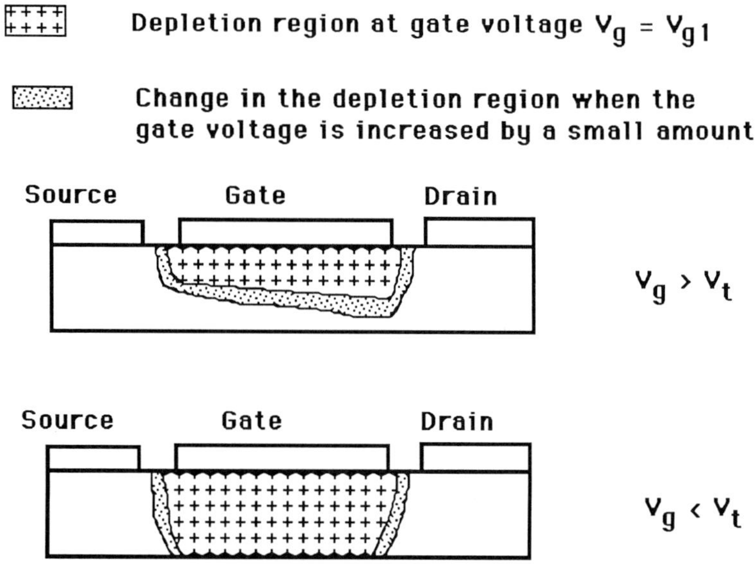

Fig. 6. Incremental changes of depletion region in GaAs MESFET.

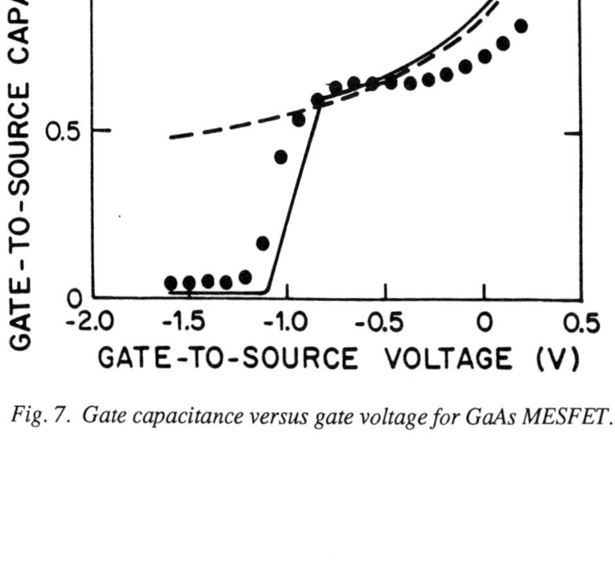

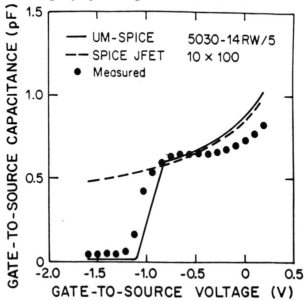

Fig. 7. Gate capacitance versus gate voltage for GaAs MESFET.

A better capacitance model was proposed by Takada et al. (1982). His model assumes that at gate voltages above the threshold the device capacitances are equal to the Schottky gate capacitances and the fringing sidewall capacitances. Below the threshold, the internal capacitances are equal to the fringing sidewall capacitances. This model is in fairly good agreemet with experimental data for low-pinch-off voltage devices ($V_{po} \leq 2V$ or so), see Fig. 7. For devices with larger pinch-off voltages, a

more sophisticated model for GaAs MESFET capacitances that takes into account non-uniformity of the doping profile, the effect of traps in a semi-insulating substrate, and a possible formation of a high field domain (a dipole layer) at the drain side of the gate was developed by Chen and Shur (1985). This model predicts the complicated dependence of the gate-to-source capacitance and gate-to-drain capacitance on the gate and drain voltages that qualitatively agree with experimental data.

The analytic model described above was implemented into a GaAs circuit simulator (called UM-SPICE) and used for the design of GaAs digital integrated circuits (see Hyun et al. (1986)). As an example, we show in Fig. 8. the dependence of the propagation delay of a GaAs MESFET ring oscillator on the load current computed using UM-SPICE.

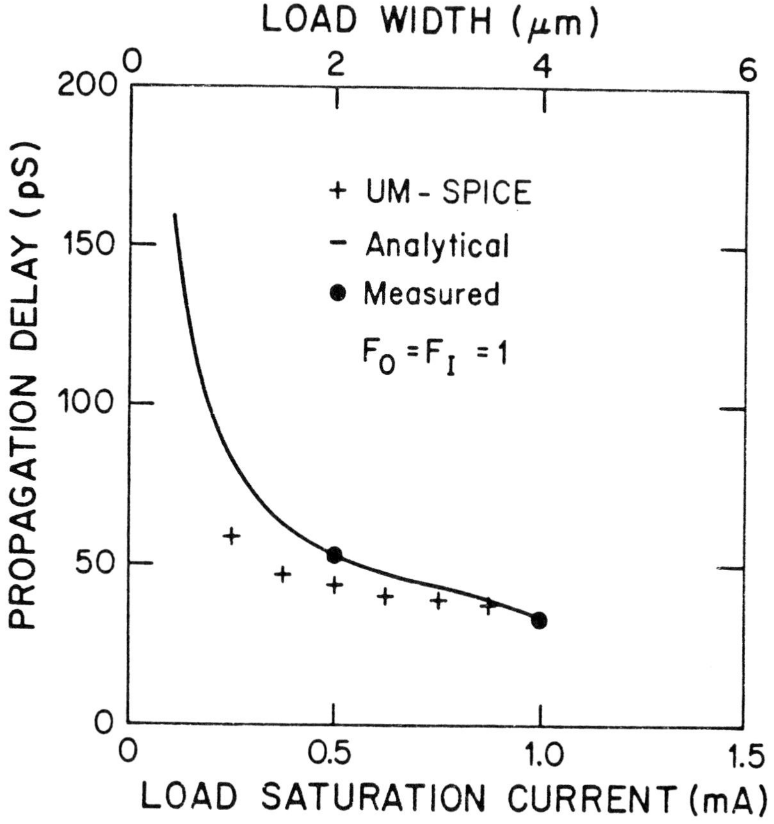

Fig. 8. Dependence of propagation delay of GaAs MESFET ring oscillator on the load current computed using UM-SPICE.

Models for Heterostructure Field Effect Transistors.

A schematic cross-section of an n-channel HFET and the band diagram of the structure are shown in Fig. 9 . In this structure, a two dimensional electron gas is formed in the unintentionally

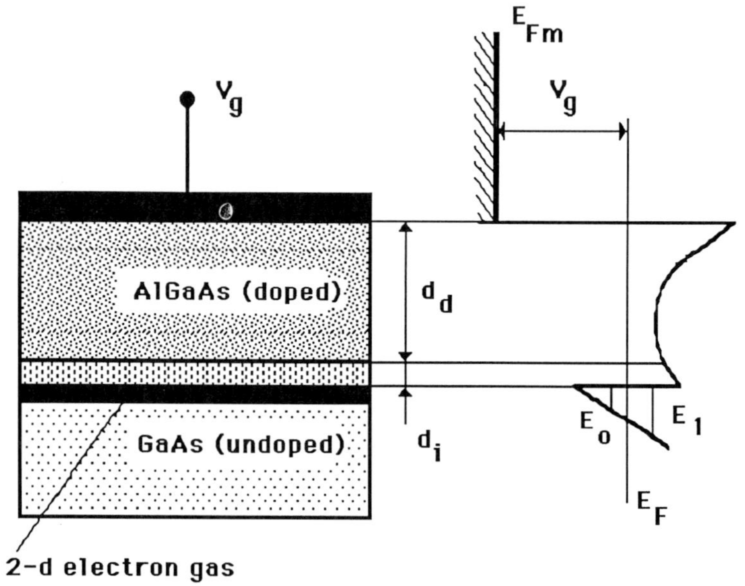

Fig. 9. Modulation doped structure

doped GaAs buffer layer at the heterointerface. More recently, complementary n- and p-channel Heterostructure Insulated Gate Field Effect Transistors were developed offering potential for high-speed low-power operation (see Cirillo et al. (1985), Daniels et al. (1986), Kiehl et al. (1987), Daniels et al. (1988), and Ruden et al. (1989a)).

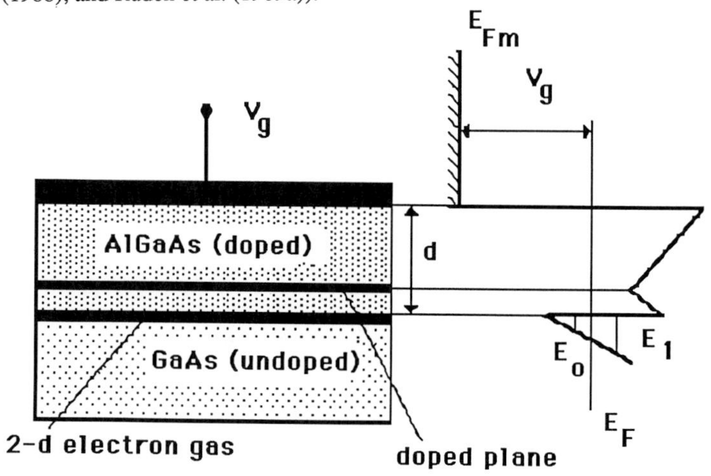

Fig. 10. Planar doped modulation doped structure.

A different type of an HFET utilizes a plane of dopants in the AlGaAs layer. The band diagram and a schematic cross-section of this device is shown in Fig. 10. This device has two distinct advantages. First of all, doping of AlGaAs in conventional HFETs leads to many undesirable effects related to traps associated with dopants (see, for example, Shur (1987), p. 583). In planar doped devices (sometimes called delta-doped devices) the effect of traps is diminished. The same goal -- trying to diminish the effects of traps in the AlGaAs layer -- led to the development of Superlattice HFETs (see Baba (1983), Shah et al. (1986), Shur et al. (1987)). In this device dopants are incorporated into several thin GaAs quantum wells that form a superlattice structure in the AlGaAs layer.

However, all these different HFETs operate in a similar way. A two-dimensional gas of carriers is induced at the heterointerface. The density of this gas and, hence, the device current is modulated by the gate voltage. The composition and doping profile of the AlGaAs layer (which is sometimes called the charge control layer) determine the device threshold voltage and gate voltage swing.

Pseudomorphic HFETs where the two-dimensional electron gas is induced into a thin quantum well of $In_YGa_{1-Y}As$ sandwiched between $Al_XGa_{1-X}As$ have demonstrated a superior performance because of higher electron mobility and velocity in $In_YGa_{1-Y}As$ compared to GaAs (see, for example, Chao et al. (1989)).

The charge density of the two-dimensional electron gas, in an n-channel device, n_s, and the charge density of the two-dimensional hole gas, p_s, versus the gate voltage may be calculated numerically by solving Poisson's equation and the Schrödinger equation in a self-consistent manner (see Stern (1974), Lee et al. (1983), Baek et al. (1987)). To a first order, the concentration of two-dimensional electron gas induced into the device channel at zero drain-to-source voltage is given by (see, for example, Drummond et al. (1982))

$$n_s \approx C_o(V_g - V_t)/q \qquad (10)$$

Here $C_o \approx \varepsilon/(d + \Delta d)$ is the gate capacitance per unit area,

$$V_t = \phi_b - \Delta E_c - (q/\varepsilon) \int_0^d N_d(x)dx \qquad (11)$$

is the threshold voltage, ϕ_b is the metal barrier height, N_d is the doping density in the $Al_XGa_{1-X}As$ layer, q is the electronic charge, d is the distance between the gate and heterointerface, Δd is the effective thickness of the two-dimensional gas, x is the space coordinate (x =0 at the boundary between the gate and AlGaAs layer), $\varepsilon = \kappa\varepsilon_o$ is the dielectric permittivity of $Al_XGa_{1-X}As$, $\varepsilon_o = 8.854\times10^{-14}$ F/cm, $\kappa \approx 13.18 - 3.12X$, X is the molar fraction of Al in $Al_XGa_{1-X}As$, $\Delta E_c = \Delta E_g - \Delta E_v$ is the conduction band discontinuity, ΔE_g is the energy gap difference. For an $Al_XGa_{1-X}As/In_YGa_{1-Y}As$ system (see Landolt-Börnstein (1986)) $\Delta E_v = 0.4\Delta E_{gg}$, ΔE_{gg} (eV) = $1.247X + 1.5Y - 0.4Y^2$, $\Delta E_g = \Delta E_{gg}$ for $X < 0.45$ and ΔE_g (eV) = $0.476 + 0.125X + 0.143X^2 + 1.5Y - 0.4Y^2$ for $X \geq 0.45$.

For a uniformly doped AlGaAs layer ($N_d(x)$ = const), the threshold voltage, found from eq. (11), is given by

$$V_t = \phi_b - \Delta E_c - qN_d d_d^2/(2\varepsilon) \qquad (12)$$

where d_d is the thickness of the doped AlGaAs layer. For a planar-doped structure, the evaluation of the integral in eq. (4-3) yields

$$V_t = \phi_b - \Delta E_c - qn_d d_d/\varepsilon \qquad (13)$$

where d_d is the distance between the metal gate and the doped plane, and n_d is the surface concentration of donors in the doped plane. The calculated dependences of V_t on d_d for uniformly doped and planar-doped devices are compared in Fig. 11.

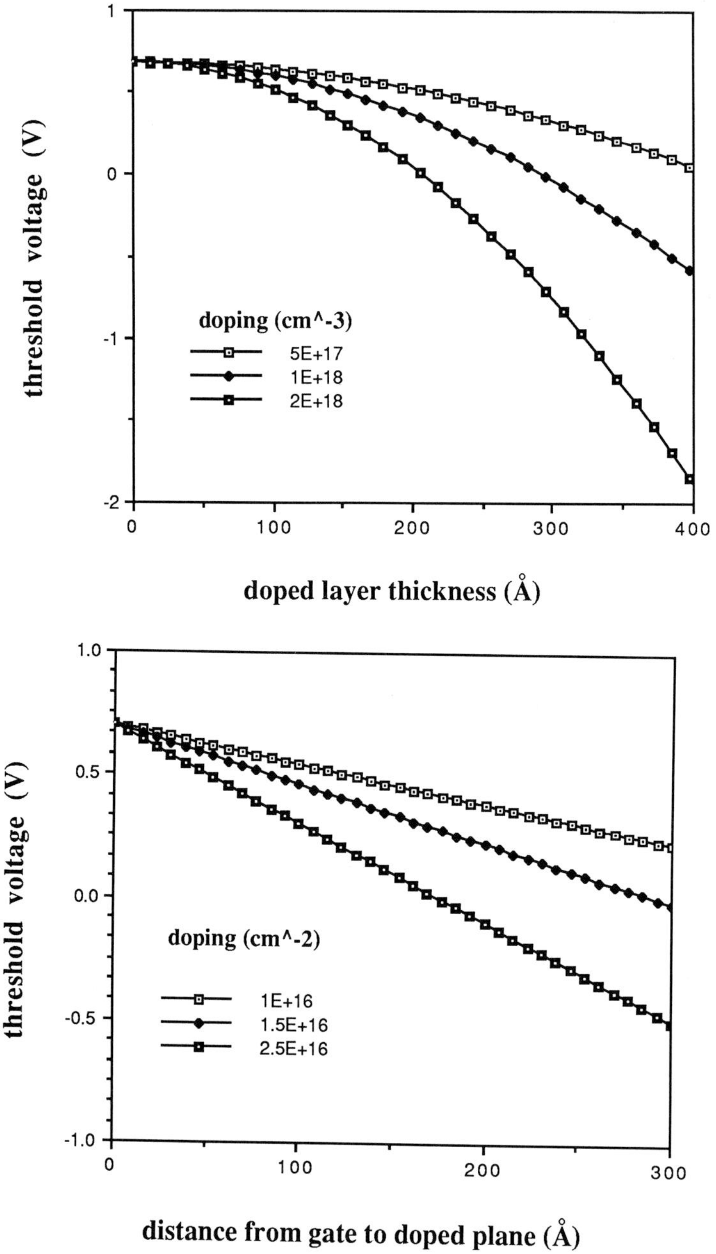

Fig. 11. Threshold voltages for conventional and planar-doped HFETs.

Eq. (11) does not take into account several effects that may be quite important, such as real space transfer of electrons into the AlGaAs layer which may play a crucial role in modulation-doped devices, non-linear dependence of the Fermi level on the carrier concentration of the 2-d electron gas, n_s, at low values of n_s, the effect of gate leakage current, etc. More accurate numerical calculations of

n_s on V_g should be based on the self-consistent solution of the Schrödinger equation and Poisson's equation that describe the electron density and potential in the vicinity of the heterointerface (see Stern et al. (1974)). The results of such calculations can be very well approximated by analytic interpolation formulas (see Baek et al. (1987)).

In Fig. 12 (from Chao et al. (1989) we show the computed dependences of n_s and n_t on the gate voltage. As can be seen from Fig. 12, at high gate voltages the concentration of electrons in the

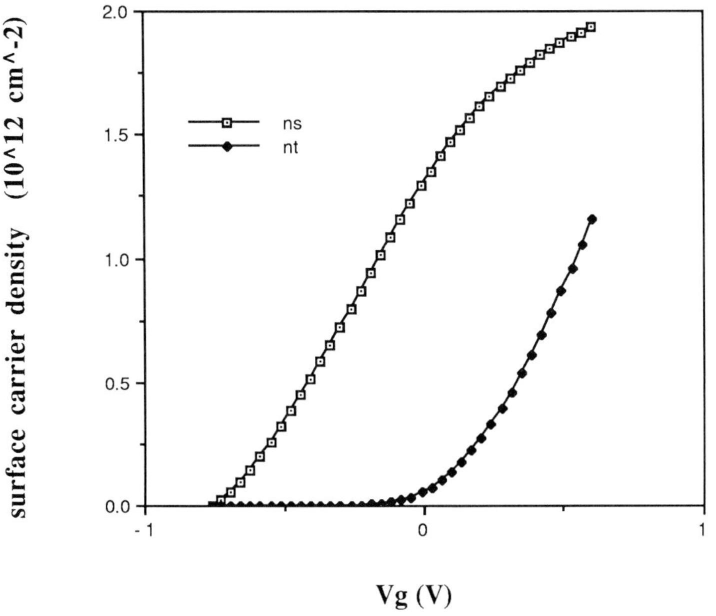

Fig. 12. Density of electrons per unit area in two-dimensional gas and in AlGaAs layer (from Chao et al. (1989)).

AlGaAs layer increases sharply and the slope of the n_s vs. V_g dependence decreases. This result can be understood by analyzing the energy band diagram of the modulation-doped structure. As an example, we show simplified band diagrams for planar-doped HEMTs for different values of the gate voltage, V_g in Fig. 13. As can be seen from this figure, at large gate voltage swings the electron quasi-Fermi level in the AlGaAs structure reaches the bottom of the conduction band so that the carriers are induced into the conduction band minimum in the AlGaAs layer.

An approximate analytical model for the description of HFET current-voltage characteristics was developed by Grinberg and Shur (1989). This model is based on the equation for drain current, valid at low drain-to-source voltages when the electric field in the channel is small and the electron velocity, v, is proportional to the electric field, F:

$$I_D \approx q\mu n_{xs}F \qquad (14)$$

where $n_{xs}(x)$ is the surface carrier density at position x of the channel, μ is the electron mobility, q is the electronic charge,and

$$F = dV/dx \qquad (15)$$

is the absolute value of the electric field. In the frame of the gradual channel approximation

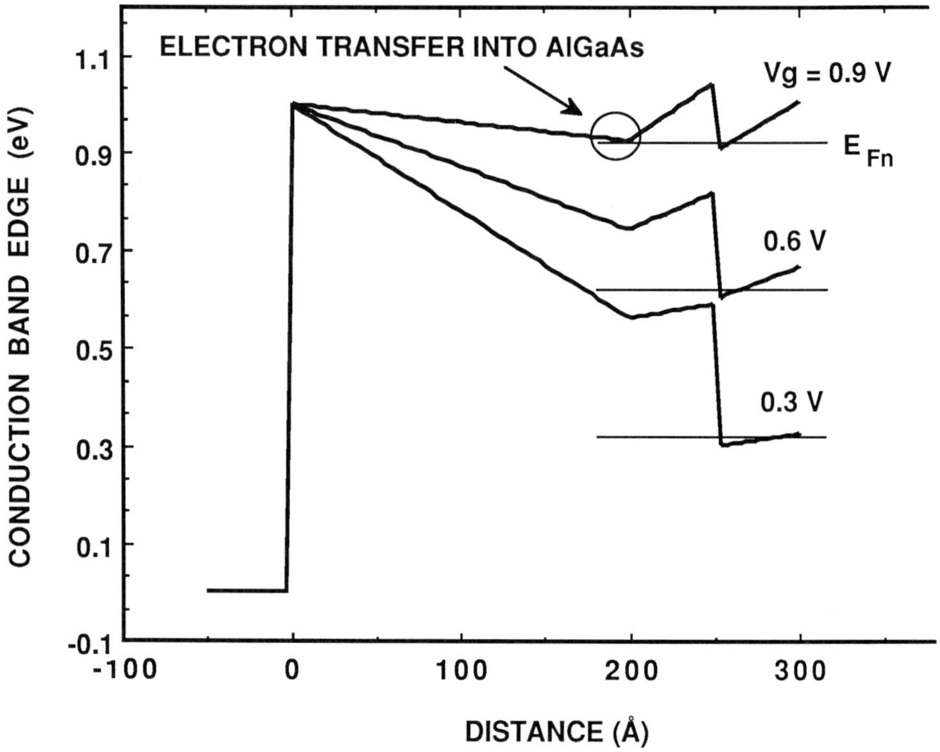

Fig. 13. Band diagrams for planar doped HEMTs at different values of gate voltage. At small gate voltages, the energy separation between the electron quasi-Fermi level and the conduction band minimum in the AlGaAs charge-control layer is large and the electron concentration in the AlGaAs layer is negligible. At large gate voltages, the electron quasi-Fermi level nearly touches the conduction band minimum in the AlGaAs charge-control layer, and the transfer of electrons into this layer becomes dominant, sharply reducing the device transconductance (after Chao et al. (1989)).

$$n_{xs} = n_s - \varepsilon V(x)/(qd) \qquad (16)$$

where ε is the dielectric permittivity of the AlGaAs layer, d is the distance between the gate and the channel, and $n_s = n_s(V_g)$ is the surface carrier density at the source (independent of V). Substituting eqs. (15) and (16) into eq. (13) and integrating from 0 to L with respect to x (where L is the gate length) and from 0 to V_D with respect to V (where V_D is the drain voltage) we obtain the following expression for the current-voltage characteristic at drain voltages smaller than the saturation voltage, V_D:

$$I_D \approx q\mu(W/L)[n_s V_D - \varepsilon V_D^2/(2qd)] \qquad (17)$$

This expression for the current-voltage characteristics is valid for an arbitrary dependence of n_s on V_g.

As we did above for GaAs MESFETs, we assume that the current-voltage characteristics in short channel devices saturate when the electric field, F, at the drain side of the channel reaches the electric field of velocity saturation, $F_s = v_s/\mu$, where v_s is electron saturation velocity. The electric field at the drain, $F = I_D/(q\mu W n_d)$ or, using eq. (16)

$$F = I_D/\{q\mu W[n_s - \varepsilon V_D/(qd)\} \qquad (18)$$

Substituting $F = F_s$ into eq. (18) and using eq. (17), we find the drain saturation voltage, V_{dsat},

$$V_{dsat} = (qdn_s/\varepsilon)[1 + a - (1 + a^2)^{1/2}] \qquad (19)$$

where $a = \varepsilon F_s L/(qn_s d)$ and n_s is a function of V_g . From eqs. (17) and (19) we find the drain saturation current (see Grinberg and Shur (1989))

$$I_{dsat} = qn_s\mu F_s W[(1 + a^2)^{1/2} - a] \qquad (20)$$

This model uses a piece-wise linear approximation for electron velocity. Based on the results of Monte Carlo simulation (see, for example, Cappy et al. (1980)) and transient calculations (see Shur (1976)), we know that v_s in short channel devices is very much different from the electron saturation velocity in a long bulk semiconductor sample. It is considerably higher and dependent on the effective gate-length, L_{eff}. One-dimensional simulations for GaAs n-i-n structures led to the following interpolation formula for v_s (see Shur and Long (1982))

$$v_s \approx (0.22 + 1.39L_{eff}) x 10^7/L_{eff} \quad \text{(cm/s)} \qquad (21)$$

where the effective gate length, L_{eff}, is in microns. Also, numerical simulations clearly show that electron velocity in short structures does not saturate in high electric fields. As a matter of fact, the velocity depends on both electric field and potential. Hence, this simple model relies on effective values of the electron velocity in high electric fields that give a general idea of electron velocities in a device channel but does not give any information about the exact shape of the velocity profile. For devices with recessed gates the effective gate length can be estimated as

$$L_{eff} \approx L + \theta(d + \Delta d) \qquad (22)$$

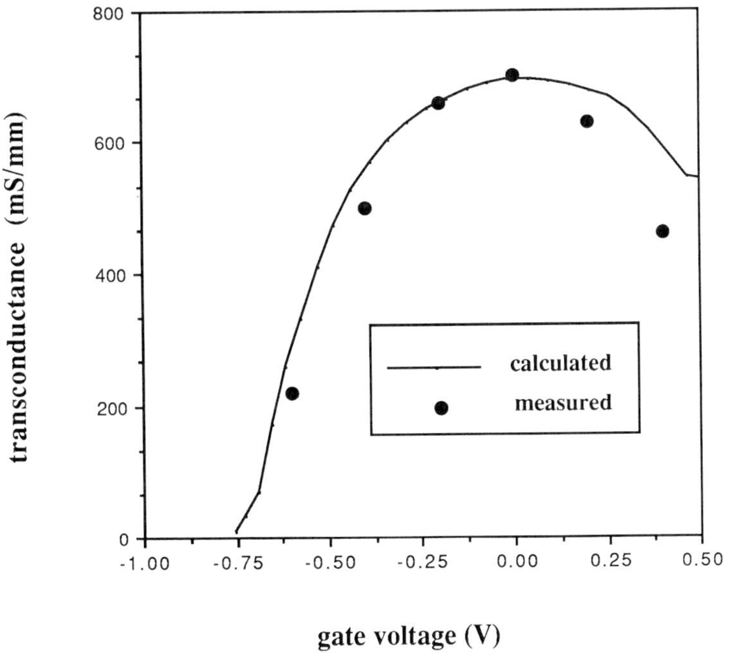

Fig. 14. Calculated and measured transconductance of 0.08 μm planar-doped pseudomorphic HFET(from Chao et al. (1989)).

where Δd is the effective thickness of the two-dimensional electron gas, and θ is a constant. The length $\theta(d+\Delta d)$ in the equation represents the total lateral depletion width and is a function of gate recess width in a heterostructure FET. Chao et al. (1989) estimated $\theta \approx 2$.

Fig. 14 (from Chao et al. (1989)) shows a comparison of this model with experimental data for device transconductance. Because of the ultra-short gate-length in this device, the maximum device transconductance is primary determined by the effective saturation velocity and only slightly affected by the low-field mobility. The effect of the gate-to-channel spacing is shown in Fig. 15 (from Chao et al. (1989)). With a decrease in d, the maximum transconductance increases if the

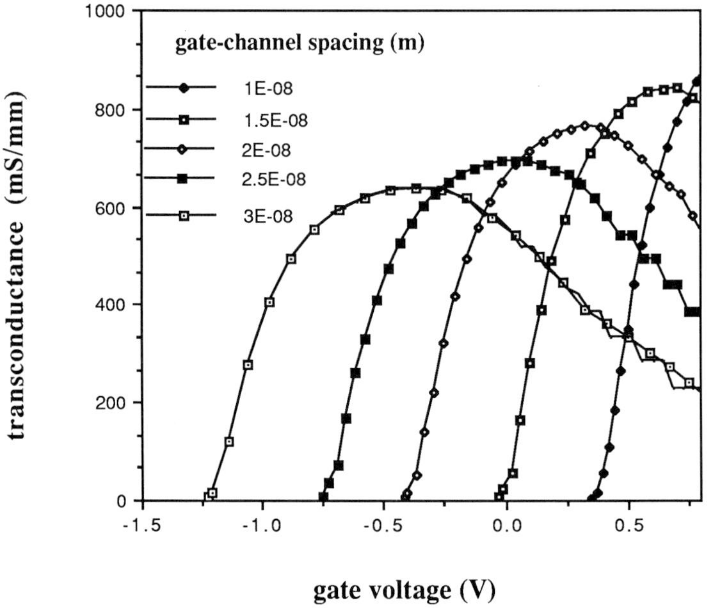

Fig. 15. Transconductance of 0.08 μm planar-doped pseudomorphic HFET calculated for different values of channel-to-gate spacing (from Chao et al. (1989)).

source series resistance remains constant (as was assumed in the calculations for Fig. 15). In fact, the source series resistance is expected to rise with the increase in the threshold voltage. This resistance plays a very important role in limiting the device transconductance. Experimentally, a gate-channel spacing of 200 to 250 Å yields maximum device transconductance. This depends, however, on the doping profile and on the shape of the gate recess.

The output current-voltage characteristics can be calculated using the same approach as for GaAs MESFETs. This model has been implemented in the GaAs circuit simulator UM–SPICE (see Hyun et al. (1986)). The equivalent circuit of an HFET used in UM-SPICE is shown in Fig. 16. This circuit takes into account the mechanism of the gate leakage current explained in Fig. 17.

As shown in Fig. 17, the gate current mechanism changes at $V_g \approx \phi_b - \Delta E_c$ where ϕ_b is the Schottky barrier height at the metal semiconductor interface and ΔE_c is the conduction band discontinuity. At smaller gate voltages the gate current is primarily determined by the Schottky barrier at the metal semiconductor interface. At larger gate voltages the gate current is limited by conduction band discontinuity or, more precisely, by the effective barrier height equal to the difference between the bottom of the conduction band in the AlGaAs layer at the heterointerface and the electron quasi-Fermi level in the two-dimensional gas. This barrier height changes very little with the gate voltage (and only as a consequence of the dependence of the quasi-Fermi level on electron concentration in the 2-d electron gas and, hence, on the gate voltage). Thus, the interpolation of the experimental dependence

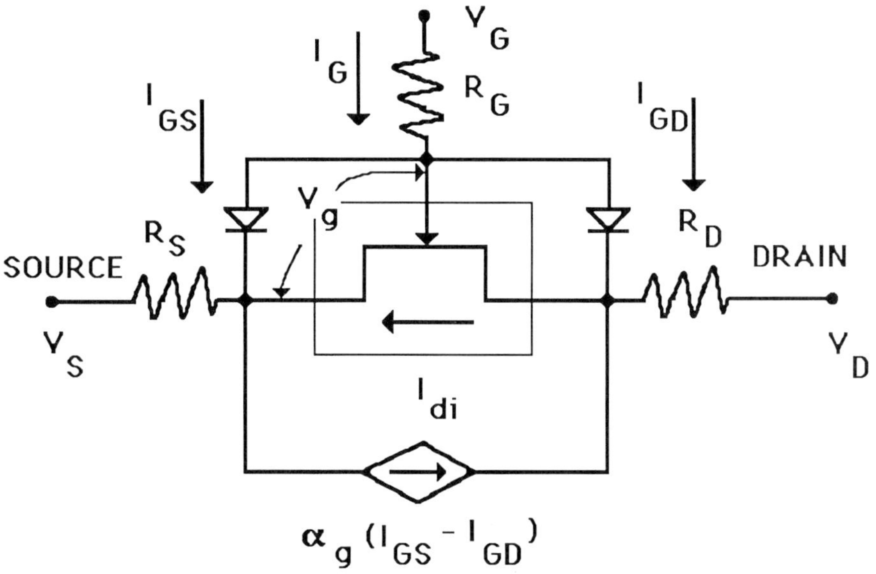

Fig. 16. HFET equivalent circuit (from Ruden et al. (1989)).

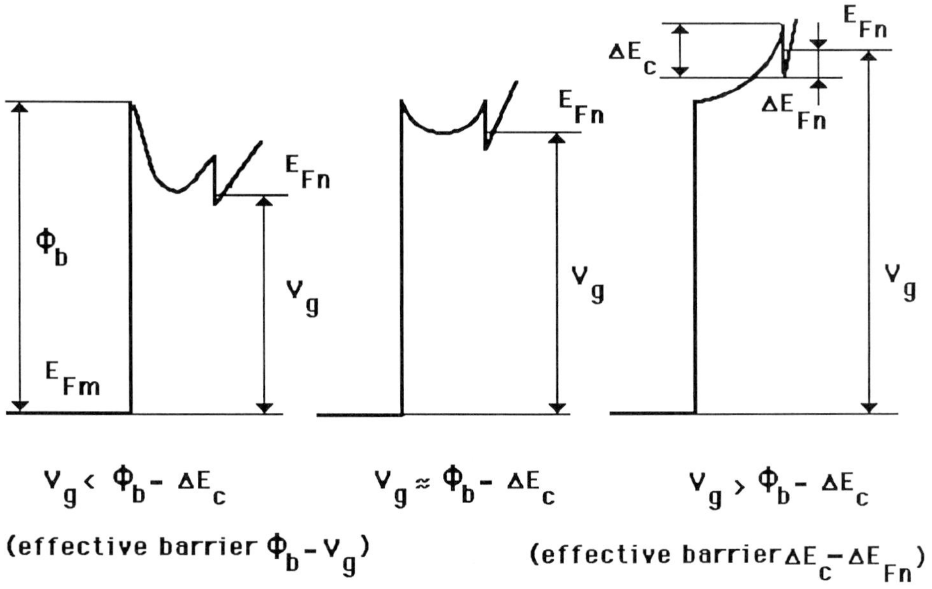

Fig. 17. Gate current mechanism in HFETs.

of the gate current on the gate voltage in this regime, given by the diode equation, has a very large ideality factor (usually from 5 to 20). Another interesting effect that is seen in the measured current-voltage characteristics at high gate voltages is a negative differential resistance. This effect may be much more pronounced (see Shur et al. (1986)) and is related to the real space transfer mechanism, first proposed by Hess et al. (1983), i.e. to the transfer of hot electrons from the channel over the barrier created by the conduction band discontinuity. A similar effect was observed in MODFETs by Y. Chen et al. (1987).

The equivalent circuit shown in Fig. 16 takes into account the effect of gate current on channel current. Indeed, the gate current is distributed along the channel, with the largest gate current density near the source side of the channel. This leads to the redistribution of electric field along the channel, with the increase in electric field near the source side of the device and a decrease in channel current. A numerical solution of coupled differential equations describing the gate and channel current distributions along the channel is given by Baek and Shur (1988). This calculation provides a justification for the new equivalent circuit shown in Fig. 16. As illustrated by Fig. 18, this new equivalent circuit allows us to obtain much better agreement with the experimental data than a conventional model.

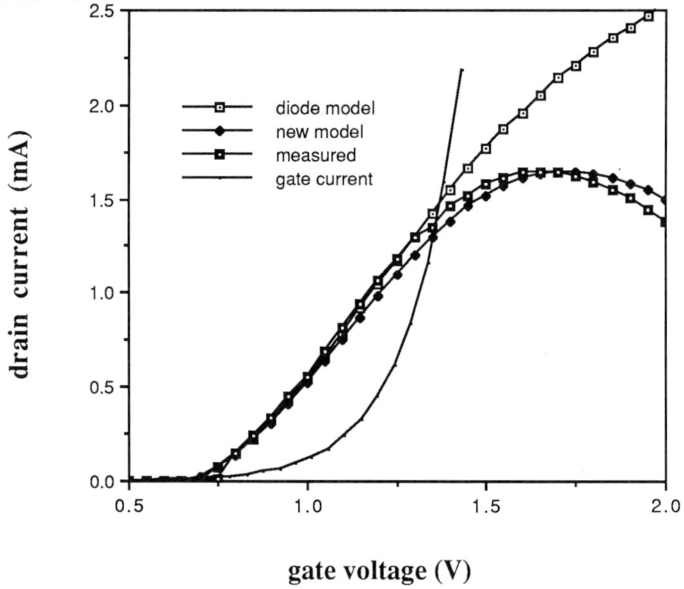

Fig. 18. Drain saturation current vs. gate voltage (from Ruden et al. (1989)).

Charge storage effects in an HFET may be represented by equivalent nonlinear capacitances $C_{gs}(V_{gs},V_{ds})$ and $C_{gd}(V_{gd},V_{ds})$. These capacitances can be estimated using a modified Meyer's model (see Meyer (1971) and Hyun et al. (1986)). Such a model has been implemented in the GaAs circuit simulator UM-SPICE. As an example we show in Fig. 19 the output waveform of an 11 stage HFET ring oscillator simulated using UM-SPICE.

Conclusion.

We described analytical models for GaAs and AlGaAs/GaAs devices that provide some insight into the complicated device physics. They have been used as design and characterization tools. A better understanding of device physics and further development of both analytical and computer models is required to help compound semiconductor technology to mature and fully realize its apparent potential.

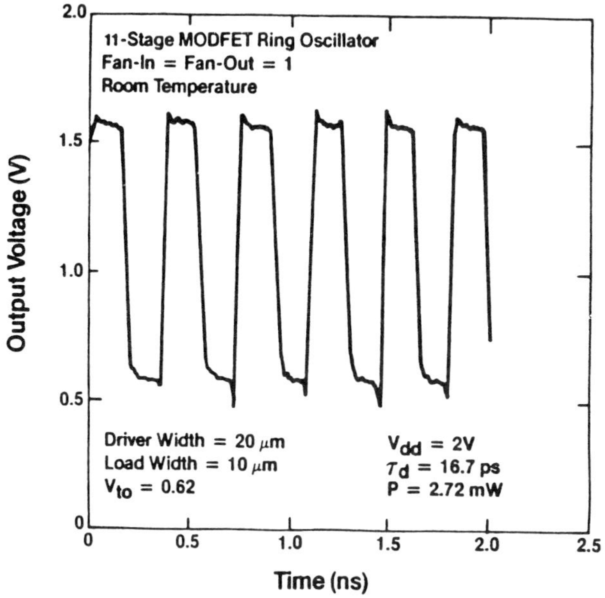

Fig. 19. Output waveform of 11 stage HFET ring oscillator simulated using UM-SPICE (from Hyun et al. (1986)

References.

P. M. Asbeck, C. P. Lee, and F. M. Chang, IEEE Trans. Electron Devices, **ED-31**, p. 1377 (1984)
T. Baba, T. Muzutani, M. Ogawa, and K. Ohata, Jpn. J. Appl. Phys., **23**, p. L654 (1983)
J. H. Baek, M. Shur, R. R. Daniels, D. K. Arch, J. K. Abrokwah, and O. N. Tufte, IEEE Trans. Electron Devices, **ED-34**, No. 8, pp. 1650-1657, August (1987)
J. H. Baek and M. Shur, Mechanism of Negative Transconductance in Heterostructure Field Effect Transistors, unpublished
S. M. Baier, M. S. Shur, K. Lee, N. C. Cirillo, and S. A. Hanka, IEEE Trans. Electron Devices, **ED-32**, No. 12, pp. 2824-2829, December 1985
A. Cappy, B. Carnes, R. Fauquembergues, G. Salmer, and E. Constant, IEEE Trans. Electron Devices, **ED-27**, pp. 2158-22168 (1980)
P. C. Chao, M. Shur, , R. C. Tiberio, K.H. G. Duh, P. M. Smith, J. M. Ballingall, P. Ho, and A. A. Jabra, DC and Microwave Characteristics of Sub-0.1μm gate-length planar-doped pseudomorphic HEMTS, IEEE Trans. Electron Devices, to be published (1989)
T. H. Chen and M. S. Shur, IEEE Trans. Electron Devices, **ED-32**, No. 5, pp. 883-91, May (1985)
C. H. Chen, M. Shur, and A. Peczalski, IEEE Trans. Electron Devices, **ED-33**, pp. 792-798 (1986)
C. H. Chen, A. Peczalski, , M. Shur, and H.K. Chung, IEEE Trans. Electron Devices, **ED-34**, No. 7, pp. 1470-1481, July 1987
Y. K. Chen, D. C. Radulescu, G. W. Wang, A. N. Lepore, P. J. Tasker, L. F. Eastman, and Eric Strid, in Proceedings of IEEE MTT Symposium, p. 871, Las Vegas, June (1987)
N. C. Cirillo, M. S. Shur, P. J. Vold, J. K. Abrokwah and O. N. Tufte, IEEE Electron Device Letters, **EDL-6**, No. 12, pp. 645-647, December (1985)
W. R. Curtice, IEEE Trans. Microwave Theory and Tech., **MTT-28** (5), pp. 448-456 (1980)
H. Dambkes, W. Brokerhoff, and K. Heime, IEDM Technical Digest, pp. 621-624(1983)
R. R. Daniels, R. Mactaggart, J. K. Abrokwah, O. N. Tufte, M. Shur, J. Baek and P. Jenkins, IEDM Technical Digest, pp. 448-449, Los Angelos CA (1986)
R. R. Daniels, P.P. Ruden, M. Shur, D. E. Grider, T. Nohava, and D. Arch, IEEE Electron Device Letters, **EDL-9**, pp. 355-357, July (1988)

G. C. Dasey and I. M. Ross, Bell Syst. Tech. J., **34**, pp. 1149 - 1189 (1955)
T. J. Drummond, H.Morkoç, K. Lee, and M. Shur, IEEE EDL., **EDL-3**, No. 11, pp.338-341, Nov.(1982)
R. W. H. Engelmann and C. A. Liehti , IEEE Trans. Electron Devices, **ED-24**, pp. 1288-1296 (1977)
T. A. Fjeldly, IEEE Trans. Electron Devices, **ED-33**, pp. 874-880 (1986)
A. A. Grinberg and M. Shur, New Analytic Model for Heterostructure Field Effect Transistors, J. Appl. Phys., accepted for publication (1989)
A. B. Grebene and S. K. Ghandhi, Solid State Electronics, **12**, pp. 573-589 (1969)
K. Hess, H. Morkoç, H. Shichijo, and B. G. Streetman, Appl. Phys. Lett., **35**, p. 459 (1979)
P. Hower and G. Bechtel, IEEE Trans. Electron Devices, **ED-20**, pp. 213-220 (1973)
C. H. Hyun, M. Shur, and A. Peczalski, IEEE Trans. CAD ICAS, **ED-33**, No.10, pp.1421-1426, April (1986)
C. H. Hyun, M. S. Shur, N. C. Cirillo, IEEE Trans. CAD ICAS, **CAD-5**, No.2,pp.284-292, April (1986a)
P. R. Jay and R. H. Wallis, IEEE Electron Device Lett., **EDL-2**, pp. 265-267 (1981)
R. A. Kiehl, D.A. Frank, S. L. Wright, and J.H.Magerlein, IEDM Technical Diest, pp. 70- 73, Washington (1987)
Landolt-Börnsten, Group III, vol. **22**, Semiconductors, Spinger-Verlag, Berlin (1982)
K. Lee, M. Shur, T. J. Drummond, and H.Morkoç, J. Appl. Physics, **54**, pp. 2093-2096 (1983)
K. Lee, M. Shur, T. J. Drummond, and H. Morkoç, IEEE Trans. on Electron Devices, **ED-30**, No. 3, pp. 207-212, March (1983)
K. W. Lee, M. S. Shur, K. Lee, T. Vu, P. Roberts, and M. Helix, IEEE Transactions on Electron Devices, **ED-32**, No. 5, pp. 987-992, May (1985)
T. J. Maloney and J. Frey, IEEE Trans. Electron Devices, **ED-22**, pp. 357-358 (1975)
J. E. Meyer, RCA Review, **32**, pp. 42-63, March (1971)
A. Peczalski, C. H. Chen, M. Shur and S. M. Baier, IEEE Trans. Electron Devices, **ED-34**, No.4, pp. 726-732 (1987)
R. A. Pucel, H. Haus, and H. Statz, in Advances in Electronics and Electron Physics, **38**, Academic Press, New York, pp. 195-205 (1975)
J. G. Ruch, IEEE Trans. Electron Devices, **ED-19**, pp. 652-654 (1972)
P.P. Ruden, M. Shur, A. I. Akinwande, and P. Jenkins, "Distributive Nature of Gate Current and Negative Transconductance in Heterostructure Field Effect Transistors", IEEE Transactions on Electron Devices, accepted for publication (1989)
P. P. Ruden, M. Shur, D. K. Arch, R. R. Daniels, D. E. Grider, T. Nohava, Quantum Well p-channel AlGaAs/InGaAs/GaAs Heterostructure Insulated Gate Field Effect Transistors, IEEE Trans. Electron Devices, submitted for publication (1989a)
N. J. Shah, S. S. Pei, and C. W. Tu, M. IEEE Trans. Electron Devices, **ED-33**, pp. 543 (1986)
M. S. Shur, Electr. Lett., **12**, No. 23, pp.615-616 (1976)
M. S. Shur, D. K. Arch, R. R. Daniels, and J. K. Abrokwah, IEEE Electron Device Letters, **EDL-7**, No. 2, pp. 78-80, February (1986)
M. Shur, GaAs Devices and Circuits, Plenum Publishing Corporation, New York (1987)
M. Shur, IEEE Trans. Electron Devices, **ED-32**, No. 1, pp. 70-72, January (1985)
M. Shur and D. Long, IEEE Electron Device Lett., **EDL-3**, p. 124 (1982)
M. Shur, J. K. Abrokwah, R. R. Daniels, and D. K. Arch, J.Appl. Phys, **61**(4), pp. 1643-1645, Feb. 15 (1987)
M. S. Shur and L.F. Eastman, IEEE Trans. on Electron Devices, **ED-25**, pp. 606-611, June (1978)
C. M. Snowden and D. Loret, IEEE Trans. Electron Devices, **ED-34**, pp. 212-223 (1987)
H. Statz, P. Newman, I. W. Smith, R. A. Pucel, and H. A. Haus, IEEE Trans. Electron Devices, **ED-34**, pp. 160-169 (1987)
F. Stern, CRC Crit. Rev. Solid State Sci., p. 499 (1974)
T. Takada, K. Yokoyama, M. Ida, and T. Sudo, IEEE Trans. Microwave Theory and Technique, **MTT-30**, pp.719-723 (1982)
R. A. Warriner, Solid State Electron Devices, **1**, p. 105 (1977)
K. Yamaguchi and H. Kodera, IEEE Trans. Electron Devices, **ED-23**, pp. 545-553 (1976)
L. Yang and S. T. Long, IEEE Electron Device Lett., **EDL-7**, pp. 75-77 (1986)

Modelling of Semiconductor Laser Diodes

Roel Baets

University of Gent - IMEC

1. INTRODUCTION

Semiconductor laser diodes play an ever increasing role in a variety of systems. The application of these light sources in optical telecommunication and in optical recording systems is well known. But also in medical systems, in optical pumping of certain laser types and in metrology does the laser diode replace other light sources. It owes its popularity to its small dimensions, high power efficiency, relatively high power output, electrical modulation capability, long lifetime and low cost if produced in large volume.
Different applications mean different specifications. Depending on the specific usage, a laser diode will be needed with a specific wavelength, a low threshold current, a good spectral purity (e.g. dynamic single longitudinal mode operation and/or a narrow linewidth), a high modulation bandwidth, special beam characteristics or high output power. Nobody will be surprised to hear that it is difficult to reconcile all good properties in one device. Therefore many different types of laser diodes have been proposed. This number is even more extensive because of the large number of semiconductor materials, especially III-V alloys, that can be used for laser action and the highly different epitaxial growth methods that have been developed to form the semiconductor layer stack typical for all laser diodes.

The behaviour of laser diodes is not simple. Both electronic and optical phenomena are present in the same device. The charge carrier concentrations (of both electron and holes) are high and various recombination processes need to be taken into account. The material is optically nonuniform and the propagation of waves and formation of cavity modes is complex.
Most properties of laser diodes can only be analysed by fairly complex models, and even then most models rely on approximate expressions for a number of physical dependencies, often based on experimental data. Due to the high complexity of laser diode behaviour and the large degree of uncertainty concerning a number of parameters, the modelling of these devices is mostly qualitative rather than quantitative. The main role of modelling is to acquire a better understanding of experimental laser behaviour so as to improve the design of the laser. Alternatively, new designs can be analysed prior to fabrication.

There are different types of models related to semiconductor lasers [1]. There are material models that describe the physical interactions in the semiconductor and there are process models that describe the device structure, given the fabrication sequence. Most attention is paid however (both in literature and in this paper) to models that describe the device behaviour, for given physical relationships and for a given device structure. In section 2 of this paper a number of general aspects of laser models will be discussed. In section 3 to 5 a number of very different device models are presented in some detail and are illustrated with examples. Finally section 6 describes two material models for the gain of III-V semiconductors.

2. GENERAL ASPECTS OF LASER DIODE MODELLING

In fig. 1 the simplified geometry of a semiconductor laser diode is shown. It consists of a stack of semiconductor layers, n-type at one side of the "active" layer, p-type at the other side. The active layer needs to have a composition, such that its bandgap is smaller and its refractive index larger than that of the surrounding layers. By applying a voltage between the top metal stripe and the bottom contact a current will flow. In the active region charge carriers, both electrons and holes, will be injected and remain confined by the action of the heterojunctions. Therefore high concentrations of charge will be present, leading to efficient recombination. In a limited wavelength range, absorption will become less probable than stimulated recombination and therefore the active layer will exhibit gain.

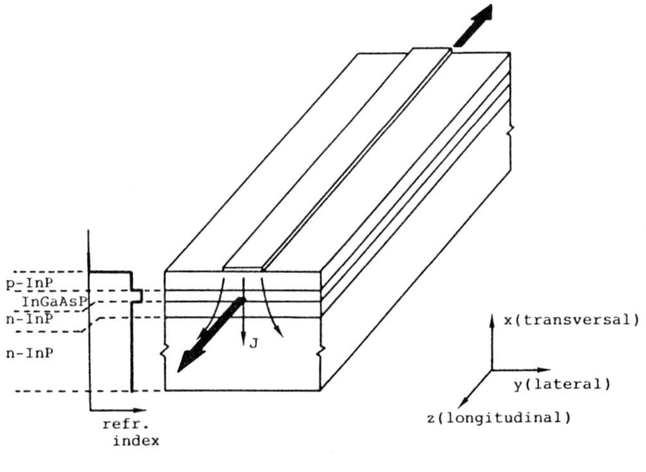

Fig. 1. Simplified geometry of a semiconductor laser diode

A cavity is formed by confining light in three dimensions. In the transverse or x-direction (see fig. 1), a relatively large index step exists at both sides of the active layer, which means that it forms a slab waveguide that can guide light. In the longitudinal or z-direction the structure is terminated by cleaved facets, with or without dielectric coatings, that provide reflection back into the device. In the lateral or y-direction, the situation is far more complex. In the simple situation as shown in the figure, there is no confinement or guiding mechanism. However the gain that is present under the stripe, can be sufficiently high, such that it compensates for diffraction loss. This diffraction loss is relatively large because the carrier injection profile causes a depression of the refractive index at the gain maximum, which is an anti-guiding effect. Numerous designs have been proposed to incorporate

a lateral waveguiding effect in the laser structure, either by having a real refractive index step in lateral direction or by having a lateral geometry variation that induces a lateral change of the effective index seen by the transverse mode.
The physical interactions are schematically shown in fig. 2. The potential distribution, carrier concentrations and current densities are linked through Poisson's equation, current equations and continuity equations. The latter need to take into account absorption and recombination phenomena. That means that the net gain (or loss) and also the power density of the optical fields appear in these equations. The net gain (and also the refractive index change) depend on the carrier density and are needed to describe the wave propagation and the cavity resonance. In some models temperature effects are taken into account. Both electrical and optical heat dissipation are then introduced into a heat diffusion model that yields the temperature increase, which affects most material parameters.

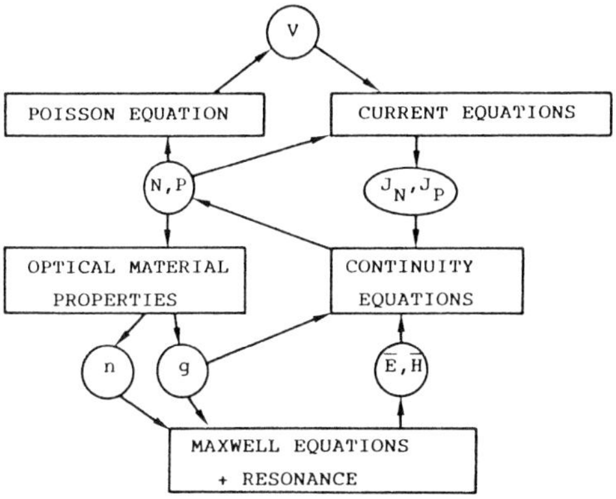

Fig. 2. Physical interactions in a laser diode

The independent variables in the laser modelling problem are the spatial coordinates x,y,z, time t (or modulation frequency f), wavelength λ and charge carrier energy E. The most important dependent variables are the voltage V, carrier concentrations N and P, the (real) refractive index n and gain g and the optical field amplitude ψ. It is not feasible to include all effects with all independent variables in one model. Such an approach would be necessary if one is interested in obtaining quantitatively accurate results. If the main goal however is to explain or anticipate certain relationships between device properties and device structure, it is much better to simplify the model in such a way as to retain the relevant elements. This leads to a wide range of different models, depending on what specific aspect of the laser is studied. In each of these a subset of the dependent and independent variables is retained.

The very simplest model for a laser diode is one in which all independent variables are left out. It is a static model with all quantities averaged over space, wavelength or energy. This basic model is governed by two equations:

$$\frac{J}{qW_a} = R_{spon}(N) + g(N)\,\Gamma\,\frac{S}{\Omega} \tag{1}$$

$$\Gamma\, g(N) - \alpha_i = \frac{1}{2L} \ln \frac{1}{R_1 R_2}$$

The first equation describes the charge continuity in the active layer. The two terms in the right hand side describe spontaneous and stimulated recombination respectively. J is the current density through the active layer with thickness W_a, S is the photon flux (sum of both propagation directions), Ω the cross-sectional area of the active layer and Γ is the confinement factor (ratio of optical power in the active layer to total optical power flux). The second equation is a resonance condition for the cavity, with R_1, R_2 and L the mirror reflectivities and cavity length respectively and α_i an additional loss term.
A number of assumptions underlie these equations. Only one carrier type is considered, assuming that the other is defined by neutrality considerations. The carrier confinement of the double heterostructure is assumed to be perfect, so that all recombination occurs within the active layer. In the resonance condition the contribution from spontaneous emission to laser light is omitted. To solve equations (1) the relationship $R_{spon}(N)$ and g(N) need to be known. From this very simple model the threshold current density and differential quantum efficiency can be deduced. The phase resonance condition is not taken into account.

The above model can be extended to include time dependencies and is then well known as rate equation model. The equations (1) transform into

$$\frac{dN}{dt} = \frac{J}{qW_a} - R_{spon}(N) - g(N)\,\Gamma\,\frac{S}{\Omega} \qquad (2)$$

$$\frac{dS}{dt} = v_g(g(N)\Gamma S + \beta_{sp} R_{spon}(N)\Omega) - \frac{S}{\tau_p}$$

The first equation is straightforward. The second equation takes on a slightly different form by incorporating all loss factors, including mirror losses into the cavity lifetime τ_p and by introducing here the spontaneous emission contribution to laser light. β_{sp} is the spontaneous emission coupling factor, v_g is the group velocity. These two simple differential equations are sufficient to derive the pulse response or small signal bandwidth of a laser diode. Typical effects such as the turn-on delay, the relaxation oscillations and the increase of bandwidth with current above threshold are readily obtained by solving these equations. A further extension of this model is to include the z-dependencies of J, N and S. Such a model is described in section 5.

A completely different extension of the basic model retains the y dependence of all variables [2], in particular of J, N, g and ψ. Here wave propagation aspects come in. It is assumed that the optical field $\psi(y)$ is an eigenmode solution of the complex refractive index profile $n(y) + j\, g(y)/2k_o$. The concept of self-consistency between carrier density and optical field is important in this type of model, as is clear from fig. 2. The carrier density determines the waveguide profile, from which the field amplitude is derived, which itself influences the carrier density again through stimulated recombination. The eigenmode solution needs to be such that the imaginary part of the complex propagation constant is zero, which means that the modal gain compensates for all the losses. This type of model will in general be iterative. It is useful in that it not only gives a more accurate solution for the threshold current and power/current curve, but that it also provides information about the near field and far field in the y-z plane, about the number of modes and about possible instabilities or kinking due to index anti-guiding. Although the x-dependence is not taken into account

explicitly, it can be included with various degrees of sophistication. For the optical eigenmode problem, most workers reduce the x-y geometry to a y-geometry by applying the effective index method. This does not increase the complexity of the model substantially. The calculation of the current density through the active layer from knowledge of the voltage across the device electrodes is much more complex but has been done [2]. Another extension of the y-model is to include the longitudinal dimension (y-z model) [3]. In such a situation one can solve for the complete cavity modes, which are not necessarily equal to the local normal modes in each lateral plane. The resonance condition now says that a lateral field distribution $\psi(y)$ at a given position z will remain identical after one complete propagation roundtrip. Such a model is particularly interesting if the laser geometry contains longitudinal variations. The optical field problem is then solved by using a beam propagation method.
The t-model and y-model can also be taken together to form a y-t model. Such an approach may be useful to study such effects as self-sustained pulsations due to oscillation of the width of the optical field.
One could imagine the combination of the z-t, y-t and y-z models into one y-z-t model but we are not aware of anybody having done so. Such a model would be relatively far-fetched and certainly computer time intensive.

A last series of models have been developed to study the spectral content of a laser diode. There are basically two aspects about the laser spectrum, one is the linewidth of the individual laser modes (for which a noise model is needed), the other is the spectral position of these modes. We will concentrate on the latter aspect. The basic model can be extended to include a number of modes with a resonance condition for each mode. In such a λ-model the gain g is assumed to be a function of carrier density and wavelength. The individual mode wavelengths are derived from a phase resonance condition. It is necessary to include the spontaneous emission in the (amplitude) resonance condition in this model, since otherwise the strongest mode would suppress all other ones. This is at least true if the thermal distribution of charge carriers over energy is not disturbed by the stimulated recombination. If such a disturbance occurs (spectral hole burning) the spectral gain distribution is strongly perturbed as well. Such effects can in principle be modelled by a combined λ-E model, where E is the charge energy.
Another extension of the λ-model is to include either the y-dependence or the z-dependence. The latter is necessary if the spectral properties of cleaved-coupled-cavity (C^3) or distributed feedback (DFB) lasers are studied. In both cases the longitudinal variations of the field amplitude and phase have a strong influence on the number and relative strength of the modes.
In the case of DFB-lasers the set of wave equations for the forward and backward beams must be replaced by a set of coupled wave equations in the description of the optical problem. For a λ-z model those coupled wave equations can be solved with the well-known propagator matrix formalism, while for a λ-y-z model an extended BPM [4] can be used. An iteration on wavelength is now indispensable due to the strong wavelength dependence of the distributed reflections. However, the carrier-photon interactions and the electronic problem remain the same as for the Fabry-Perot type laser.
A combined λ-y-z model has been applied to the C^3-laser [5] and may be useful in a number of special cases. A combined λ-z-t model would be very useful to assess dynamic single longitudinal mode operation or wavelength chirping effects. Such a model is very complex however and, to our knowledge, has not been presented.

In fig. 3 most of the models mentioned so far are schematically brought together. The models described in the following sections are underlined.

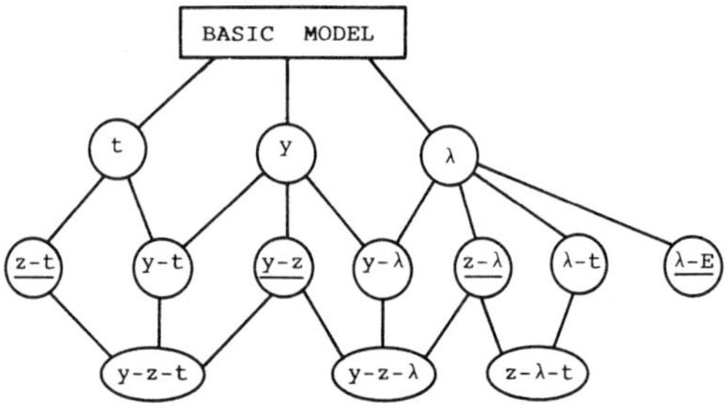

Fig. 3. General classification of the different models by the independent variables involved.

3. STATIC (y,z)-MODELS

3.1 Electronic Modelling

The current flow in a semiconductor laser consists of different parts, namely drift and diffusion. Carriers recombine for instance at interfaces or in the active region, this either spontaneously or through stimulated emission.
The current carried by the electrons ($\bar{J}_N$) and the holes ($\bar{J}_P$) in the different layers of the laser structure is related to the electron and hole concentration (N and P) and to their respective quasi Fermi levels (E_{FN} and E_{FP}).

$$\begin{aligned}
\bar{J}_N &= \mu_N\, N\, \nabla E_{FN} \\
\bar{J}_P &= \mu_P\, P\, \nabla E_{FP} \\
\nabla . \bar{J}_N &= q(R_{spon} + R_{stim}) \\
\nabla . \bar{J}_P &= -q(R_{spon} + R_{stim})
\end{aligned} \tag{3}$$

R_{spon} and R_{stim} represent the spontaneous and stimulated recombination. In the active layer the equations (3) reduce to an equivalent diffusion equation determining either the electron or the hole concentration [2]. It is generally assumed that the carrier density is uniform over the thickness of the active layer. In the passive layers on the other hand, the equations (3) reduce to a Poisson problem (fig.4), such that we obtain two coupled problems:

Active layer

$$\frac{1}{W_a(y)} \frac{\partial}{\partial y} \left(W_a(y) \cdot D_e \frac{\partial N}{\partial y} \right) = \frac{\eta\, J(y)}{eW_a} + R_{spon} + R_{stim} \tag{4}$$

$$R_{spon} = \frac{N}{\tau} + BN^2$$

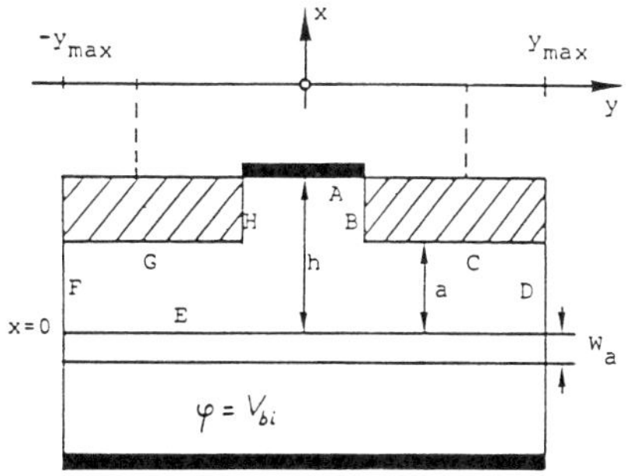

Fig. 4. Schematic view of the simplified laser cross-section.

Passive layer

$$\nabla^2 V = 0$$

$$J(y) = \sigma(\frac{\partial V}{\partial x}) \quad \text{at interface with active layer } (x=0)$$

W_a is the active layer thickness, D_e is an effective diffusion coefficient, which is, in the most general situation, concentration dependent. If proper boundary conditions are specified, the coupled set of equations (4) can be solved. The coupling of both problems arises from the fact that the boundary condition V(x=o,y) is found from the electron concentration in the active layer: $V(x=0,y) = V_a(N(y))$ (eq.(14)). One can either solve (4) self-consistently [2] or one can determine J(y) analytically [6], such that both problems become independent of one another.

3.2 Electromagnetic Modelling

The three-dimensional, vectorial electromagnetic problem is far too complex to be solved entirely for a semiconductor laser. In most cases it can be shown that the electromagnetic field consists mainly of a TE-wave, which satisfies a three-dimensional Helmholtz equation. By means of an effective index method [6] this equation can be reduced to a two-dimensional Helmholtz equation, in which the complex refractive index is dependent on the optical field :

$$\nabla^2_{yz}\psi + k_0^2 \, n^2_{eff}(y,z)\psi = 0 \qquad (5)$$

The complex effective refractive index is found from a perturbation method where only linear terms in the electron concentration are retained in the expression for n^2_{eff}.
The non linearity in (5) now arises from the fact that the stimulated recombination depends on F:

$$R_{stim} = \sum_i \frac{g_i \Gamma}{W_a (h\nu)_i} \left(|\psi_{Fi}|^2 + |\psi_{Bi}|^2 \right) \tag{6}$$

$$\psi_i = \psi_{Fi} + \psi_{Bi}$$

ψ_{Fi} and ψ_{Bi} denote the forward and backward travelling waves belonging to the lateral mode i and g_i is the gain in the active layer. This gain depends of course on the electron density.

3.3 Iteration Procedure And Examples

The iteration proceeds as follows: for a given electromagnetic field ψ the problem (4), (6) is solved and a new value of n_{eff} is determined. Subsequently we solve (5) with appropriate boundary conditions, to determine a better approximation F. This procedure is repeated until self-consistency is achieved.

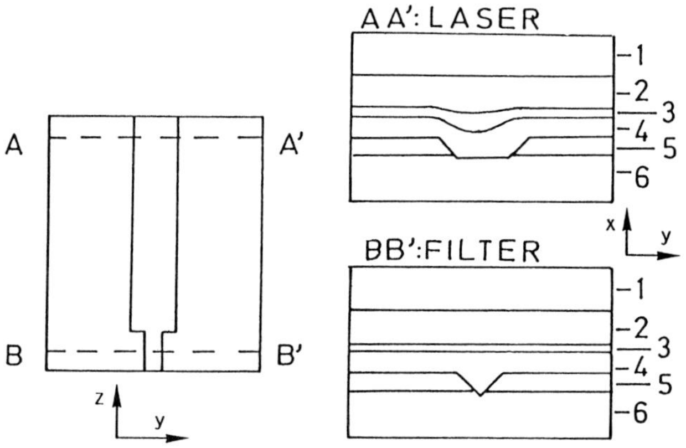

Fig. 5. Schematic view of a laser with filter section (1:top layer, 2:cladding layer, 3:active layer, 4:cladding layer, 5:current blocking layer, 6:substrate).

We will discuss an example for which the injected current J(y) is given by the analytic expression derived by Yonezu [7]. The laser consists of two sections: a 7µm wide laser section and a 3-4µm wide filter section (fig. 5). A current blocking layer is used to establish the current confinement. The purpose of the filter section is to suppress the higher order lateral modes. In fig.6 we have depicted the calculated light-current characteristics for different lengths of the filter section (0,20,40,60µm) with a constant laser section length (125µm). The light current characteristics are calculated up to the onset of a second order cavity mode. From fig. 6 it is seen that the output power from the laser section can be doubled by means of a filter section, before a higher order mode appears.

The price which is paid, is a 7mA higher threshold current for the laser with the longest filter section. From fig. 6 it is also seen that the output power from the filter section shows a saturating behaviour as

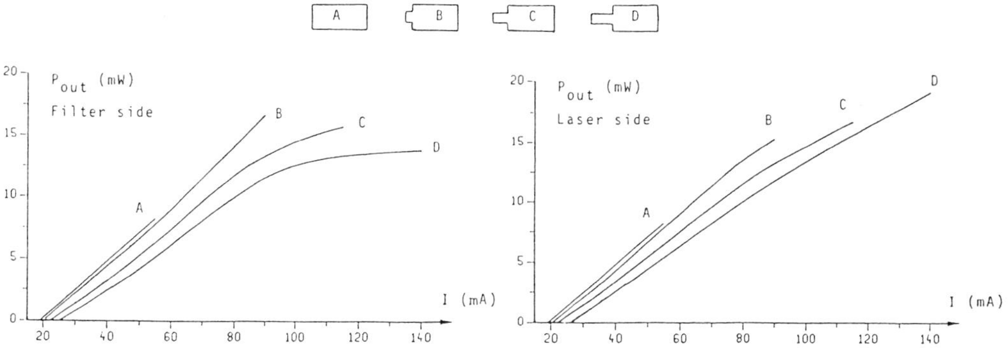

Fig. 6. Light-current characteristics of lasers with different filter section dimensions, varying in length between 0 and 60 µm.

a function of current. This saturating behaviour originates from the fact that local higher order modes (generated at the interface between the laser section and the filter section) radiate away in the unpumped region [8], [9].

4. STATIC (z,λ)-MODELS

Starting from the y-z-models, it is quite easy to make an extra dimension reduction, by averaging over the lateral direction [10]. As explained in [10], [11] we can furthermore reduce the order of the differential equation determining the electromagnetic field. Indeed, if we put

$$\psi_{F,B,i}(y,z) = Y_{F,B,i}(y) \cdot e^{-j\beta_{F,B,i}z} \tag{7}$$

where $\beta_{F,B,i}$ denotes the complex propagation constant of the forward (F) or backward (B) propagating wave of the i-th longitudinal mode, we obtain ([10],[11]) a first order differential equation for the forward propagating power P_{Fi}:

$$\frac{dP_{Fi}}{dz} = 2\mathrm{Im}(\beta_{Fi})\, P_{Fi} + \gamma_i$$

$$P_{Fi}(z) = \int_{-\infty}^{+\infty} |\psi_{Fi}(y,z)|^2 \, dy \tag{8}$$

In (8) we have neglected the (slight) z-dependence of β_{Fi}, γ_i denotes a source term, arising from spontaneous emission. Similar relations hold for the backward propagating power and the phases of the fields. An expression for the complex propagation constant can be obtained from a variational principle [12], [10], [11]. It turns out that β_{Fi} and β_{Bi} are linearly perturbed by $N_t(z)$, given by

$$N_t(z) = \int_{-\infty}^{+\infty} N(y,z)dy \tag{9}$$

An expression for $N_t(z)$ can be found from (4) by simple integration [10]. It is evident that the gain and refractive index determining β_{Fi} and β_{Bi} are intrinsically wavelength dependent.
This method can be extended to multi-section lasers and external cavity lasers [10], [11], although this requires special precautions to guarantee convergence.

As an example we calculated the spectrum of two gain guided lasers, one $\lambda/4$-coated and one normally cleaved laser. The output spectra for these two lasers have been drawn in fig. 7a and fig. 7b for nearly equal (and high) output powers of about 60mW.

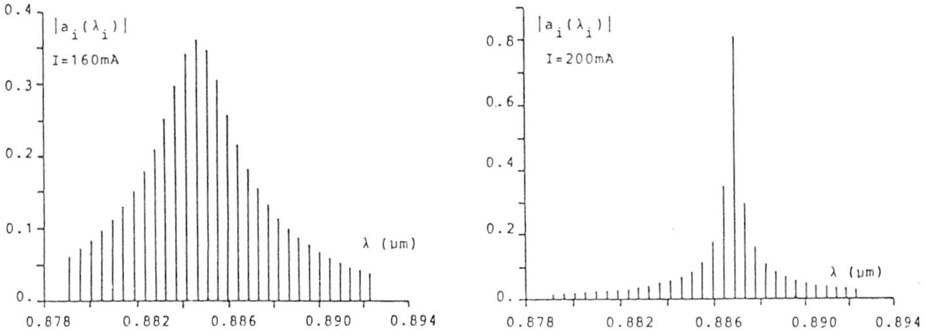

Fig. 7. Output spectrum of laser diodes at a given output power. a. Normally cleaved laser, b. Laser with $\lambda/4$ coating.

The spectrum of the high power, $\lambda/4$-coated laser is very broad as observed experimentally. This is due to the small mirror reflectivity at one of the mirrors (2%) and due to the strong longitudinal variations inside the cavity. At the coated mirror the power becomes very large and consequently, the electron concentration is depleted strongly. At the normally cleaved mirror the power is much lower and the electron concentration is much higher. This results in an appreciable variation of the gain along the laser axis. As a consequence the gain characteristic as a function of wavelength broadens and the laser becomes multimodal. Another slight difference can be observed between fig. 7a and 7b, being the fact that the central wavelength is slightly smaller for the $\lambda/4$-coated mirror. This is a consequence of the higher threshold current and gain, which causes the gain to shift towards higher photon energies.

5. DYNAMIC (z,t)-MODEL

A time dependent, longitudinal model is presented, in which only the amplitude resonance is taken into account. Phase resonance will not be considered. This means that the wavelength of each longitudinal mode (λ_i) has to be specified. For a Fabry-Perot(FP)-like device this invokes an uncertainty on λ_i equal to the modespacing. This is no problem since the exact knowledge of the λ_i's has hardly any influence on the laser behavior in such FP-devices. The dynamic wavelength shift will however be defined. Furthermore the model is restricted to strongly index-guided lasers with electron confinement in both lateral and transverse directions. Fundamental lateral and transverse mode operation is assumed.

A longitudinal dynamic model becomes particularly interesting when studying devices in which strong spatial unhomogeneities exist. This is the case for devices with unhomogeneous current injection (multi-segment single cavity devices). Also for low reflectivity devices (high power at the mirrors) or lasers with strongly asymmetric mirrors, the classical models [13] are not entirely satisfactory [14]. These models consider the photon loss through the end facets as a distributive process. In certain circumstances multi-segment lasers show bistable [15] or unstable (Self-Sustained Pulsations) [16] behaviour. Gain-switched lasers with an absorptive region have shown ultra-short light pulse generation [17].

5.1 Description Of The Equations Of The Model.

Using the slowly varying amplitude approximation [18], one can derive from Maxwell's equations an intensity and phase equation for each longitudinal mode of the laser cavity :

$$\frac{\partial I_{i,F,B}}{\partial z} + \frac{1}{v_g}\frac{\partial I_{i,F,B}}{\partial t} = -\alpha_1 I_{i,F,B} + \Gamma\, g_i(N)\, I_{i,F,B} + \frac{1}{2}\beta_{sp,i}\, BN^2$$

$$\frac{\partial \phi_{i,F,B}}{\partial z} + \frac{1}{v_g}\frac{\partial \phi_{i,F,B}}{\partial t} = \frac{1}{2}\Gamma\,\alpha\, A_i(N-N_{th}) \qquad (10)$$

These equations are called travelling wave rate equations. Here $I_{i,F,B}(z,t)$ and $\phi_{i,F,B}(z,t)$ are the phase and intensity of the complex, slowly varying amplitude of the ith mode. The subscripts F and B refer to respectively the forward and backward beams. We assume that the z-axis is always in the direction of propagation. For the backward wave we can thus use the same equation as for the forward wave. The lateral electrical laser field component for the ith longitudinal mode is then given by:

$$E_{y,i}(x,y,z,t) = I_i(z,t)\, e^{j\phi_i(z,t)}\, e^{-j(\omega_i t - k_i z)}\, \Phi(x,y)$$

Where $\Phi(x,y)$ is the transverse /lateral field distribution of the laser waveguide, assumed independent of time. Phase and frequency modulation around the reference frequency ω_i are expressed by $\phi_i(z,t)$. Furthermore v_g is the groupvelocity in the laserguide, α_1 represents the internal cavity losses, Γ is the two-dimensional filling factor, N(z,t) is the electron density, $g_i(N) = A_iN-B_i$ is the wavelength dependent gain, $\beta_{sp,i}$ expresses the fraction of spontaneous emission coupled into the mode i, α is the linewidth enhancement factor and N_{th} is the electron density at threshold.

The boundary conditions for the intensities at the two facets are:

$$I_{i,F}(0,t) = R_1\ I_{i,B}(0,t)$$
$$I_{i,B}(L,t) = R_2\ I_{i,F}(L,t) \qquad (11)$$

Where L is the laser length and R_1 and R_2 are the power reflectivities at facets 1 and 2. As we consider no phase resonance for ease of convergence, we will only use the phase equation to obtain the dynamic frequency shift $\Delta\omega_i(t)$. This is possible because we know at each instant the refractive index along the whole cavity. As long as the speed of time dependent phenomena is such that phase resonance is maintained, the wavelength is determined. The boundaries for ϕ_i can be chosen as:

$$\phi_{i,F}(0,t) = 0$$
$$\phi_{i,B}(L,t) = \phi_{i,F}(L,t) \qquad (12)$$

The electron density N(z,t) which determines the gain along the cavity is determined by the following diffusion equation :

$$\frac{\partial N}{\partial t} = D\frac{\partial^2 N}{\partial z^2} + \frac{J(z,t)}{eW_a} - \frac{N}{\tau} - B\ N^2 - C\ N^3 - \frac{1}{v_g}\sum_i g_i(N)\ (I_{i,F}+I_{i,B}) \qquad (13)$$

With D being the diffusion constant, W_a the active layer thickness, τ the linear recombination lifetime, B the bimolecular recombination constant, C the Auger recombination constant and J(z,t) the current density injected in the active layer. This current density is determined by the potential problem in the top cladding layer, between the top contact with given voltage and the active layer of which the voltage is defined by [19]:

$$V_a(N(z,t)) = \frac{2kT}{q}\ln(1+\frac{N}{I_s}) + C_1N + C_2N^2$$
$$I_s = \sqrt{N_cN_v}\ \exp(-E_g/2kt) \qquad (14)$$

with C_1, C_2, N_c, N_v being material constants and E_g the bandgap. We assumed here a perfectly conductive substrate and neutrality in the active layer. The combination of all these equations allows us to describe the dynamic behaviour of a laser in detail. In the next section we shall discuss how these equations can be solved in a self-consistent way.

5.2 Solution Methods For The Travelling Wave Rate Equations.

Due to its complexity and relatively large CPU-time consuming character, only a few authors [20], [21], [17], have solved the travelling wave equations, defined in the former section. Two methods are generally met.

The first method is based on a propagation technique [21]. Therefore the travelling wave equations are transformed to forward difference equations in time and position. The roundtrip time is divided into a number of time steps. The time step (t_h) and position step (z_h) are taken interdependent: the position step is chosen equal to the distance the light travels in a time period equal to one time step, thus $z_h=v_gt_h$. Mathematically this means:

$$F(z+z_h, t+t_h) \approx F(z,t) + \frac{\partial F}{\partial z}(z,t)\, z_h + \frac{\partial F}{\partial t}(z,t)\, t_h$$
$$\approx F(z,t) + z_h\left(\frac{\partial F}{\partial z} + \frac{1}{v_g}\frac{\partial F}{\partial t} \right)$$

The derivatives in (10) are thus replaced by $(F(z+z_h, t+t_h)-F(z,t))/z_h$. The beam is now propagated through the device for several roundtrip times, until a stable field pattern is obtained.

In the second technique [20], [17] the spatial variable is discretised so that the set of partial differential equations (10), (13), results in a system of ordinary non-linear differential equations in time. This system can then be integrated in time, using for instance Euler's method [22]. An alternative is to consider the ordinary differential equations as network equations, in which the field and carrier quantities are represented by electrical quantities. Such an implementation [20] has the advantage that network simulators can be used for solving the numerical problem. The network simulator should however be able to handle non-linear elements (voltage dependent current sources, etc.). In this case the potential problem, determining the injection current, can be replaced by a resistive network, while in the other methods a separate Poisson problem must be considered. Furthermore the network implementation offers the important possibility to easily simulate the laser diode in combination with parasitic elements as well as with its steering and controlling electronics.

In conclusion it is noticed that the (z,t)-model can be extended to include coherent light injection by adapting the boundary condition at the injected facet. In this way travelling-wave (single pass) and Fabry-Perot (multiple pass) amplifiers can be simulated in case of intensity or phase modulated input beams. Also injection locked lasers become then a possible subject of investigation. A further and more complicated extension is the study of external cavity lasers, this in order to analyse mode-locking [23] as well as dynamic wavelength shift reduction.

5.3 Calculated Example

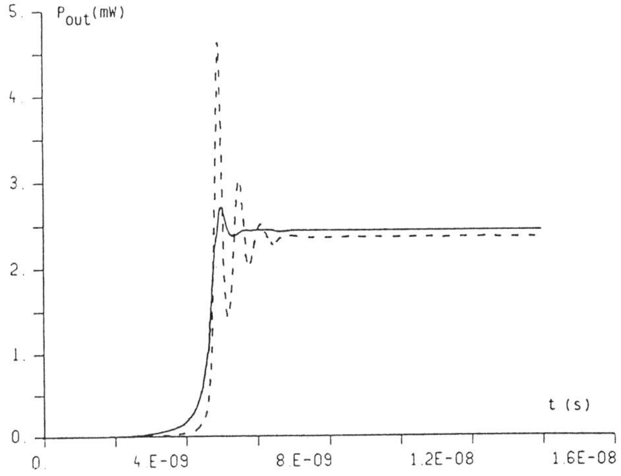

Fig. 8. Transient response of a laser with a 1 reflectivity.

Fig. 8 depicts the transient behaviour when a step current is applied to a laser diode with 1% power reflectivity at both mirrors. The full line is obtained with a longitudinal model while the dashed line results from a classical rate equation model (t-model). Both models have been implemented on a networksimulator. For the longitudinal model the laser length was divided into 35 sections of equal length. One clearly observes the strong suppression of the transient relaxation in the z-t-model. This is due to the fact that the longitudinal model takes superluminescence [14] into account. For low mirror reflectivities, thus high mirror losses, the gain above threshold is high. Therefore spontaneous emission is strongly amplified and a lot of spontaneous emission is coupled into the modes, causing a strong damping effect [24]. To obtain the same effect in the classical model,one should introduce an effective spontaneous emission factor β_{eff} in the photon rate equation.

6. MODELS FOR THE GAIN OF INVERTED III-V BULK MATERIALS

Under high carrier injection III-V direct materials show a considerable amount of optical gain due to population inversion causing stimulated emission. There are two main types of models describing the gain as a function of carrier concentration and photon energy.

6.1 Gain Models Based On The Einstein Relations

These models usually start from the Einstein relations for systems with two interacting energy bands. The relations give the net effect of absorption and stimulated emission under light injection of frequency $h\nu$. The net number of electrons per unit of volume, time and incident light power going from an energy level E_2 in the conduction band to an energy level E_1 in the valence band, with $E_2-E_1=h\nu$, is given by [25]:

$$g_s = B_{12}(h\nu)\ (\ f_2(E_2) - f_1(E_1)\ )\ /\ v_g$$
$$f_i = 1\ /\ [\ 1 + \exp\ (\ (\ E_i - F_i)\ /\ kT\ )\ ]\quad i=1,2 \tag{15}$$

Here f_1 and f_2 are the probabilities that the levels E_1 and E_2 will be occupied with an electron, v_g is the group velocity, $B_{12}(h\nu)$ is the transition probability and F_1 and F_2 are the Fermi levels for holes and electrons respectively. Up to now we considered only two discrete energy states in g_s. Therefore g_s should be extended to the case of transitions between two continuous sets of energy levels. Usually the one-electron approximation is used [26]. Transitions between two energy levels E_1 and E_2 can occur between every pair of states belonging to those energies. Therefore g_s must be multiplied with $\rho_c(E_2)\rho_v(E_1)$, $\rho_c(E_2)$ and $\rho_v(E_1)$ being the density of states at E_2 and E_1. Furthermore the transition with a photon of energy $h\nu$ can happen between several sets of energies E_1 and E_2, as long as $E_2-E_1=h\nu$. The gain g_s for a single set of energies can thus be extended by integration over the energy bands [27]. This gives:

$$g(h\nu) = \frac{1}{v_g} \int B_{12}(h\nu)\ (f_2 - f_1)\ \rho_v(E_1)\ \rho_c(E_2)\ \delta(E_2 - E_1 - h\nu)\ dE_2 \tag{16}$$

Solution of (16) is as follows [25]. Given a certain impurity concentration, an electron density is chosen for which the gain will be calculated as a function of photon energy. Assuming that the distributions of the density of states of electrons and holes are known one can now derive the position of the Fermi-levels relative to the band edges. In the case of lowly doped materials a parabolic band structure can be taken and the calculation of the Fermi-levels is straightforward using the inverse of the Fermi-Dirac integrals. For the strongly doped case energy tails are introduced in both energy bands [28]. In this way the low energy tail found in experimental absorption spectra can be justified. The Fermi levels must be determined iteratively because the density of states is now concentration dependent. For parabolic bands the transition probability is a simple expression of the form $B_{12} \sim \delta(\bar{k}_c - \bar{k}_v)$ [29], where the Dirac function expresses the "k-selection rule" [25]. For the bandtail models k-selection is however no longer valid and B_{12} becomes a complex function of $h\nu$. Several methods [30], [31], with increasing perfection, have been proposed for the calculation of B_{12}. The combination of a Gaussian fit to the Halperin-Lax Band Tails and the B_{12} calculated by Stern [31] was the first model [32] to give an acceptable order of magnitude for the gain in the strongly doped case.

6.2 Gain Models Based On Relaxation Broadening Theory

The former models need the assumption of high doping to explain the low energy tail observed experimentally. Furthermore they do not take into account the effect of spectral hole burning, which is a decrease of the gain in the neighbourhood of those wavelengths at which a strong optical field exists. This optical field causes carriers to recombine at certain energy levels by stimulated emission. Therefore these levels become depleted and they will be refilled in a small but finite time by intraband relaxation so that a quasi-equilibrium will be established. Intraband relaxation is a process in which energy is redistributed over the carriers through collisions such as electron-phonon scattering, etc. In models based on such a process [33], absorption tails can be explained by relaxation broadening, this means that carriers can relax out of the bands before recombining.

Recently Asada and Suematsu [34] have proposed an improved model. Gain is attributed to a non-linear electric susceptibility χ_{nl} caused by electron-hole pairs. The optical field E and the polarisation $P=\chi_{nl}E$ are decomposed in their different spectral components:

$$E = \Sigma\, E_p(t)\, \exp(j\omega_p t)$$
$$P = \Sigma\, P_p(t)\, \exp(j\omega_p t) \qquad (17)$$

In this way the gain at an optical frequency ω_p can be written as

$$g(\omega_p) = \frac{\omega_p}{\epsilon_o n_r^2}\ \mathrm{Im}\ (\ P_p\ /\ E_p) \qquad (18)$$

in which n_r is the refractive index of the material.

The macroscopic polarisation P is interpreted as $P = N\,\langle\bar{R}\rangle$ [34], with N the total number of electrons in both conduction and valence bands and $\langle\bar{R}\rangle$ the dipole moment per electron, averaged over the ensemble of N electrons. This dipole moment associated with an electron is formed by the combination of the electron (negative) and the rest of the atom (the hole being positive). An incident optical field drives the electron and hole in opposite directions and the dipole tends to direct itself parallel with the field. This rotation involves a change in potential energy $h\nu$, which is either stimulated emission or absorption.

Using density-matrix theory [35], the macroscopic polarisation can be written as

$$P = N \sum_{n} (\rho_{mn} R_{nm} + R_{mn} \rho_{nm}) \tag{19}$$

ρ_{mn} are the components of the density matrix ρ, which satisfies a rate equation of the form [34]:

$$\frac{\delta\rho}{\delta t} = \Lambda + \frac{j}{\hbar} [\rho, H] - (\rho - \rho_o) / \tau_r \tag{20}$$

in which the Λ represents carrier injection, $H = H - RE$ is the Hamiltonian operator, with H_o occurring at equilibrium and $-RE$ the energy of the dipole R in a field E, and $[\rho, H]$ is the commutator between ρ and H. The second term of the right hand side represents the effect of collisions and the associated effect of relaxation, τ_r being the relaxation time and ρ_o the density matrix at equilibrium. R_{nm} is the matrix representation of the dipole operator and for parabolic bands its derivation is simple [29], but beyond our scope.

Theoretically the problem is now completely determined. Solving (20) gives us ρ_{nm} and then P_{nl} can be found with (19). For practical calculations ρ is expanded in a power series of the optical field E

$$\rho = \rho^{(o)} + \rho^{(1)} E + \rho^{(2)} E^2 + \ldots \tag{21}$$

so that in a second order approximation the gain $g_p \sim \mathrm{Im}\,(P_p / E_p)$ becomes:

$$g_p(\omega_p) = g_p^{(1)} - \sum_q g_{pq}^{(3)} \, |E_q|^2$$

with $g_p^{(1)}$ the linear gain at ω_p and $g_{pq}^{(3)}$ the coefficient describing gain suppression at ω_p due to an optical field at ω_q. The combination of (21), (17) and (18) give us g_p. In fig. 9 this model is applied to 1.3µm $In_{0.74}Ga_{0.26}As_{0.57}P_{0.43}$.

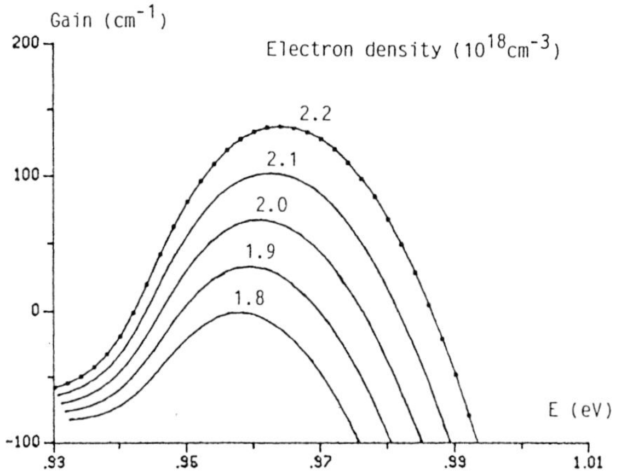

Fig. 9. Gain spectra of InGaAsP (1.3 µm material).

7. CONCLUSIONS

Laser models are fairly complex, leading to unpractical computer times if too extended models are used. Therefore one should always try to minimize the number of independent variables and physical interactions taken into account in the model. Of course, if not done with care, this could substantially reduce the simulation possibilities and accuracy. This means that a compromise must be sought between simplicity and sufficient rigour in modelling the physical processes. It also implies that, depending on the device type to be simulated, specific and optimised models must be developed for it. Therefore prior to any modelling effort, the model designer should anticipate what independent variables and which physical phenomena have a dominant influence on the behaviour of the specific laser he wants to study.

8. REFERENCES

[1] Buus J. (1985), "Principles of semiconductor laser modelling", IEE Proc., Vol.132, Pt.J, no.1

[2] Van de Capelle J.P., Vankwikelberge P., Baets R. (1986), "Lateral Current Spreading in DH lasers above threshold", IEE Proceedings, Part J, UK, 133, nr.2, pp.143-148

[3] Baets R., Lagasse P.E. (1984), "Longitudinal static-field model for DH lasers", Electr.Lett., vol.20, pp.41-42

[4] J.Van Roey, P.E.Lagasse (1982), 'Coupled-beam analysis of integrated-optic Bragg reflectors', J.Opt.Soc.Am., Vol.72, nr.3, pp.337-342

[5] J.P.Van de Capelle, R.Baets, P.E.Lagasse (1987), 'Twodimensional model for C^3- and external cavity lasers', ESSDERC conference proceedings, Italy, pp.1025-1028.

[6] Baets R., Van de Capelle J.P., Lagasse P.E. (1985), "Longitudinal analysis of semiconductor lasers with low Reflectivity Facets", IEEE J. Quantum Electronics, 21, nr.6, pp.693-699

[7] Yonezu H., Sabuma I., Kobayashi K., Kamejima T., Veno M., Nannichi Y. (1973), "A GaAs-$Al_xGa_{1-x}As$. Double Heterostructure Planar Stripe Laser", Jap. J. Appl. Phys., 12, nr.10, pp.1585-1592

[8] Verbeek B.H., Opschoor J., Vankwikelberge P., Van de Capelle J.P., Baets R. (1986), "Analysis of index-guided AlGaAs lasers with mode filter", El. Lett., 22, nr.19, pp.1022-1023

[9] Vankwikelberge P., Van de Capelle J.P., Baets R., Verbeek B.M., Opschoor J. (1987), "Local Normal Mode Analysis of Index-Guided AlGaAs Lasers with Mode Filter", IEEE J. Quantum Electronics, 23, nr.6, pp.730-737

[10] Van de Capelle J.P., Baets R., Lagasse P.E. (1987), "Multilongitudinal mode model for cleaved cavity lasers", IEE Proceedings Part J, 134, nr.1, pp.55-64

[11]Van de Capelle J.P., Baets R., Lagasse P.E. (1987), "Multilongitudinal mode model for cleaved coupled cavity lasers", IEE Proceedings Part J., 134, nr.4, pp.232-248

[12]Wilt D.P., Yariv A. (1981), "A self-consistent static model for a double heterostructure laser", IEEE J. Quantum Electron., 17, nr.9, pp.1941-1949

[13]Tucker R.S. (1985), "High-speed modulation of semiconductor lasers", IEEE Trans. on Electron Devices, vol.ED-32, no.12, pp.2572-2584

[14]Lau K.Y., Yariv A. (1982), "Effect of superluminescence on the modulation response of semiconductor lasers", Appl.Phys.Lett., vol.40, no.6

[15]Kawaguchi H. (1982), "Optical bistable-switching operation in semiconductor lasers with inhomogeneous excitation", IEE Proc., Vol.129, Pt.I, no.4

[16]Ueno M., Lang R. (1985), "Conditions for self-sustained pulsation and bistability in semiconductor lasers", J.Appl.Phys, vol.58, no.4

[17]Wong Y.L., Carroll J.E. (1987), " A travelling wave rate equation analysis for semiconductor lasers", Solid-state Electronics, vol.30, no.1, pp.13-19

[18]Schubert M., Wilhelmi B. (1986), "Nonlinear optics and quantum electronics", Chapt.1, John Wiley & Sons, New York

[19]Joyce W.B. (1980), "Current-crowded carrier confinement in double-heterostructure lasers", J.Appl.Phys, vol.51, no.5, pp.2394-2401

[20]Habermayer I. (1981), "Nonlinear circuit model for semiconductor lasers", Optical and Quantum Electronics, vol.13, pp.461-468

[21]Demokan M.S. (1986), "The dynamics of diode lasers at microwave frequencies", GEC Journal of Research, vol.4, no.1, pp.15-27

[22]Hildebrand F.B. (1968), "Finite-difference equations and simulations", Chapt.2, Prentice-Hall, Inc., Englewood Cliffs, N.J.

[23]Demokan M.S. (1986), "A model of a diode laser actively mode-locked by gain modulation", Int.J.Electronics, vol.60, no.1, pp.67-85

[24]Boers P.M., Vlaardingerbroek M.T., Danielsen M. (1975), "Dynamic behaviour of semiconductor lasers", Electr.Lett., vol.11, no.10, pp.206-208

[25]Casey H.C., Panish M.B. (1978), "Heterostructure lasers", Part A, chpt.3, Academic Press, New York

[26]McKelvey J.P. (1966), "Solid state and Semiconductor Physics", Harper, New York

[27]Lasher G., Stern F. (1964), "Spontaneous and stimulated recombination radiation in semiconductors", Phys. Rev. A, vol.133a, pp.553-562

[28]Halperin B.I., Lax M. (1966), "Impurity-band tails in the high-density limit. I. Minimum counting methods", Phys.Rev., vol.148, pp.722-740

[29]Kane E.O. (1957), "Band structure of Indium Antimonide, J.Phys. Chem.Solids, vol.1, pp.249-261

[30]Hwang C.J. (1970), "Properties of spontaneous and stimulated emission in GaAs junction lasers. II. Temperature dependence of threshold current and excitation dependence of superradiance spectra", Phys.Rev.B, vol.2, pp.4126-4134

[31]Stern F. (1971), "Band-tail model for optical absorption and for the mobility edge in amorphous silicon", Phys.Rev.B, vol.3, pp.2636-2645

[32]Casey H.C, Stern F. (1976), "Concentration-dependent absorption and spontaneous emission of heavily doped GaAs", J.Appl.Phys., vol.47, no.2, pp.631-642

[33]Yamada M., Suematsu Y. (1981), "Analysis of gain suppression in undoped injection lasers", J.Appl.Phys., vol.52, no.4, pp.2653-2664

[34]Asada M., Suematsu Y. (1985), "Density-matrix theory of semiconductor lasers with relaxation broadening model - Gain and gain-suppression in semiconductor lasers", IEEE J.Quantum Electron., vol.QE-21, no.5, pp.434-442

[35]Marcuse D. (1980), "Principles of Quantum Electronics", Chapt.7, Academic Press, New York

9. ACKNOWLEDGEMENT

This paper was published in "Annales des Télécommunications", 43, no. 7-8, pp. 423-433, 1988. The author thanks the editors of this journal for permission to republish.

The author would also like to thank J.P. Van de Capelle, P. Vankwikelberge and D. Botteldooren for contributions to this overview of laser modelling.

Equivalent Circuit Models for Silicon Devices

Margaret E Clarke

University of Manchester

1. INTRODUCTION

The equivalent circuit modelling of semiconductor devices was on the scene long before numerical device simulation became sufficiently developed to form a useful design tool. In spite of the rise of more glamorous forms of device modelling, the familiar equivalent circuit models of undergraduate text-books are still the workhorse of circuit design engineers. They are used to model the latest sub-micron devices and incorporate the many new effects introduced by recent technological advances.

It must be admitted that equivalent circuit models taken on their own are not very interesting; they are rather like a map of a country without the associated history, geography and sociology. In this session therefore we will be looking at not just the equivalent circuit models themselves but also at their context and use, their links with analytical and numerical device modelling and their application to circuit design and development.

First we will consider what is meant by an equivalent circuit model and what types of circuit are commonly used. Then, rather than attempt to catalogue a whole range of different device models and variations, I have chosen to discuss the equivalent circuit representations of two particularly well known low power silicon devices, the bipolar transistor and the MOSFET. It is interesting to see how these models stand up to recent developments in technology. Will we still be using them in ten years time?

In addition, the vertical DMOSFET and the Lateral Insulated Gate Transistor (LIGT) will be used to illustrate the different types of problem that are now being encountered with high power device equivalent circuit models.

Some practical aspects of equivalent circuit modelling will also be considered. Sooner or later, if a model is to be of practical use, numerical values must be assigned to the elements. Thus, parameter extraction, which is the science of obtaining values for the components of equivalent circuit models, forms an important topic.

Finally we will look at applications of the models; as steps in a complete device and process optimisation, by analogue circuit designers and by digital design engineers. Some examples and some recent developments will be discussed.

2. WHAT IS AN EQUIVALENT CIRCUIT ?

The need for equivalent circuit device models arises out of the need to understand the behaviour of devices in circuits and to simulate and validate circuit designs before committing them to silicon. It is therefore pertinent to say a few words about circuit simulation.

The breadboarding methods of the past cannot simulate accurately the devices and the chip environment in which the devices will be embedded. Since the mid seventies, chip designers have come to rely instead on the present generation of circuit simulators. Although several different circuit simulators were developed during the seventies, the most famous and most widely used is SPICE (Nagel 1975). Its tremendously widespread use has no doubt been helped by the fact that it was available to anyone for the cost of a magnetic tape. Having become hooked on circuit simulators, many are now prepared to pay for commercial versions of SPICE with pre- and post-processing enhancements which make them much more convenient to use. These are the so-called "Alphabet" versions: HSPICE, MSPICE, PSPICE etc.

References to SPICE are to be taken as shorthand for any SPICE-type circuit simulator, this particular version having acquired a name in the field of simulation similar to that of Hoover in the domestic arena.

A device model in a circuit simulator is expressed in terms of lumped circuit elements. These elements include independent and controlled current and voltage sources, resistors, capacitors and inductors. The elements are described by analytical expressions which are evaluated by the simulator. The equivalent circuit model is required to represent an approximation to the device behaviour which is sufficiently accurate for a particular application.

An example which can be used to illustrate the basic principles of equivalent circuit models is the pn junction diode. Figure 1 shows five different representations of a diode.

The DC behaviour is simply modelled by the equation for the diode current

$$I_d = I_S \left[e^{qV_d/kT} - 1 \right] \tag{1}$$

where the saturation current I_S is the only parameter to be provided. A series resistor is added to account for the resistance of the semiconductor and the contacts (Fig. 1b).

For transient analysis two capacitances are added to the DC circuit, to model the currents due to charge flow. Capacitance is defined as $C = dQ/dV$.

C_j , in Fig. 1c , models the capacitance due to the junction depletion layer and C_D the stored charge which is equal, but of opposite sign, to the charge of the minority diffusion carriers. The variation of these two capacitances with voltage is well understood. Expressions for these form part of the circuit simulator device model.

In a transient analysis the simulator calculates the circuit's initial bias point and then calculates the circuit's time response at a series of time intervals, or steps, using the full non-linear device equations.

The linearised small signal AC model is shown in Fig. 1d . The capacitances and resistors are the same as in the transient model but the non linear current represented by equation (1) is replaced by a diode conductance g_d. In AC calculations we are concerned with small changes in current ΔI caused by small changes in voltage ΔV and so require the differential of the I-V characteristic evaluated at the bias point of the diode. Thus for the diode:-

$$g_d = \frac{qI_d}{kT} \qquad (2)$$

One further circuit representation of the diode is shown in Fig. 1e . This circuit models the AC noise response of the diode. As this subject will be covered elsewhere in the course I will simply note here that the circuit is derived by adding an equivalent noise current diode and an equivalent noise current series resistor in parallel with the diode and resistance components.

This simple example demonstrates the different types of equivalent circuit used in various types of analysis.

3. TWO LOW POWER SILICON MODELS

Most SPICE users will be familiar with one or other of the equivalent circuit models of two more complex devices, the bipolar transistor and the MOSFET (metal-oxide-silicon field effect transistor).

Figure 2 shows a cross section of a typical planar bipolar transistor together with the DC/transient and AC small signal equivalent circuit models as used in SPICE. The linearised small signal model is the well known hybrid-π circuit. This circuit gives a good approximation to the circuit behaviour of the bipolar transistor at frequencies for which the transistor has useful gain for anlalogue and digital applications.

The model shown in Fig. 2b has only ten components, some of which appear, reassuringly, to be resistors. It therefore comes as a bit of a shock to some first time users of SPICE to find a list of more than forty parameters for the model. The origin of the equivalent circuit model and the meaning of the parameters is described in many text-books, but the book by Getreu (1976) gives a particularly thorough treatment which is helpful for those who need to understand the significance of all the parameters.

A number of the parameters are fairly easily understood, the ohmic resistances for example (although the derivation of suitable values is another matter). The junction capacitances are modelled by an equation which quickly becomes familiar and the storage/diffusion capacitances are proportional to the transit times τ_F and τ_R.

Equations for the forward and reverse currents model high current, low current, basewidth modulation and base widening effects. Use of default parameters for second order effects gives a representation equivalent to the simpler Ebers-Moll models, EM1 and EM2, while the full modelling of all the effects corresponds to the Gummel-Poon model (Getreu 1976). Temperature effects are modelled by the built in temperature dependencies of some parameters and by the inclusion of temperature coefficients for others.

This model has been remarkably successful over many years and has been used to model low power bipolar transistors made by a wide variety of processes. It does not even seem to have run out of steam in the sub-micron era. A Gummel-Poon model was used to simulate the 30ps gate delay of the "super-selfaligned transistor" (Ichino et al 1984). In this case a macro model of the transistor was made up for the 5 emitter stripe transistor. Similar techniques have been used to model distributed base and emitter edge effects. Avalanche breakdown can be modelled by adding a current-dependent current source between the collector and base terminals, though this can cause convergence problems in some programs. In fact whenever a new complication crops up it always seems to be possible to model it by combining elemental bipolar models into a macro model for the device together with other lumped equivalent components. However, as we shall see later, the model does run into problems for high power transistors.

The second of the models I would like to discuss is the silicon MOSFET. Figure 3 shows the equivalent circuit models used in the SPICE circuit simulator. Again there are a large number of parameters associated with the SPICE model, most of them occurring in the expressions which model the current source I_{ds}.

The best known analytical models used with these equivalent circuits are the levels 1, 2 and 3 models which form part of SPICE2. Level 1 is the Schichman-Hodges model in which I_{ds} as a function of V_{ds} is modelled in the linear and saturation regions in terms of the the gate-source voltage V_{gs} and the threshold voltage V_t together with some geometric factors. This first order model is used for channel lengths greater than about ten microns and when a simple model is required for hand calculations.

Level 2 is an analytical one-dimensional model which incorporates most of the second order effects of small devices (Vladimirescu and Liu, 1980). These second order effects become important for channel lengths of two microns or less. They include the channel length modulation in saturation which gives rise to a finite output conductance, velocity saturation effects, the effects of short and narrow channels on the threshold voltage and other effects. Level 3 models the same effects as level 2 but is a semi-empirical model. It has fewer parameters and needs less computing time (Vladimirescu and Liu, 1980).

In contrast to the Gummel-Poon bipolar transistor model the MOS SPICE models are not usually found to be completely satisfactory by different manufacturers and many users substitute their own models in place of the generally available ones. Thus some newer versions of SPICE and in-house circuit simulators include options for adding one's own model equations. Examples of recently developed short channel MOSFET models are given in papers by Sheu et al (1987) and Lee and Rennick (1988). The book by Tsividis (1988) provides a comprehensive treatment of all aspects of the analytical modelling of MOS transistors.

The gate capacitances in the models shown in Fig. 3 are often replaced nowadays by a charge model. Ward and Dutton (1978) showed that the use of expressions for capacitance rather than expressions for charge led to errors in simulations of particular types of circuit. This is the phenomenon known as charge nonconservation. Much has been written about this problem; Yang et al (1984) give a clear account.

The charges associated with each of the four MOSFET terminals are affected by voltage changes at all four terminals. There are then sixteen possible capacitances; but it can be shown that only nine of these are independent. Also, they are not in general reciprocal eg $C_{dg} \neq C_{gd}$ in the saturation region.

Thus for applications where the simple capacitances of Fig. 3 do not give an accurate representation for MOSFET simulation, models contain expressions for the four terminal charges in each of the different operating regions, from which the necessary charge derivatives can be calculated.

It seems questionable to me whether an equivalent circuit representation of nine capacitances, which could include several nonreciprocal pairs, each capacitance varying differently in different regions of operation, serves any useful purpose. Equivalent circuit models can certainly be derived. For example Tsividis (1988) shows how five out of nine capacitances provide a suitable model at low and medium frequencies. Four more "transcapacitances" are defined which can be added to the low frequency equivalent circuit model for accurate high frequency modelling. But a circuit simulator would derive its data from expressions of charge, and the equivalent circuit has now become so complex that it cannot be easy for a designer to use it to provide a simple understanding of the device operation in terms of circuit elements.

Will the intrinsic MOS transistor model disappear, perhaps, into a black box in the equivalent circuit model, surrounded by the extrinsic transistor components ?

4) HIGH POWER SILICON DEVICE MODELS

The producer of high voltage discrete devices is in a different situation from that of an integrated circuit manufacturer. Even when the integrated circuit circuit designer is carrying out semi custom design remote from the manufacturer, the CAD tools and device models are usually supplied by the manufacturer. Circuit simulation incorporating high voltage devices, on the other hand, is carried out by many different purchasers who only have access to the manufacturers' data sheets. The challenge then is to find an equivalent circuit representation which will provide an adequate behavioural description of the device and which can be used with standard versions of SPICE.

An example of a high voltage device which illustrates this problem is the vertical DMOSFET. Figure 4 shows an N-type enhancement device in which the channel is formed at the surface of a diffused P-well.

There are several reasons why the $I_{ds}-V_{ds}$ behaviour of this device may not be accurately described by the low power MOS equations: the so-called bulk region in which the channel is formed is not uniformly doped but is graded along the channel length. Also, at high currents the increased device temperature can significantly affect the electron saturation velocity to the extent that, in devices without any form of compensation, the conductance, dI_{ds}/dV_{ds}, can be negative for high values of V_{ds} and V_{gs}. Thus it is unlikely that the same set of parameters can be used to model the $I_{ds}-V_{ds}$ curves for both high and low values of I_{ds}.

The drain of the DMOSFET is a particularly difficult region to model for switching applications. When the device is turned off, the large potential drop across the p-n bulk-drain diode results in a wide depletion region in the high resistivity epi-drain so that it is possible for the drain region between two neighbouring transistors to be almost completely pinched off. This phenomenon is largely responsible for the highly non linear behaviour of the gate-drain capacitance which can vary by over two orders of magnitude during a switching cycle.

While these problems may provide fertile ground for the device theorists, the company selling the devices is faced with the problem of providing simple behavioural models for the end user in the form of equivalent circuit models which can be interfaced in the standard manner to SPICE simulators. Various approaches have been adopted for the modelling of the critical gate-drain capacitance: a polynomial approximation for C_{gd} has been used (Hancock 1987), a cascode JFET is included in series with the MOS transistor (Wheatley et al 1985) and a piecewise modelling of the various capacitances during switching is proposed by Castro Simas et al (1987). So this is an area where new work is being carried out on the development of equivalent circuit approximations.

It can be seen that in the short term this type of equivalent circuit modelling is necessary, but it is apparent that in order to model device behaviour adequately the equivalent circuits are becoming increasingly complex, and artificial means of switching in different circuit components are contrived to account for non linear components.

The time would appear to be ripe for a new approach to the whole problem, in which the equivalent circuit representation is abandoned in favour of a more direct physical model which can still be interfaced to a circuit simulator.

Just such a new approach is proposed by McDonald and Fossum (1988). They take as an example the lateral insulated gate transistor (LIGT) which is a high voltage switching transistor comprising merged bipolar and MOS sructures (Fig. 5a). The built in SPICE models do not properly account for many of the features of the device behaviour, eg. the low BJT gain caused by carrier recombinaton in the wide base, base region conductivity modulation and latch up.

McDonald and Fossum propose the replacement of the equivalent circuit model by a set of User Defined Controlled Sources (UDCSs) each of which is defined by an equation in terms of the nodal voltages. The steady state behaviour is described by three current equations and a voltage equation (I_{CN},I_{cp},I_{mos} and V_{epi} in Fig. 5b) and the transient behaviour by two charge based UDCSs of the form dQ/dt. It is stressed that the "circuit" model shown in Fig. 5b is not an equivalent circuit. Computer subroutines are used to solve the necessary equations for the evaluation of the UDCSs at the required nodal voltages.

This approach is intuitively satisfying because it provides a more direct physical description of the device, and the model is interfaced directly to SPICE. It is also akin to the methods now used for the charge based modelling of the low power MOSFET.

There are however a number of reasons why it may be some time before new methods like this become widely used:-

1) The computing times required for the UDCS models are substantially longer than the times needed for the built-in SPICE models.
2) Some versions of SPICE are not designed for the easy interfacing of new types of model. Many users do not have the resources and expertise to implement new SPICE models. The manufacturer of discrete devices in particular is tied to using models which can be employed by non-expert users of standard versions of SPICE.
3) The biggest problem with the UDCS model is that of obtaining the necessary parameters for the equations. McDonald and Fossum derive most of the parameters from a knowledge of the structure and doping densities of the device, but for real life applications it is important to be able to derive all the essential parameters from devices or test structures made by the same process as the target devices.

5. PARAMETER EXTRACTION

Parameter extraction is an extremely important aspect of device modelling for the purposes of circuit simulation. A model used for circuit simulation is only as good as the numerical parameters that are available for it. An elegant and potentially accurate analytical description of a device is no good to the end user if reliable numerical values cannot be obtained for the parameters.

Many of the parameters for the model current equations are derived from a series of current-voltage measurements. Sophisticated numerical methods are required in some cases to fit a number of different parameters to several measured curves. Fortunately for those who do not have readily available resources to develop these techniques, software packages can now be bought which, together with measurement equipment, provide an automatic parameter extraction system. TECAP, by Hewlett Packard, which is used by a number of companies in the UK, is a good example. It is usual to make measurements on several transistors of different shapes and sizes to obtain the information which is necssary to model the geometry dependencies of the parameters.

Measurements of junction and overlap capacitance on the actual devices to be modelled are rarely feasible, because the capacitances of the active components are very small compared with the parasitic capacitances associated with the connections to the device necessary for testing, pads for example. The usual method is to incorporate large capacitors of different edge-length to area ratios for all the important types of capacitance on a purpose designed parameter evaluation chip. Measurements made as a function of voltage allow the extraction of the pn junction capacitance parameters to be derived. Many junctions are neither true "step" junctions or true "graded" junctions, but it is usually possible to find values of C_0, φ and m in the equation

$$C = \frac{C_0}{(1 + V/\varphi)^m} \tag{3}$$

which fit the variation of capacitance reasonably accurately over the range of voltages which will be used.

The diffusion capacitance of bipolar transistors is found by deriving τ_F from s-parameter measurements made on individual devices. Considerable care is needed to de-embed the device characteristics from those of the parasitic pads and package.

Contact resistance and the sheet resistance of various diffusions can be measured using a set of test structures, but spreading resistances, where the current does not flow in parallel lines in a 2-D plane, can present a problem. Spreading resistances cannot be accurately derived from sheet resistance measurements without the aid of numerical simulation, preferably 3-D.

In the bipolar transistor the measurable effects of base resistance are not easily disentangled from those of other parameters. An additional complication is that the effective base spreading resistance depends on the region of operation of the transistor. The current flow during DC operation is quite different from that during transient or AC operation when base current is flowing to charge the junction and diffusion capcitances. The base resistance in reverse operation is also different from that during forward operation because of the asymmetric geometries of the emitter and collector.

The DC base resistance can be measured by fitting a pair of values for emitter and base resistance to the high current region of the I_c and I_b vs V_{eb} curves. Second order effects which might allow the separation of the two resistances are too small to give reliable results so an independent estimate of the emitter resistance is required. This is not as easy as the text-books would suggest because the classic method of measuring emitter resistance, in which the base current is measured as a function of collector-emitter voltage with an open-circuited collector, gives incorrect values for planar transistors. However a good estimate of the emitter resistance can be made from a knowledge of the emitter contact resistance, diffusion sheet resistance and junction depth. In fact, as with the empirical fit of parameters to equation (3) it does not matter whether values for the two resistors are accurately apportioned as long as the model predicts the correct transistor I-V behaviour.

The measurement of AC base resistance presents similar problems. In principle, and according to the textbooks, s-parameters measurements can be used. In practice the measurements are very difficult to interpret because of the effects of comparatively large parasitic pad capacitances.

I have gone into the example of base resistance because it is not always appreciated that some of the apparently simple parameters are not easy to measure and that methods that might quite well have suited the type of transistor made in the sixties can be grossly inaccurate for small scale planar transistors.

The spreading resistance due to current flow in the bulk silicon, which can be important for the modelling of latchup in MOS transistors and for the calculation of the resistance of contacts to the back of the chip, can be difficult to estimate. Some recent papers (Chen and Wu 1986, Deferm et al 1988) show how numerical methods can be used to calculate these parameters from a knowledge of the resistivities of the different layers in the wafer.

It should not be forgotten that on VLSI chips, the equivalent circuit model parameters of passive components external to the transistors, interconnects and resistors for example, can be as important as the transistor parameters in circuit simulations. Parameter evaluation chips will therefore usually include structures for the measurement of the passive component parameters.

6. THE APPLICATIONS AND USE OF EQUIVALENT CIRCUIT MODELS

Equivalent circuit models are used in two very different modelling worlds which hardly come into contact. These could be called the "real" world of measurable, useable devices on silicon and the theoretical simulation world.
The interests and needs of the users of equivalent circuit models in these two worlds are quite different. Theoreticians may spend many happy years studying the effects of emitter high doping density effects on bipolar transistor gain, but to the circuit designer and device producer the gain is a simple parameter over which the process engineer has a good deal of control and which can be very simply measured on the finished device. The circuit designer is far more interested in what the theoretician would regard as "the boring bits", spreading resistance and junction capacitance for example. As in much of electrical engineering the product RC is of supreme importance and the designer wants to know, within say 10%, the mid range values and production spreads of critical resistances and capacitances. A large error in one of these estimates can mean that a design completely fails to meet a specification.

Numerical simulators cannot supply accurate values for real devices even if they are sufficiently accurate in themselves. To do so they need access to accurate three dimensional data on the doping density distributions in manufactured devices. Without wanting to go into the details of the problems of process simulation and silicon analysis techniques, I would say that the present tools are a long way from being able to predict the doping distributions in sub micron devices with sufficient accuracy.

Thus circuit designers are inclined to regard numerical device simulation as an esoteric waste of time. But the point is that the empirical modelling of the real world and the computer modelling of the simulation world serve totally different purposes. The theoretical simulations are used for "what if" and sensitivity analysis, understanding of the effects of changes in process and device structure on circuit design, optimisation of device structure to produce desired electrical characteristics etc – all very important in the development of new devices and processes.

Until recently only very large companies have had the resources to develop complete integrated simulation environments with good interfaces between all the various simulators. Such a system would include process simulators (SUPREM, SUPRA, etc), device simulators (PISCES, MINIMOS, BIPOLE etc) and circuit simulators, all linked by interpolation and parameter extraction tools to pass the parameters from one simulator to the next. A system of this type is illustrated in Fig. 6. However it is now possible to buy sets of linked simulation packages; for example TMA (Technology Modelling Associates) market software which includes much of the well known semiconductor modelling work produced by the University of Stanford together with the necessary linking software.

The equivalent circuit model sits at one of the interfaces in these integrated systems. The practice has been to extract the circuit model parameters from a numerical device simulator by producing simulated device measurements and using the same parameter extraction methods as for experimentally produced curves. This requires a large number of simulated "measurements". A more elegant approach is described by Wan et al (1988) who show how they can obtain almost all the parameters for a level 2 MOSFET model using only six biasing conditions. Their software MOSGEN provides the interface between the two-dimensional device simulator PISCES and the SPICE circuit simulator.

Is it possible perhaps to dispense altogether with the equivalent circuit model as a stepping stone between device and circuit simulators? Several people have now implemented mixed-mode simulators in which device simulators are interfaced directly with circuit simulators (eg Engl and Dirks 1981, Axness et al 1987). So far only specific problems for very small circuits have been tackled using two-dimensional device simulation. A recently published example is by Rollins and Choma (1988) who have written a computer program to allow the simultaneous solution of an electrical network containing both nonlinear circuit elements and two-dimensional solid-state models. They point out that this type of modelling is necessary for some problems such as the "single event upset" where the interaction between several solid state devices must be modelled. The development, over the next few years, of powerful and cheap computers with parallel processing, could make this type of seamless modelling a practical proposition for larger problems.

It could be said that the equivalent circuit model is really only a "model of convenience" in the total simulation world, a sort of D/A converter; but in the real analogue world it is still a very important conceptual and analytic tool.

A classic example of the use of equivalent circuit models for digital circuit design is in the analysis of gate delay. The question might be asked "Given a process for which we know the unit junction capacitances and basic resistivities, which of a single base contact and a double base contact bipolar transistor will give the shorter gate delay?" The problem amounts to finding whether the effect of the decreased base resistance of the double base transistor on the gate delay is greater than the effect of the associated increased collector-base capacitance for any given power dissipation.

An analysis of the equivalent circuit model of the gate, either by hand using first order circuit theory approximations or using a circuit simulator, yields an expression for the gate delay in terms of the lumped equivalent values of all the capacitances and resistances associated with the gate, (see for example Tang and Solomon 1979). Gate delays can then be plotted as a function of power dissipation for each of the two transistors.

The calculation of RC delay times is also one of the main applications of equivalent circuit models for MOS VLSI circuit design where relatively simple models are usually adequate. The full complex models are more likely to be needed by the designer of new transistors and cells.

Analogue designers need both the complex device models for circuit simulation and much simpler models which are suitable for hand calculations. In the early stages of design simple models are used as an aid to understanding circuit operation and configuration. Detailed calculations can then be performed using SPICE once the circuit configuration has been determined. Even then, the fullest complex models may not be needed because the designs have to be tolerant of quite large spreads in parameters due to batch-to-batch and manufacture-to-manufacturer variations. It is therefore not worth modelling some of the smaller effects.

We have seen that developments in the processing and design of silicon devices, such as the scaling to submicron geometries and the merging of bipolar and MOS technologies, lead to increasingly complex analytical device models for which the equivalent circuit models are either very complex or are replaced by representations which do not use the standard lumped element approximations. It is worth considering how these developments affect the circuit designers; the people who play a crucial role in the transformation of process and device developments into the working devices which bring in the money and so keep the whole industry going.

If the circuit designer is to exploit the many recent improvements in device design, he or she still needs models which give a simple understanding of a device in terms of its circuit behaviour. Blind use of of circuit simulators using "black box" models is inefficient and does not provide the insight needed for good design. I would therefore like to make a plea for better communications between engineers working in different areas of semiconductor design. Device physicists need to appreciate that circuit designers, experts in their own field, cannot be expected to keep up with all the developments in the modelling field in anything like the same detail as themselves. Clearly written guides are needed which enable the circuit designer to select the simplest model needed for any particular application and to appreciate which of the plethora of newly hatched parameters are likely to be significant. Is this perhaps an area where some form of Expert System, used in conjunction with circuit simulators, could play a useful role?

7. CONCLUSIONS

In this chapter we have touched on many different aspects of the development and use of equivalent circuit models, showing that this humble approximation has proved to be an essential and powerful tool in a complex technological scene.

However many of the recent developments suggest that the days of the equivalent circuit device model may be numbered. If this is so, there is a lot more that will have to change if the key role that equivalent circuits now play in the modelling world is to be filled by another type of model. There will be a need for new parameter extraction packages, and circuit simulators will have to be adapted to interface to the new types of model. But most importantly, new models must retain the essential property of equivalent circuit models, which is to provide behavioural models of devices which are meaningful to circuit designers.

REFERENCES

Axness CL, Weaver HT, Giddings AE, Shafer BD (1987) Single levent upset in CMOS static RAM and latches. In: Miller JJH (ed) Proceedings NASECODE-II Conf., Boole Press, Dublin

Castro Simas MI, Simoes Piedade M, Costa Freire J (1987) A new method for the evaluation of internal capacitances of power MOSFETs. EPE'87, Grenoble, France.

Chen M-J, Wu C-Y (1986) A new analytical three-dimensional model for substrate resistance in CMOS latchup structures. IEEE Trans on Electron Devices ED-33: 489-493

Deferm L, Claeys C, Declerck GJ (1988) Two and three-dimensional calculation of substrate resistance. IEEE Trans on Electron Devices ED-35: 339-351

Engl WL, Dirks H (1981) Functional device simulation by merging numerical building blocks. In: Browne BT, Miller JJH (eds) Proceedings NASECODE-II Conf., Boole Press, Dublin

Getreu I, (1976) Modeling the bipolar transistor, Tektronix Inc, Beaverton

Hancock JM (1987) A MOSFET simulation model for use with microcomputer SPICE circuit analysis. Proceedings PCI September 1987

Ichino H, Suzuki M, Hagimoto K, Konaka S (1984) Si bipolar multi-Gbit/s logic family using super self-aligned process technology. ICSSDM Dig. Tech. Papers, Kobe, Japan

McDonald RJ, Fossum JG (1988) High-voltage device modeling for SPICE simulation of HVIC's. IEEE Trans on Computer- Aided Design CAD-7:425-432

Nagel LW (1975) SPICE2: A computer program to simulate semiconductor circuits. Electronics Res.Lab. Memo ERL-M520, Univ. of California, Berkeley

Rollins JG, Choma J (1988) Mixed-Mode PISCES-SPICE coupled circuit and device solver. IEEE Trans on Computer-Aided Design CAD-7: 862-867

Sheu BJ, Sharfetter DL, Ko PK, Jeng M-C (1987) BSIM Berkeley short-channel IGFET model for MOS transistors. IEEE J Solid-State Circuits SC-22: 558-565

Tang DD, Solomon PM (1979) Bipolar transistor design for optimized power-delay logic circuits. IEEE J Solid-State Circuits SC-14: 679-684

Tsividis YP (1988) Operation and modeling of the MOS transistor. McGraw-Hill

Vladimirescu A, Liu S (1980) The simulation of MOS integrated circuits using SPICE2. Electronics Res. Lab. Memo ERL-M80/7, Univ. of California, Berkeley

Wan C-P, Sheu BJ, Lu S-L (1988) Device and circuit simulation interface for an integrated VLSI design environment. IEEE Trans on Computer-Aided Design CAD-7: 998-1004

Ward DE, Dutton RW (1978) A charge oriented model for MOS transient capacitances. IEEE J Solid State Circuits SC-13: 703-707

Wheatley CF Jr, Ronan HR Jr, Dolny GM (1985) Spicing up SPICE II software for power MOSFET modelling. RCA application note, 1985.

Yang P, Epler BD, Chatterjee PK (1983) An investigation of the charge conservation problem for MOSFET circuit simulation. IEEE J Solid State Circuits SC-18: 128-138

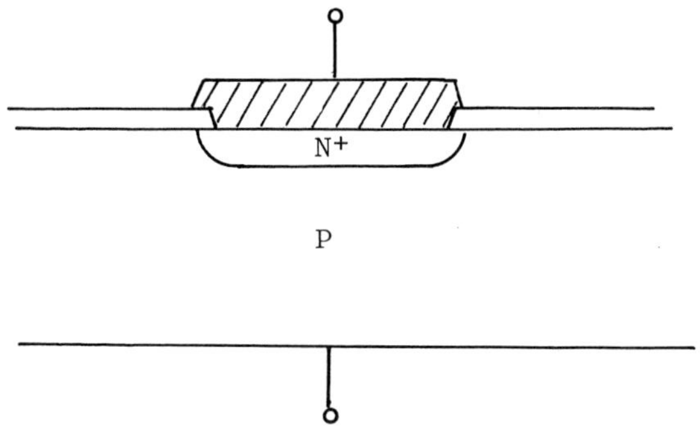

a) Schematic cross section of a planar diode

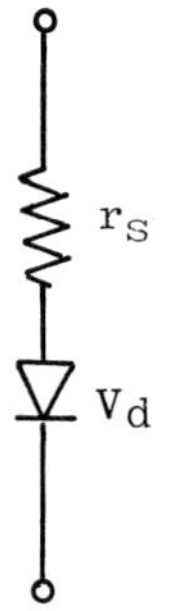

b) DC analysis

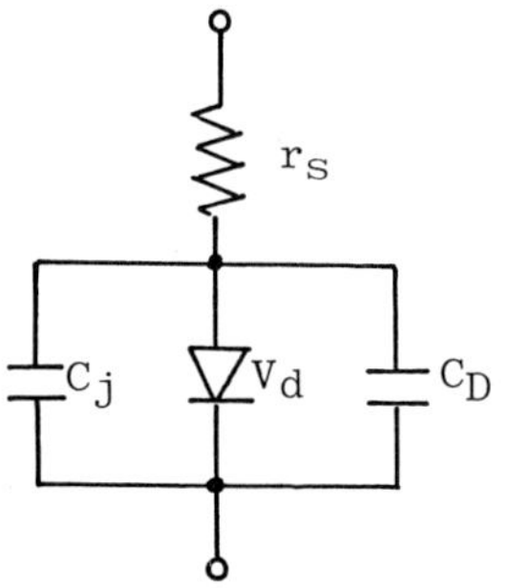

c) Transient analysis

Where:

$$I_D = I_S \left[e^{qV_d/kT} - 1 \right]$$

r_S = series resistance
C_D = diffusion capacitance
C_j = junction capacitance

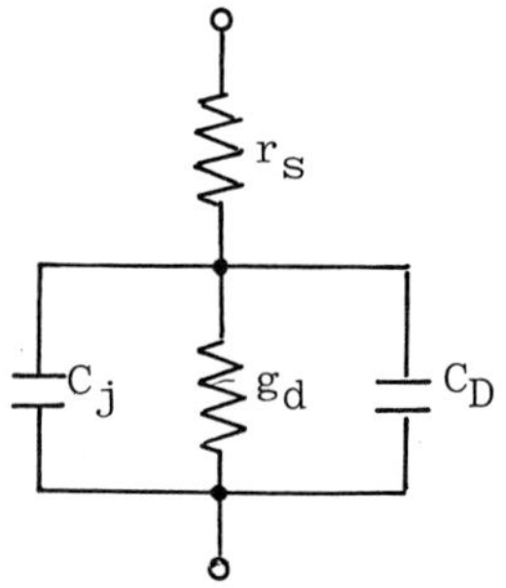

d) AC analysis

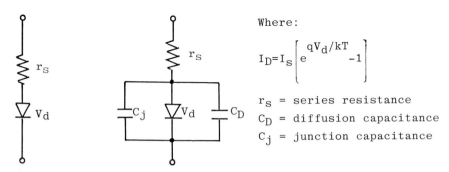

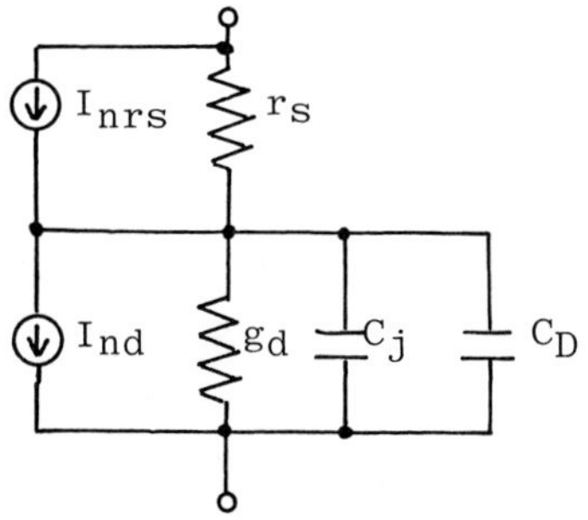

e) AC noise analysis

Where:

g_d = diode conductance
I_{nd} = equivalent noise current diode
I_{nrs} = equivalent noise current series resistor

Fig. 1 Equivalent circuit models for the pn junction diode

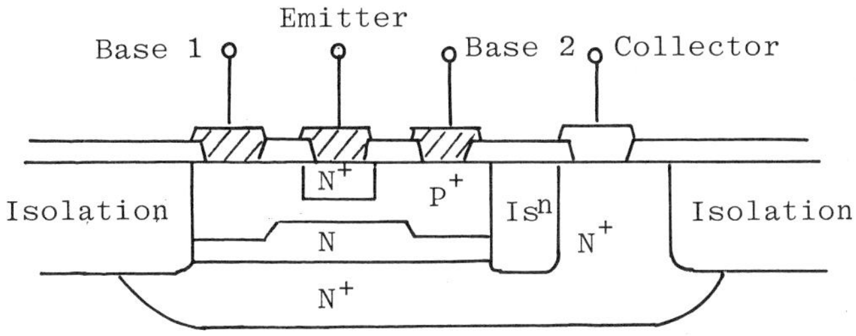

a) Schematic cross section of a planar bipolar transistor

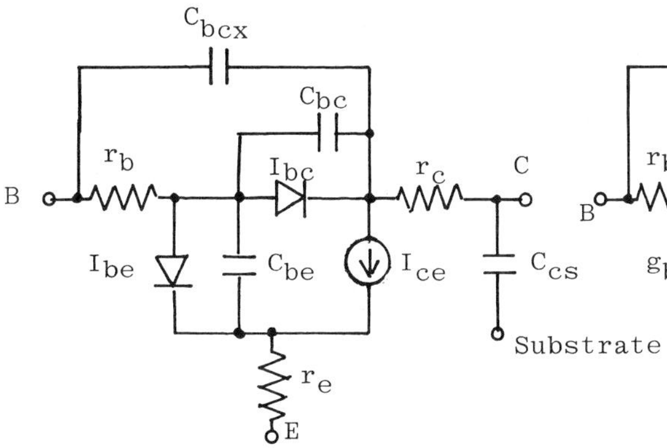

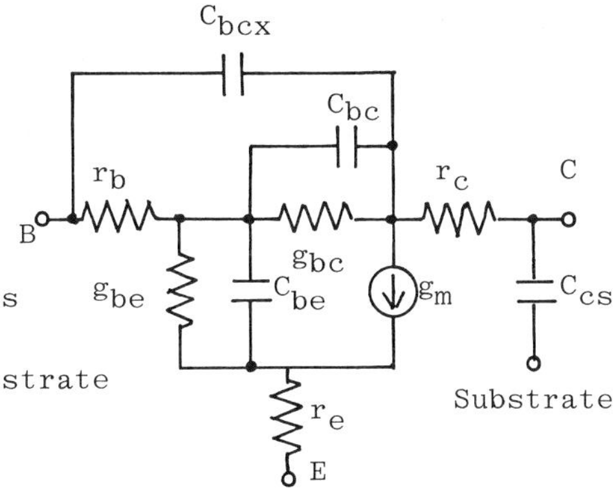

b) For DC and transient analysis c) For AC analysis

Where:

r_b, r_c, r_e are the internal base resistance, the emitter resistance and the collector resistance.

C_{bc} and C_{bcx} are the base to collector capacitances for the internal and the external bases.

C_{be} is the base to emitter capacitance, including both the diffusion and the depletion capacitances.

C_{cs} is the collector to substrate capacitance.

I_{be} and I_{cs} are the base to emitter and base to collector DC currents.

g_{be} and g_{bc} are the forward and reverse base conductances.

g_m is the transconductance.

Fig. 2 Equivalent circuit models for the bipolar junction transistor.

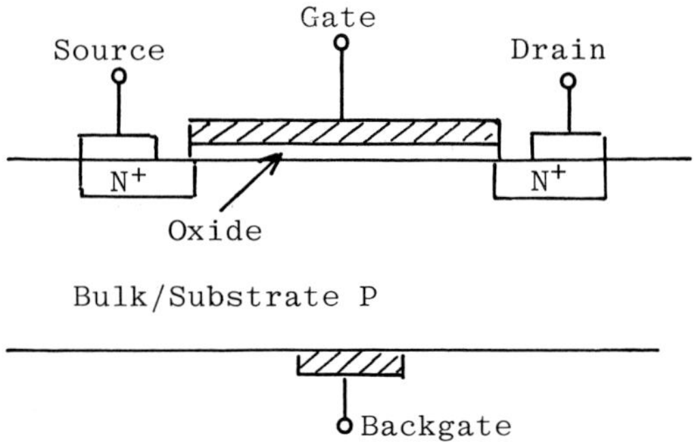

a) Schematic cross section of a MOSFET

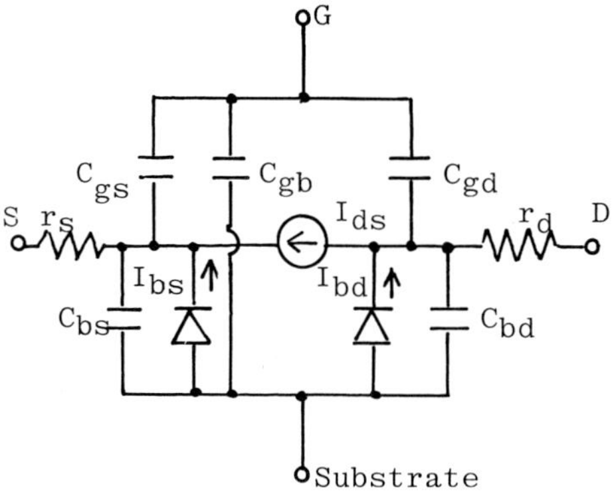

b) For DC and transient analysis

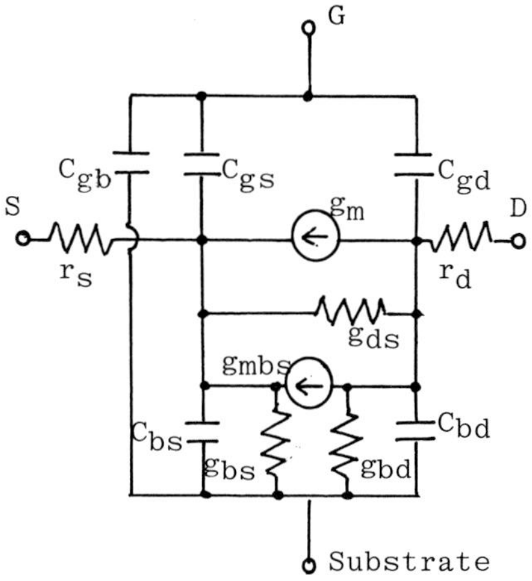

c) For AC analysis

Where:

C_{bd} and C_{bs} are the bulk to drain and bulk to source capacitances.

C_{gb}, C_{gd} and C_{gs} are the gate to bulk, gate to drain and gate to source capacitances.

I_{ds}, I_{bs} and I_{bd} are the drain to source, bulk to source and bulk to drain DC currents.

r_d and r_s are the drain and source resistances.

g_m is the AC transconductance controlled by V_{ds}.

g_{mbs} is the AC bulk transconductance drain to source controlled by V_{bs}.

g_{bd} and g_{bs} are the bulk to drain and bulk to source conductances.

g_{ds} is the AC conductance drain to source controlled by V_{ds}.

Fig. 3 Equivalent circuit models for the MOSFET

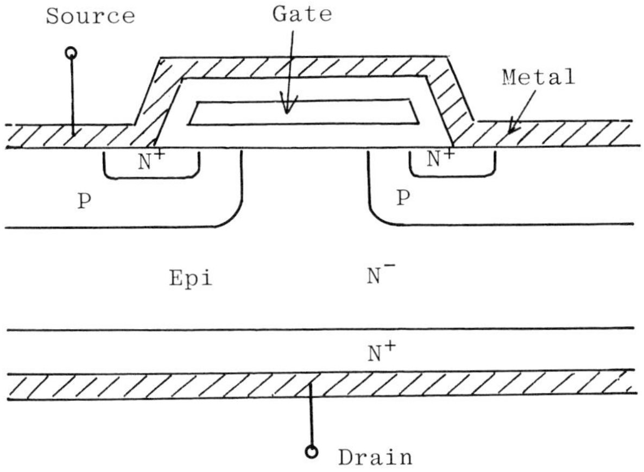

Fig. 4 Schematic cross section of an n-channel DMOS power FET.

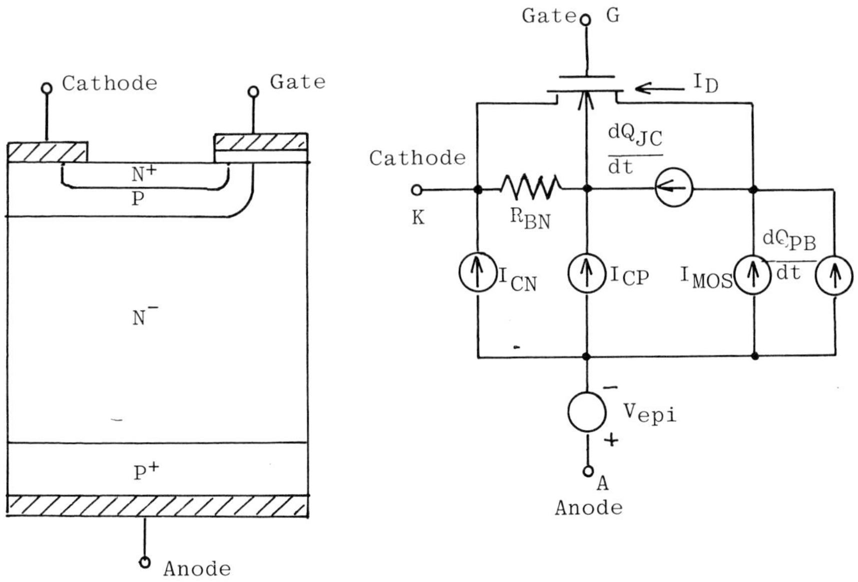

a) The basic LIGT model b) UDCS representation of the LIGT

Fig. 5 The UDCS model of the high voltage insulated gate transistor (McDonald and Fossum 1988).

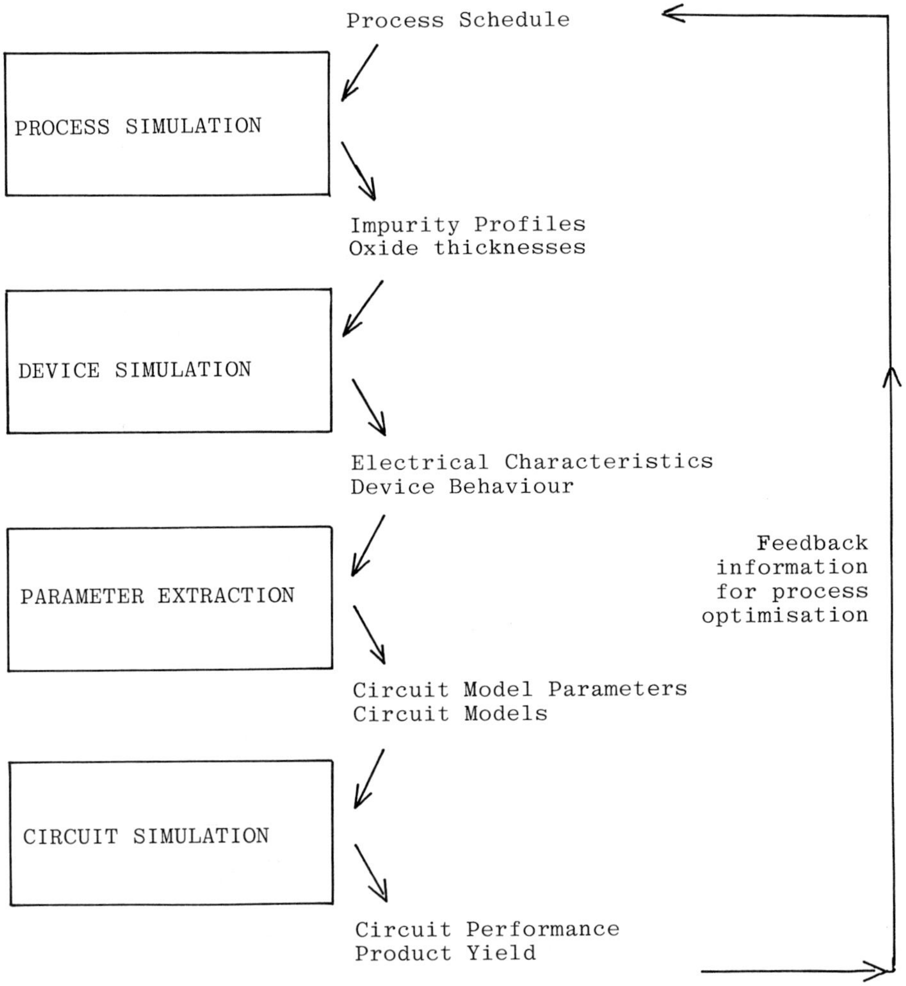

Fig. 6 An integrated system for the simulation of the entire design sequence from unprocessed silicon to final circuit performance and yield.

High Frequency Equivalent Circuit Models

Michael J. Howes
University of Leeds

Introduction

The following notes form the basis for a discussion on equivalent circuit modelling of solid-state devices which are used in present high frequency electronic subsystem designs.

There are a number of approaches to equivalent circuit modelling. The most satisfying, in an intellectual sense at least, is to construct the equivalent circuit topology from a detailed knowledge of the device geometry and other physical parameters, and then to derive mathematical expressions (ideally in closed form) for the circuit element values from the internal physics and geometry of the device in question.

An alternative approach is to determine the equivalent circuit element values directly from measurements on the device. [Z], [Y] or [S] parameter measurements are made, the voltage dependencies of the non-linear elements being determined by making measurements at different bias voltages. The procedure is simple in principle, and involves calculating the [Z], [Y] or [S] parameters of the equivalent circuit and varying the equivalent circuit element values until adequate agreement is reached. Inevitably optimisation procedures are required for problems which are multi-variable in nature. Consequently there are no unique solutions and satisfactory agreement with measurement can often be obtained with equivalent circuit models containing elements which have realistic values, and with models which have element values which are non-physical.

Assuming that the models are accurate over the required bandwidth under small signal and large signal conditions, it is then possible to include the device models in computer aided design packages and thereby simplify the design procedure considerably.

The disadvantages of using equivalent circuit models are that it is almost always difficult to determine accurate element values under small-signed conditions and that it is almost always impossible to do so under large- signal conditions.

Schottky Diodes [1]

The Schottky diode is a majority carrier device and largely immune to minority carrier effects. Consequently the junction capacitance and current change almost instantaneously with junction voltage and the d.c. values for these parameters are valid up to very high frequencies. The equivalent circuit model is shown in Fig 1.

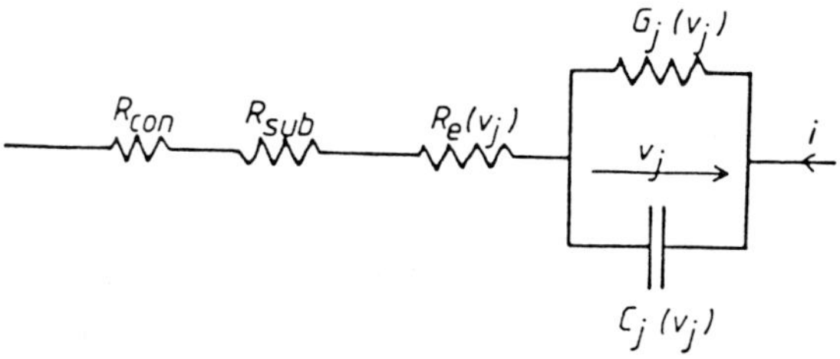

Fig. 1
Equivalent Circuit Model of Schottky barrier diode

The model simply consists of a voltage dependent conductance and capacitance for the metal-semiconductor junction and a voltage dependent series resistance. The mathematical form of these voltage dependencies is determined by the physical properties of the diode and is normally available in closed form. Under high injection levels minority carrier effects can dominate the behaviour of Schottky barrier devices even to the extent that the equivalent circuit model becomes largely inductive.

A small signal circuit model for the Schottky barrier may be derived by studying the physical structure of the device. The junction capacitance, C_j, has a voltage dependence given by

$$C_j = \varepsilon A \left[\frac{(m+2)\varepsilon\phi}{e\,\alpha} \right]^{\frac{-1}{m+2}} \left\{ 1 - \frac{v_j}{\phi} \right\}^{\frac{-1}{m+2}} \tag{1}$$

where the doping is given by $N_d(X) = \alpha X^m$, and the zero bias capacitance is

$$C_o = \frac{\varepsilon A}{d_o} = \varepsilon A \left[\frac{(m+2)\varepsilon\phi}{e\alpha} \right]^{\frac{-1}{m+2}} \tag{2}$$

For a flat doping profile (m=0)

$$C_o = \varepsilon A \left[\frac{2\varepsilon\phi}{eN_d} \right]^{-\frac{1}{2}} \tag{3}$$

Note that C_j is a function of the junction voltage and not the total device voltage.

The junction conductance, G_j, is obtained by differentiating the expression for the junction current. If the ideality factor n is nearly 1 and the bias dependence of the barrier height is small, then the junction current may be expressed as

$$J = A^{**} T^2 \exp\left\{ \frac{-e\phi_b}{kT} \right\} \left[\exp\left\{ \frac{ev_j}{kT} \right\} - 1 \right] \tag{4}$$

So $\quad G_j = \dfrac{eI_s}{kT} \exp\left\{ \dfrac{ev_j}{kT} \right\}$

where $\quad I_S = A^{**} T^2 \exp\left\{ \dfrac{-e\phi_b}{kT} \right\}$

R_e is the resistance of that part of the epi-layer which is outside the depletion region and is given by

$$R_e\ (v_j) = \frac{1}{A} \int_d^W \rho(x).dx \tag{5}$$

where W is the total width of the epi-layer, and ρ is the resistivity of the epi-layer.

Since $\rho = (N_d\, e\mu)^{-1}$ and $N_d = \alpha X^m$, substituting into (5) and integrating gives for a flat doping profile (m=0)

$$R_e(v_j) = \frac{\rho}{A}\ [W-d] \tag{6}$$

$$= \frac{\rho W}{A} - \frac{\rho\varepsilon}{C_o} \left\{ 1 - \frac{v_j}{\phi} \right\}^{\frac{1}{2}} \tag{7}$$

These formulae assume a 1-dimensional current flow perpendicular to the contact plane. This is valid for the mesa-etched device but not necessarily for the planar device. For a circular contact with the current spreading into an infinite area, the epi-layer or spreading resistance is given approximately by

$$Res = \frac{\rho}{2\pi r} \tan^{-1}\left\{ \frac{2t}{r} \right\} \tag{8}$$

where r is the contact radius, and t is the layer thickness (= W-d).

By expanding $\tan^{-1}X$ as a power series, (8) may be written as:

$$R_{es} \simeq \frac{\rho}{2\pi r}\left\{\frac{2t}{r} - \frac{8t^3}{3r^3} + \ldots\right\}$$

$$= R_e\left\{1 - \frac{4t^2}{3r} + \ldots\right\} \tag{9}$$

It can be seen from (9) that when $t \ll r$, $R_{es} \rightarrow$Re.

In calculating the substrate resistance R_{sb}, both spreading resistance and skin depth must be taken into account. For a cylindrical substrate with a circular top contact

$$R_{sb} = \frac{\rho}{2\pi r_1}\tan^{-1}\frac{r_2}{r_1} + \frac{\rho}{2\pi\delta}\ln\frac{r_2}{r_1} + \frac{\rho t}{2\pi\delta r_2} \tag{10}$$

The contact resistance R_{con} can be estimated from measurements of the specific contact resistivity.

Metal Semiconductor Field Effect Transistors [3]

The small signal equivalent circuit model for a GaAs MESFET operating in the saturated current region and in the common source configuration is shown in Fig 2.

The model is only valid up to about 12 GHz and at higher frequencies the channel region of the device should be modelled by a distributed RC network.

Although a great deal of effort has been applied to the problem of deriving a large signal equivalent circuit for the GaAs MESFET an accurate model has so far alluded research workers. This is largely due to the complicated nature of the velocity field characteristic of GaAs.

The basic circuit representation of a MESFET is readily obtained. Fig 3 shows the equivalent circuit of the intrinsic MESFET. This intrinsic model is only applicable at low microwave frequencies and for higher frequencies proper account must be taken of the parasitic elements and packaging. The model given in Fig 2 is straightforward to use for small signals but for large signals the component values become voltage dependent resulting in a non linear circuit analysis.

Under small signal conditions the component values in a model would first need to be optimised to a particular device. After optimisation the S-parameters are calculated and the network representation can then be treated by standard techniques to determine matching circuits and

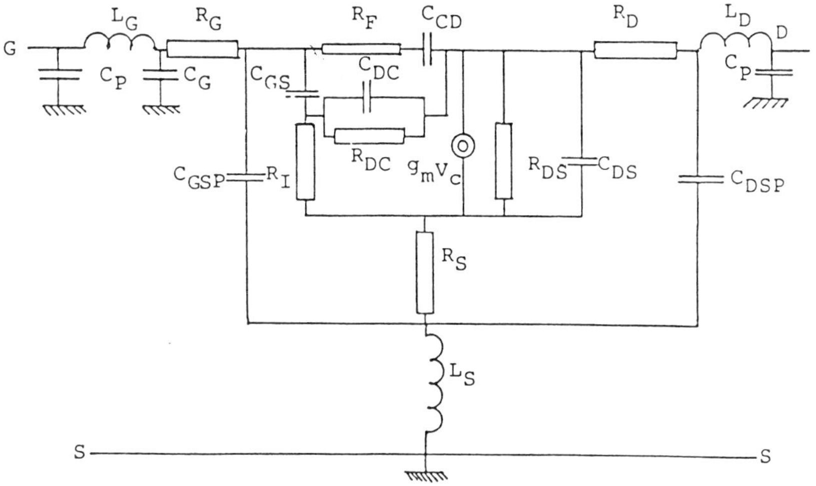

Fig. 2
Equivalent circuit for GaAs MESFET in common source configuration

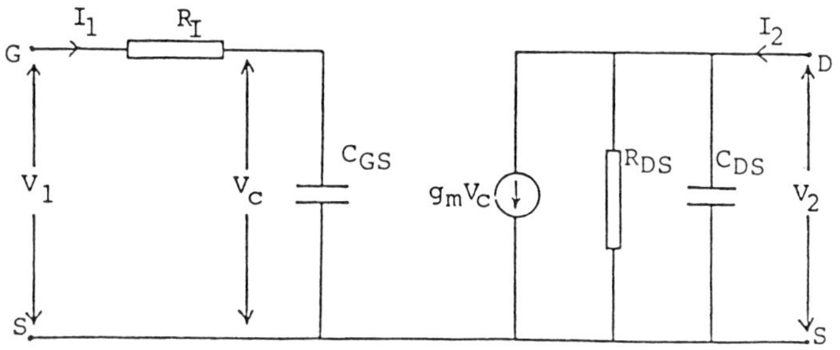

Fig. 3
Intrinsic model of a MESFET

power gains etc.. The full equivalent circuit obviously needs computer techniques for any practical treatment but the principles can be illustrated using the intrinsic model.

Rather than calculate the S-parameters directly it is easier to calculate the y-parameters of the intrinsic MESFET model. Applying these conditions gives

$$
\begin{aligned}
y_{11} &= \frac{1}{R_I + \dfrac{1}{j\omega C_{GS}}} \\
y_{12} &= 0 \\
y_{21} &= \frac{g_m}{1 + j\omega R_I C_{GS}} \\
y_{22} &= \frac{1}{R_{DS}} + j\omega C_{DS}
\end{aligned}
\tag{11}
$$

$S_{11}, S_{12}, S_{21},$ and S_{22} can be calculated using the standard conversion formula from y to S-parameters. Before proceeding let us return to the circuit and calculate the short circuit current gain $I_2/I_1 \mid_{V_2=0}$

$$I_2 = g_m V_c \tag{12}$$

and

$$I_1 = \frac{V_1}{R_I + 1/j\omega C_{GS}} \,. \tag{13}$$

Now

$$V_1 = (1 + j\,\omega R_I C_{GS}) V_c \tag{14}$$

$$\therefore \left| \frac{I_2}{I_1} \right| = \frac{g_m}{2\pi f C_{GS}} \tag{15}$$

when $f = \dfrac{g_m}{2\pi C_{GS}}$ the current gain is unity.

One of the advantages of GaAs MESFETs is their excellent noise properties. However noise is produced in the device and these noise processes limit the amount of amplification that can be achieved.

Low frequency noise is due mainly to flicker ($\frac{1}{f}$) noise and generation/recombination noise associated with localised energy levels in the energy gap of the semiconductor at different parts of the device. The levels thought to be most important are those in the depletion region under the gate and at the air/semiconductor or passivation layer/semiconductor interface. The amplitudes of the excess voltages due to these noise sources vary significantly from device to device and depend on the fabrication process. For small signal operation at microwave frequencies low frequency noise may not be important but in large signal amplifiers, oscillators and mixers this noise can be upconverted to higher frequencies.

The important sources of thermal noise in GaAs MESFETs are the conducting channel and series parasitic resistances associated with the source, gate and drain. Noise is also induced in the gate by capacitative coupling from the channel. Under normal operating conditions at microwave frequencies the thermal noise is a few orders of magnitude greater than the low frequency noise. At lower frequencies the thermal noise decreases and the $\frac{1}{f}$ noise increases resulting in a noise corner frequency where they are equal. This corner frequency is at about 100 MHz resulting in a poorer noise performance at UHF and VHF.

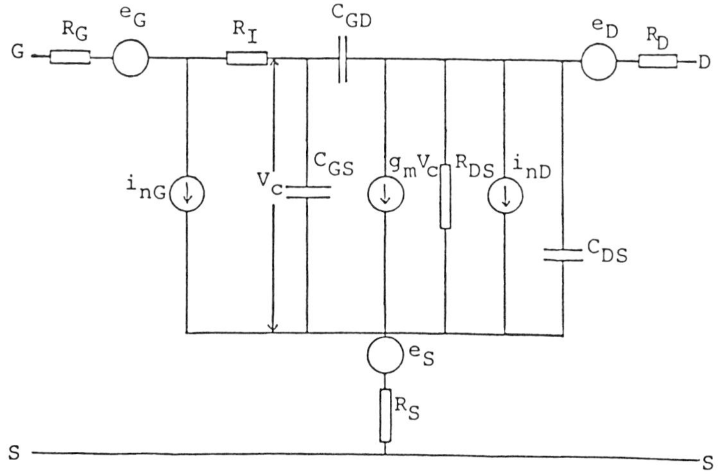

Figure 4. Simplified model of a MESFET showing the location of the noise sources

The noise sources mentioned above have been incorporated into a slightly simplified FET model in Fig 4 where e_S, e_G and e_D are the noise voltages associated with the lead resistances, i_{nG} is the induced gate noise current and i_{nD} the channel noise current. A semi empirical treatment of this model by Fukui gives the following expression for the optimal noise figure F_o

$$F_o = 1 + KLf[g_m(R_G + R_S)]^{1/2} \tag{16}$$

Where K is a filling factor and L the gate length. (The noise figure is the ratio of the available noise power at the output to the available noise power at the input.) This equation is found to agree well with measurement and shows the double dependence on the gate from the L and R_G terms. F_o is reduced for short gate lengths and low gate resistance. There is an element of compromise here because a short gate reduces the cross sectional area of the gate metalisation and increases R_G . This effect can be reduced by plating up the gate to increase its thickness. In terms of the gate length the design for a low noise figure is the same as that for a high cut-off frequency.

Dual Gate Metal Semiconductor Field Effect Transistors [4, 5, 6]

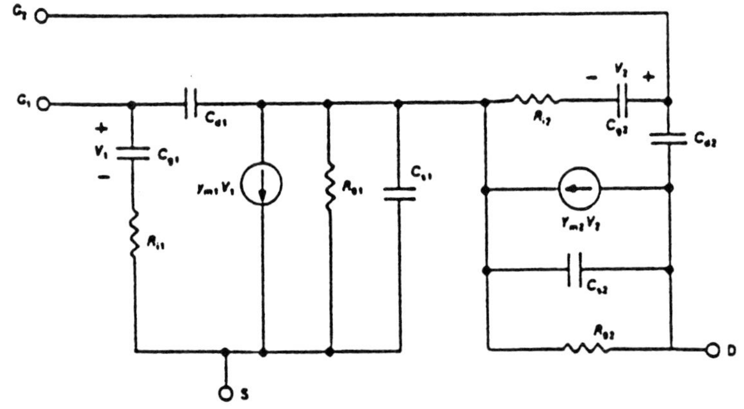

Fig. 5
Dual-Gate MESFET Model

DGMESFETs are usually modelled as two single-gate FETs in series. Since this configuration is similar to that of the cascode amplifier the dual gate FET is often described in terms of this classical configuration of two discrete devices. However, if one FET is to control the transconductance of the other, the lower device must be operated in its linear region whereas in the cascode amplifier both devices are operated in the saturated current region.

Equivalent circuit modelling is the only practical modelling technique available for dual gate FETs at the present time. Fig 5 shows the result of substituting the single gate equivalent circuit into the two FET equivalent of the DGMESFET.

Bipolar Transistors [2]

The equivalent circuit of an npn silicon planar bipolar transistor is shown in Fig 6.

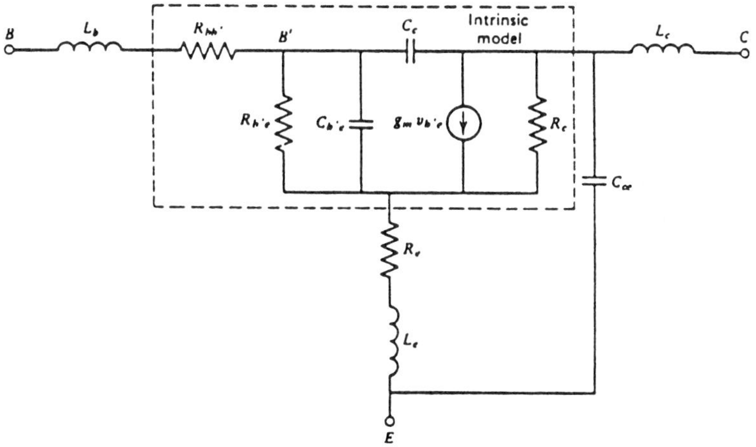

Fig. 6
Equivalent circuit for npn bipolar
transistor in common emitter configuration

Reasonably accurate estimates of the values of the circuit elements can be made under small signal and large signal conditions largely due to the rather well behaved velocity field characteristic of silicon.

Transferred Electron Devices [7] and IMPATTs [8]

Two terminal devices such as TEDs and IMPATTs may be used as the basis of reflection amplifiers or oscillators. In either case both small signal and large signal models are essential to the design procedures. Equivalent circuit models which contain accurate element values with some physical basis have proven to be allusive for both TEDs and IMPATTs. However, both of these devices may be modelled by a generalized admittance as indicated in Fig 7.

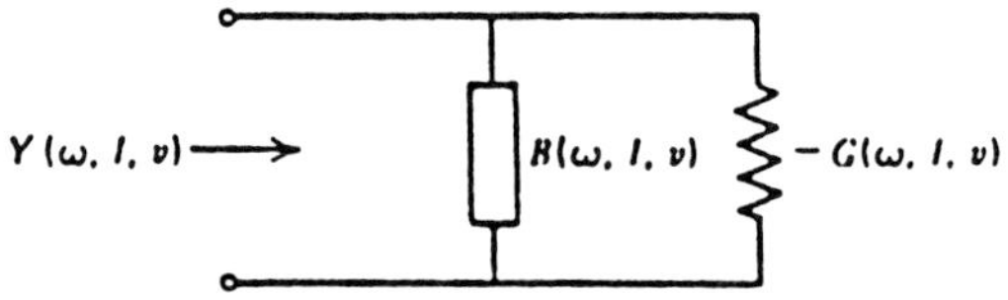

Fig. 7
Generalized model for two terminal devices such as TEDs and IMPATTs

Under large signal conditions accurate equivalent circuit models are not available at the present time and the large signal admittance of the devices must be obtained via direct measurement [9] or computer simulation techniques [10]. However closed form expressions for the real and imaginary parts of the admittance are available under small signal conditions and in order to illustrate the basic method a small signal and large signal analysis of the IMPATT diode will be described.

Small Signal Admittance of the IMPATT Diode

The basic structure of the device is shown in Fig 8. For small sinusoidal perturbations of the maximum electric field E_m, the avalanche region inductance may be calculated to be

$$L_A = \frac{\tau_i}{I_{dc}\,\overline{\alpha}'} \tag{17}$$

where I_{dc} is the d.c. device current and $\overline{\alpha}'$, is an effective ionisation rate derivative which is given by

$$\overline{\alpha}' = \frac{1}{W_A}\,\frac{\partial}{\partial E_m}\left[\int_o^W \alpha(E)dx - 1\right] \tag{18}$$

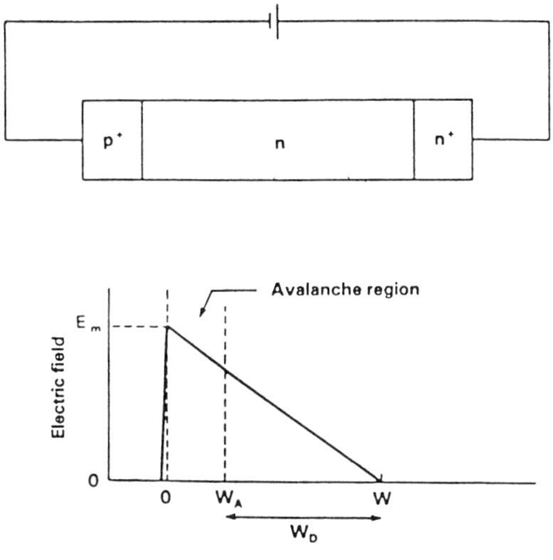

Fig. 8
Schematic diagram of an IMPATT diode

For the simple p-n junction considered in the previous sections

$$\overline{\alpha}' = \frac{\alpha_{max}\varepsilon\varepsilon_o}{qNW_A} \tag{19}$$

In addition to this conduction current the avalanche region also sustains a displacement current due to its capacitance, C_A, where for a device area A

$$C_A = \frac{\varepsilon\varepsilon_o A}{W_A} \tag{20}$$

These two components will be shunt resonant at a frequence f_a, known as the avalanche resonance frequence, which is given by

$$\omega_a^2 = \frac{\overline{\alpha}' W_A J_{dc}}{\varepsilon\varepsilon_o \tau_i} \tag{21}$$

Completing the analysis by considering the conduction and displacement currents throughout the drift region gives the following expression for the small signal device impedance, Z_D

$$Z_D = \frac{1}{j\omega C_T} \frac{1 - \dfrac{\omega_a^2 C_T}{\omega^2 C_D}(1+F)}{1 - \dfrac{\omega_a^2}{\omega^2}} \tag{22}$$

where C_T is the total depletion layer capacitance and F is the drift region transport factor

$$F = \frac{e^{-j\theta} - 1}{j\theta} \tag{23}$$

Separating real and imaginary parts gives the following expressions for the device resistance and reactance

$$R = \frac{W_D^2}{2v_s\varepsilon\varepsilon_o A} \frac{1}{1 - \frac{\omega^2}{\omega_a^2}} \frac{1 - \cos\theta}{\frac{\theta^2}{2}} \tag{24}$$

$$X = \frac{1}{j\omega C_D}\left\{1 - \frac{\sin\theta}{\theta} + \left[\frac{\sin\theta}{\theta} + \frac{W_A}{W_D}\right] \Big/ \left[1 - \frac{\omega_a^2}{\omega^2}\right]\right\} \tag{25}$$

This expression shows several significant features. It is only at frequencies greater than the avalanche resonance frequency that the device exhibits negative resistance and that the device impedance changes from inductive to capacitive at resonance. At high frequencies ($\omega \gg \omega_a$) the device can be modelled quite simply as the total depletion layer capacitance C_T in series with a negative resistance of magnitude

$$\frac{W_D\omega_a^2}{\omega^3\varepsilon\varepsilon_o A} \frac{1 - \cos\theta}{\theta} \tag{26}$$

The frequency dependence of this negative resistance is largely determined by the second term which is shown in Figure 9. It can be seen that this maximises at a drift region phase angle of $\simeq 0.74\pi$, *i.e.* 74% of the Read frequency.

At low frequencies the resistance is real and equal to the space charge resistance R_{sc} of an avalanching p-n junction which is given by

$$R_{sc} = \frac{W_D^2}{2v_s\varepsilon\varepsilon_o A} \tag{27}$$

The development of the negative resistance in J and K bands as the device current is increased is shown in Figure 10, for a device designed for operation at 20 GHz, where many of the features given above are observable.

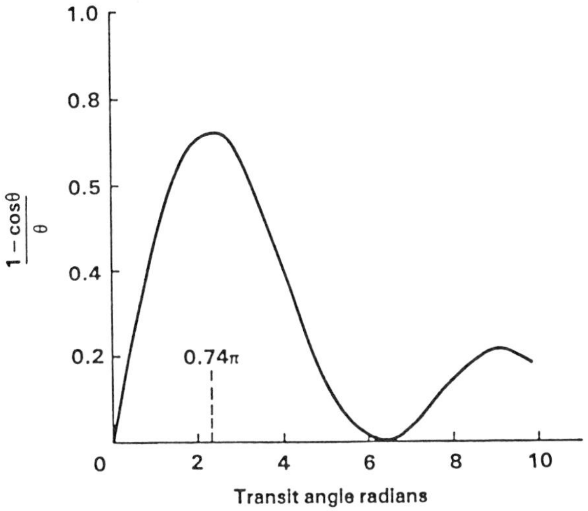

Figure 9. The real part of the drift zone transport factor

Large Signal Model

As an approximate solution for large signals the following method has been employed in which a finite sinusoidal variation in the maximum field E_m, of amplitude E_a, is included in the Read Equation which then becomes

$$\frac{dJ_a(t)}{dt} = \frac{\overline{\alpha}' W_A E_a}{\tau_i} J_a(t) \sin \omega t \tag{28}$$

A solution to this Equation can be found by the use of an expansion of the sinusoidal terms in modified Bessel functions $I_0(B)$ and $I_1(B)$, where B is a normalised electric field amplitude

$$B = \frac{\overline{\alpha}' W_A E_a}{\omega \tau_i} \tag{29}$$

The final expression for the diode impedance is identical to Equation (22) except that the avalanche resonance frequency, ω_a, is replaced by a modified large signal resonance frequency ω_R, which is related to ω_a and the normalised field parameter, B, by

$$\frac{\omega_R^2}{\omega_a^2} = h(B) = \frac{2}{B} \frac{I_1(B)}{I_0(B)} \tag{30}$$

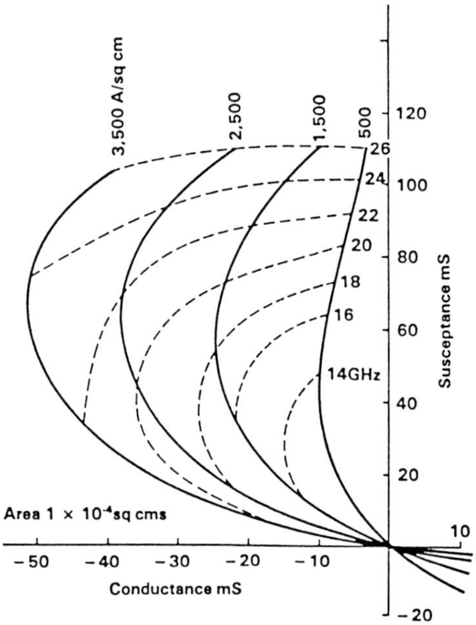

Figure 10. The small signal impedance of an Impatt diode

Thus in this model all the large signal effects are determined by the parameter B which is a function of both frequency and the a.c. electric field amplitude. The behaviour of $h(B)$ is shown in Fig 11.

At small signal levels $h(B)$ tends to unity and falls as the signal level is increased, lowering the avalanche resonance frequency which in turn leads to a reduction in the magnitude of the negative resistance at a given operating frequency. This effect is demonstrated in Fig 12 which presents the large signal admittance levels, for the same d.c. operating conditions as used for Fig 10, for a ratio of a.c. voltage to bias voltage of 0.6. There is a 2-3 fold reduction in peak negative conductance but a considerable improvement in efficiency as the output power is given by

$$P_o = \frac{1}{2}|G| V_{rf}^2 \tag{31}$$

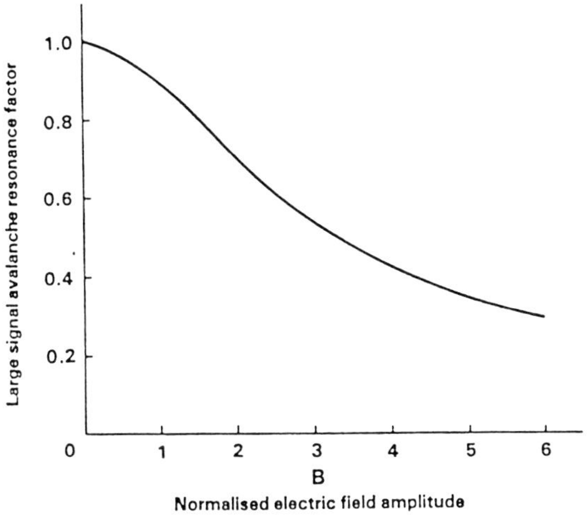

Figure 11. The large signal avalanche resonance factor

In the simplified analysis presented earlier it was shown that the width of the injected current pulse, θ_w should be small in order to attain high efficiencies and this large signal analysis shows that the ratio of the magnitude of the fundamental component of the a.c. conduction current to the d.c. bias current is

$$\frac{J_{ac}}{J_{dc}} = 2\,\frac{I_1(B)}{I_0(B)} \tag{32}$$

The function $I_1(B)/I_0(B)$ is given in Fig 13 where the value 1 represents the ideal 'sharp pulse' which is only achieved for a normalised field parameter greater than 6. For lower values of B the broadening of the avalanche current pulse leads to a reduction in power and efficiency. Examination of Equations (19) and (29) shows that large values of B are obtained by operating at large signal levels, low frequencies and using structures which have narrow avalanche regions. The frequency dependence arises from the term $\omega\tau_i$ which says that for high efficiency operation the avalanche build up time should be short compared to an r.f. cycle. The use of narrow avalanche regions has a double benefit in that not only is the field parameter increased, but the voltage drop in the avalanche region is reduced. It is these factors which are the driving force for the study of the more complex device structures.

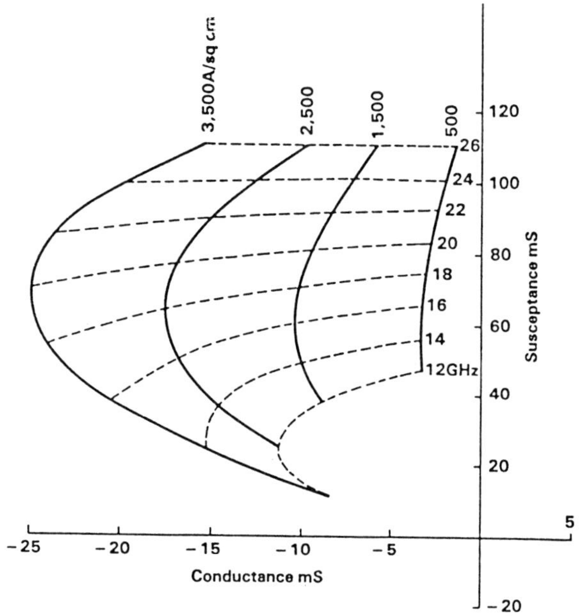

Figure 12. The large signal impedance of the same diode as Figure 10 (excluding space charge effects)

Under large signal conditions Equation (24) may be modified to give the following expression for the negative resistance

$$R_D = \frac{-2}{\omega^2 C_D^2} \frac{W}{W_D} \frac{I_{dc}}{V_{rf}} \frac{I_1(B)}{I_0(B)} \frac{1-\cos\theta}{\theta} \tag{33}$$

and since in all practical devices the device Q will be sufficiently high that the conductance may be written as R_D/X_D^2 we obtain

$$G_D = -2(1-\frac{W_A}{W}) \frac{I_{dc}}{V_{rf}} \frac{I_1(B)}{I_0(B)} \frac{1-\cos\theta}{\theta} \tag{34}$$

The device negative resistance and conductance are proportional to the bias current and decrease with increasing signal level. This latter result being essential for self starting and

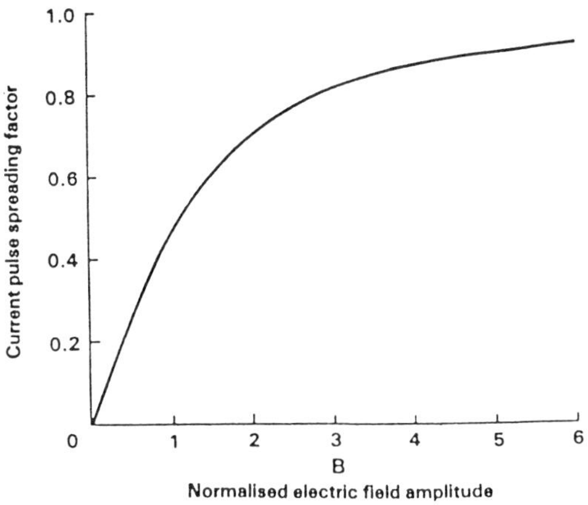

Figure 13. The current modulation factor in the avalanche zone

stability in an oscillator.

References

1. Maas, S.A., 'Microwave Mixers', Artech House, 1986.
2 Morgan, D.V., Howes, M.J. and Pollard, R.D., 'Microwave Solid-State Component and Subsystem Design', Leeds University Press, 1983.
3. Dilorenzo, J.V. and Khandelwal, D.D., 'GaAs FET Principles and Technology', Artech House, 1982.
4. Scott, J.R. and Minasian, R.A., 'A Simplified Microwave Model of the GaAs Dual-Gate MESFET,' IEEE Trans. Microwave Theory Tech., MTT-32, 1984, p.243.
5. Tsironis, C., Meirer, R. and Stahlman, R., 'Dual-Gate MESFET Mixers', IEEE Trans. Microwave Theory Tech., MTT-32, 1984, p.248.
6. Kim, B., Quen Tsung, H. and Saunier, P., 'GaAs Dual Gate FET for Operation to K-Band', IEEE Trans. Microwave Theory Tech., MTT-32, 1984, p.256.
7. Bulman, P.J., Hobson, G.S. and Taylor, B.C., 'Transferred Electron Devices', Academic Press, 1972.

8. Morgan, D.V., Howes, M.J. and Pollard, R.D., 'Microwave Solid-State Component and Subsystem Design', Leeds University Press, 1983.

9. McBretney, J., 'Galvanomagnetic Phenomena in Transferred Electron Devices', Ph.D. Thesis, Leeds University, 1977.

10. Dyer, G.R., 'Circuit Modelling of Varactor Diodes in VCOs', Ph.D. Thesis, Leeds University, 1981.

Modelling of Noise Processes

Alain Cappy

Universite des Sciences et Techniques de Lille

Contents

1. Basic mathematical concepts

1.1. Averages - correlation

In calculations about noise in electrical systems, one must often calculate the average $\overline{x^m}$ of the random variable x(t). $\overline{x^m}$ is calculated using the probability density function or distribution function f(x) defined by

$$\text{Prob}\ (x_0 < x(t) \leq x_0 + dx) = f(x_0)\, dx \tag{1.1}$$

We can now define ensemble averages as follows :

$$\overline{x^m} = \int x^m f(x)\, dx \tag{1.2}$$

The most important averages are $\overline{x}$ and $\overline{x^2}$; In addition, the noise processes encountered in physics and engineering are practically always ergodic. That means :

$$<x> = \lim_{T \to \infty} \frac{1}{T} \int_0^T x(t)\, dt = \overline{x} \tag{1.3}$$

For two continuous variables x(t) and y(t) one can evaluate the probability dP that x has a value between x and x + dx and y has a value between y and y + dy

$$dP = f(x, y, t)\, dx\, dy \tag{1.4}$$

The averages are defined in the same manner as for single variables. In practice, the most important average is $\overline{xy}$. If $\overline{xy} = 0$ the two variables are said to be uncorrelated ; if $\overline{xy} = 0$ the two variables are said to be correlated. One can define a correlation coefficient C as :

$$C = \frac{\overline{xy}}{(\overline{x^2}\ \overline{y^2})^{\frac{1}{2}}} \tag{1.5}$$

It can be shown that $|C| \leqq 1$. If $|C| = 1$ the two variables are fully correlated. If $|C| < 1$ the two variables are partially correlated. In that case one can split y into a part $\alpha.x$ that is fully correlated with x and a part z that is uncorrelated with x : $y = \alpha x + z$. α and z are given by :

$$\alpha = \frac{\overline{xy}}{\overline{x^2}} \qquad \overline{z^2} = \overline{y^2}\ (1 - c^2) \tag{1.6}$$

These formulas will be useful in the calculation of the noise figure of a two port.

1.2. Fourrier analysis - spectral intensity

Let X(t) be a continous variable in the time interval $0 < t < T$ - X(t) can be developed in Fourrier series :

$$X(t) = \sum_{-\infty}^{+\infty} a_n \exp(j\omega_n t) \quad \text{with} \quad a_n = \frac{1}{T}\int_0^T X(t)\exp(-j\omega_n t)\,dt \tag{1.7}$$

The Fourrier component of frequency $\omega_n = 2\pi n/T$ is :

$$x_n = a_n \exp(j\omega_n t) + a_{-n} \exp(-j\omega_n t)$$

the spectral intensity $S_x(f)$ is then defined as :

$$\overline{x_n^2} = 2\,\overline{|a_n|^2} = S_x(f)\,\Delta f \tag{1.8}$$

$S_x(f)$ can be calculated using the Wiener-Khintchine theorem that can be expressed as :

$$S_x(f) = 2\int_{-\infty}^{+\infty} \overline{X(u)\,X(u+\tau)}\,\exp(j\omega\tau)\,d\tau \tag{1.9}$$

So, $S_x(f)$ is twice the Fourier transform of the autocorrelation function $\overline{X(t)\,X(t+\tau)}$. Figure 1 shows the physical meaning of $S_x(f)$.

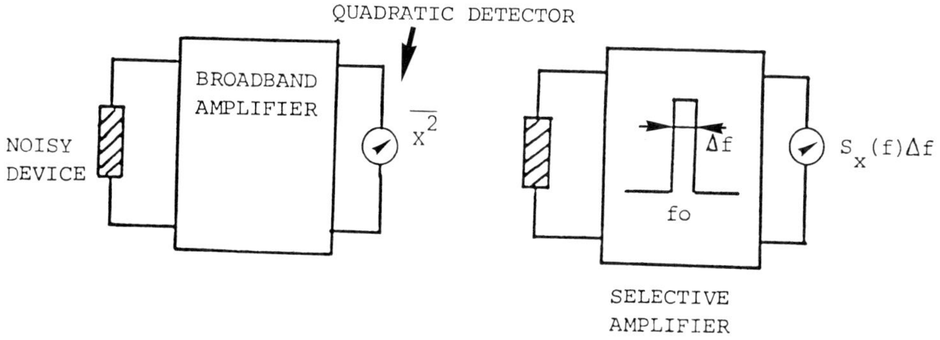

Figure 1

2. The noise sources

The current density is defined as : $j = qnv$. In this expression, n and v represent average values that can be obtained from macroscopic laws (Electrokinetic equations, Poisson's equation etc...). In fact, these quantities are fluctuating quantities and the purpose of the noise study is to characterize these fluctuations. In practice four different noise processes will be encountered.

2.1. Shot noise

It occurs whenever a noise phenomenon can be considered as a series of independent events occurring at random (for example the emission of electrons by a cathode). The spectral intensity of the fluctuating current I(t) of average value $\overline{I}$ is (Schottky's theorem) :

$$Si(f) = 2q\overline{I} \qquad (2.1)$$

2.2. Generation-recombination noise

Due to generation and recombination (band to band, deep levels, shallow levels...) the number of free carriers in a semiconductor fluctuates. For a given G.R. process, the spectral intensity of the fluctuating current I(t) is (Van der Ziel 1954) :

$$S_I(f) = \frac{4\beta \bar{I}^2}{N} \frac{\tau}{1 + \omega^2 \tau^2} \qquad (2.2)$$

where $\overline{I}$ is the average current, N the average total number of carriers, β and τ are characteristic of the G.R. process.

2.3. Flicker noise

The physical origin of Flicker noise is not very well known as yet (Van der Ziel 1988). It occurs in all physical or biological systems and is characterized by a spectral intensity of the form :

$$S_I(f) = \frac{\bar{I}^2}{N} \frac{\alpha_H}{f^\beta} \qquad (2.3)$$

where α_H is the Hooge constant and β is close to unity.

2.4. Diffusion noise

This noise is due to the random motion of carriers in any conductors. At thermal equilibrium it is called "thermal noise". This noise is dominant in the most electrical devices, at least in the high frequency range.

For a device slice of section S(x) and thickness Δx, the spectral intensity of diffusion noise is expressed by :

$$S_I = 4q^2\, S(x)\, n(x)\, D(x)/\Delta x \tag{2.4}$$

where n(x) is the carrier density and D(x) the diffusion coefficient. At thermal equilibrium, the diffusion coefficient obeys Einstein's relation : $D = kT\,\mu/q$ and therefore :

$$S_i = 4kT\, S(x)\, \sigma(x)/\Delta x \tag{2.5}$$

where σ(x) is the conductivity. This relation is the well known Nyquist law.

The noise sources that have been described in this part are local noise sources and are present at each point of any device. The purpose of the modelling of noise processes is to answer the following question : what are the effects of all the local noise sources on the external electrodes ? In order to answer this question, some basic concepts will be introduced and a general method for the external noise source calculation will be given.

3. Noise in devices

3.1. Noise in one port

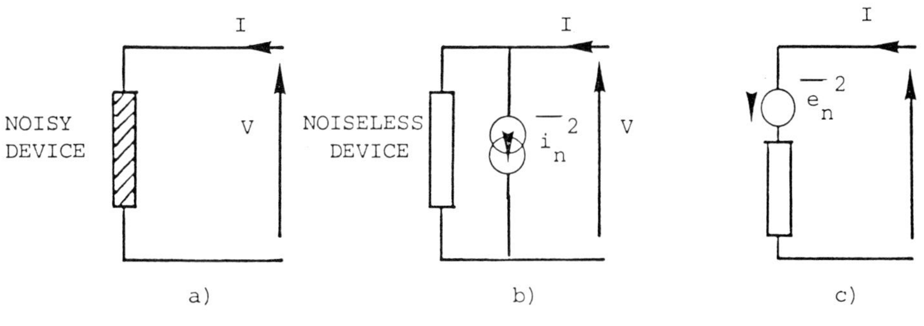

Figure 2

Figure 2a shows a noisy two terminal device. Assuming that a noiseless voltage v is applied to this device, the current i flowing in the external circuit is given by

$$i = i_n + Y.v \tag{3.1}$$

where Y is the device admittance and i_n the noise current that represents, in the external circuit, the effects of all the noise sources located in the device. The current i_n is called "equivalent noise source". So, the noisy device can be considered as a noiseless device bridged by a noise current source i_n (Fig. 2b) or in series with a voltage noise source $e_n = i_n/Y$ (Fig. 2c).

It is important to note that i_n (resp. e_n) can be used in circuit calculations as a conventional current source (resp. voltage source). Nevertheless, only the mean square value $\overline{i_n^2}$ (resp. $\overline{e_n^2}$) [related to the spectral intensity by $\overline{i_n^2} = Si(f)\,\Delta f$ (resp. $\overline{e_n^2} = Sv(f)\,\Delta f$)] represents a physical quantity and has to appear in a final result.

Remark : For an admittance $Y_s = g_s + jb_s$ at thermal equilibrium, the noise source $\overline{i_n^2}$ is given by Nyquist's law :

$$\overline{i_n^2} = 4\,kT\,g_s\,\Delta f \tag{3.2}$$

3.2. The noise in two port

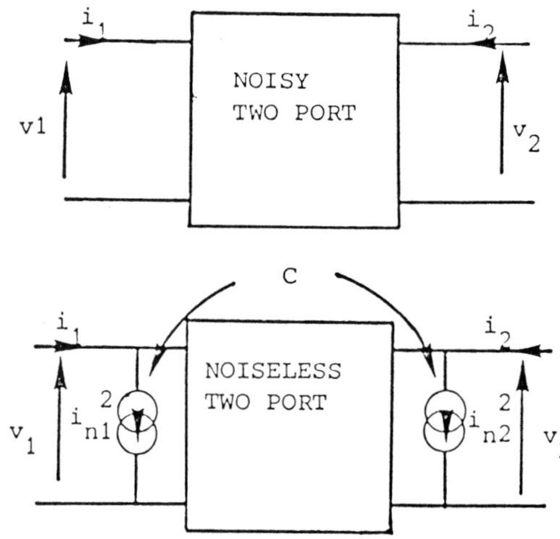

Figure 3

As in the case of the two-terminal device, a noisy two-port will be characterized by two different noise current sources i_{n1} and i_{n2} defined by (Fig. 3)

$$\begin{vmatrix} i_1 \\ i_2 \end{vmatrix} = \begin{vmatrix} Y_{ij} \end{vmatrix} \begin{vmatrix} v_1 \\ v_2 \end{vmatrix} + \begin{vmatrix} i_{n1} \\ i_{n2} \end{vmatrix} \quad (3.3)$$

these noise sources represent the noise currents flowing in the external circuit and resulting from the internal noise sources. Since the same internal sources give rise to i_{n1} and i_{n2}, these two noise sources are correlated. So, a noisy two-port is characterized by three noise parameters, $\overline{i_{n_1}^2}$, $\overline{i_{n_2}^2}$ and a (complex) correlation coefficient $C = \overline{i_{n1}.i_{n2}^*}/(\overline{i_{n1}^2}.\overline{i_{n2}^2})^{1/2}$.

3.3. The noise figure

The noise figure of a two-port characterizes the degradation of the signal-to-noise ratio at a reference temperature $T_o = 290$ K.

$$F = \left. \frac{(S/N)_{in}}{(S/N)_{out}} \right|_{T_o = 290\,K} \quad (3.4)$$

an equivalent definition is :

$$F = \left. \frac{N_a + N_{in}\,Ga}{N_{in}\,G_a} \right|_{T_o = 290\,K} \quad (3.5)$$

where N_{in} is the noise power that goes into the two-port, G_a the available gain and N_a the noise power produced by the two correlated noise sources i_{n1} and i_{n2}.
It is also possible to define the equivalent noise temperature T_e of the two-port as :

$$F = 1 + \frac{T_e}{T_o} \qquad \text{with } T_o = 290\text{ K} \quad (3.6)$$

3.4. How to calculate the noise figure

The noise figure calculation is performed in several steps (Rothe and Dahlke 1956). The required parameters are :

(i) the admittance matrix (or any equivalent two-port description)
(ii) the noise sources $\overline{i_{n1}^2}$, $\overline{i_{n2}^2}$ and their correlation coefficient C (Fig. 3b).

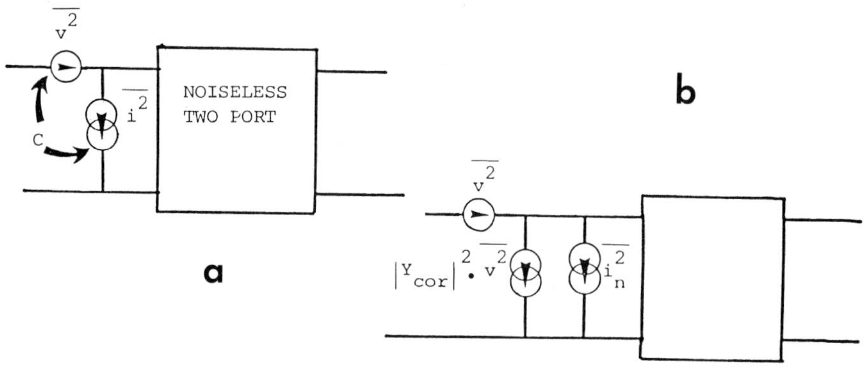

Figure 4

In a first step, i_{n1} and i_{n2} are transformed in two correlated noise sources v and i located at the device input (Fig. 4a). A simple circuit calculation yields :

$$v = i_{n2}/Y_{21}$$
$$i = i_{n1} - Y_{11}/Y_{21}\, i_{n2} \qquad (3.7)$$

It should be emphasized that i and v are correlated even if i_{n1} and i_{n2} were not. At this step i is expressed as (Fig. 4b).

$$i = i_n + Ycor\, v \qquad \text{where } \overline{i_n v^*} = 0 \qquad (3.8)$$

As shown before we have :

$$Ycor = \frac{\overline{iv^*}}{\overline{v^2}} \quad \text{and } \overline{i_n^2} = \overline{i^2}(1 - |Co|^2) \qquad (3.9)$$

$$\text{with } Co = \frac{\overline{iv^*}}{(\overline{i^2}\ \overline{v^2})^{\frac{1}{2}}}$$

At this step, the noise figure calculation is straighforward (Pucel 1974). A noisy generator with an internal admittance $Y_s = g_s + jb_s$ is introduced and the noise figure can be writen :

$$F = 1 + \frac{|Y_{cor} - Y_s|^2 R_n + G_n}{g_s} \qquad (3.10)$$

with $Rn = \overline{v^2}/4kT\Delta f$ and $Gn = \overline{i_n^2}/4kT\,\Delta f$

The noise figure becomes minimal for a generator admittance Ysopt given by :

$$Y_{sopt} = \sqrt{g_{cor}^2 + \frac{G_n}{R_n}} + jb_{cor} = g_{sopt} + j\,b_{sopt} \qquad (3.11)$$

In that case F becomes :

$$F_{min} = 1 + 2R_n\,(g_{sopt} - g_{cor}) \qquad (3.12)$$

From a practical point of view, this value represents the measured noise figure when the input is tuned for minimum noise figure. For any generator admittance, F can be written :

$$F = F_{min} + \frac{R_n}{g_s}\left(\left|Y_s - Y_{sopt}\right|^2\right) \qquad (3.13)$$

This general expression shows that four parameters determine the noise behavior of a linear two-port : F_{min}, R_n, g_{sopt} and b_{sopt}. Therefore the main problem is to calculate the value of these parameters from the knowledge of the internal noise sources.

4. The impedance field method

The purpose of the impedance field method (Shockley et al, 1966) is to calculate the effect of a noise source, located in the device, on a given electrode. For the sake of simplicity, a one dimensional formulation of this method will be presented. A more general formulation can be found in the litterature (see for example Van Vliet et al, 1975).

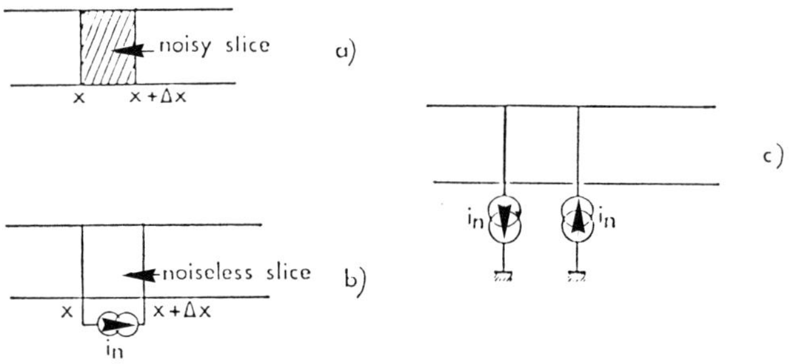

Figure 5

The impedance field method principle is presented Fig. 5. A noisy slice (x, x + Δx) of the device is firstly transformed into a noiseless slice bridged by a noise current source i_n. In a second step, the noise current source is transformed in two current sources. The noise current in is introduced between x + Δx and ground and this current is removed between x and ground.

When a small ac current $i_n(x)$ = io(x) exp(jωt) is introduced between x + Δx and ground, the resulting voltage fluctuation $\delta V(L)$ at abscissa x = L is expressed by :

$$\delta V(L) = Z\,(x + \Delta x, 0, \omega)\, i_o(x) \qquad (4.1)$$

where Z(x + Δx, 0, ω) is the impedance between 0 and x + Δx. When the ac current is removed at x, the resulting voltage fluctuation is :

$$\delta V(L) = -\, Z(x, 0, \omega)\, i_o(x) \qquad (4.2)$$

Therefore, the voltage fluctuation $\delta V(L)$ resulting from the noise source located between x and x + Δx is express as :

$$\delta V(L) = [Z(x + \Delta x, o, \omega) - Z(x, 0, \omega)]\, i_o(x) = \frac{\partial Z}{\partial x}(x, o, \omega)\, i_o(x).\Delta x \qquad (4.3)$$

the partial derivative of Z with respect to x is called the "impedance field".

The total voltage fluctuation is consequently given by a summation performed over the device length.

$$\delta V^{tot}(L) = \int_0^L \frac{\partial Z}{\partial x}\, i_o(x)\, dx \qquad (4.4)$$

and the device impedance can be expressed as :

$$Z(O, L, \omega) = \int_0^L \frac{\partial Z}{\partial x}\,(0, x, \omega)\, dx \qquad (4.5)$$

If we now consider i_n as a noise current source of spectral intensity $S_i(f)$, the resulting spectral intensity of voltage fluctuations will be expressed as :

$$Sv = \left| \frac{\partial Z}{\partial x} \right|^2 .\, Si\, .\, \Delta x^2 \qquad (4.6)$$

This expression is valid for any noise process (shot noise, G.R., 1/f, diffusion noise...). For instance, in the case of diffusion noise

$$Si = 4q^2\, S(x)\, D(x)\, n(x)/\Delta x \qquad (4.7)$$

and Hence

$$Sv = 4q^2 S(x) D(x) n(x) \left| \frac{\partial Z}{\partial x} \right|^2 \Delta x \tag{4.8}$$

Assuming that two different local noise sources corresponding to two different device slices are uncorrelated, the whole equivalent noise voltage corresponding to all the noisy slices of the device can be calculated by a simple summation performed over the whole device length :

$$Sv^{tot} = \int_0^L 4q^2 S(x) D(x) n(x) \left| \frac{\partial Z}{\partial x} \right|^2 dx \tag{4.9}$$

Application

For a simple resistance (Length L, section S doping level Nd, carrier mobility μ_0) we have :

$$Z(O, x, \omega) = \frac{1}{q\,\mu o\,Nd} \frac{x}{S} \tag{4.10}$$

and Hence :

$$\frac{\partial Z}{\partial X} = \frac{1}{q\,\mu o\,Nd\,S} \tag{4.11}$$

Introducing (4.11) in (4.9) yields (Nyquist's formula)

$$Sv = 4kTR \tag{4.12}$$

To summarize, it has been shown that the determination of the equivalent noise sources can be performed from the knowledge of the local spectral intensity of the studied noise process and of the impedance field. Some methods to determine the impedance field will be now given.

4.1. Impedance field using the local electric field

In some cases (Nougier 1985, 1987) the transport equations can be combined to give a non linear relation of type :

$$I = f(E, x) \tag{4.13}$$

where E represents the local electric field

a) the solution of this equation gives a relation between the electric field Eo(x) in the structure and the DC current Io(x) (Usually, Io is constant but it can depend on x in two-port).

b) a small ac current $\delta I(x) \exp(j\omega t)$ is introduced. As a consequence, the electric field becomes $E(x) = Eo(x) + \delta E(x) \exp(j\omega t)$. A first order development of equation (4.13) yields :

$$L\left|\delta E(x)\right| = \delta I(x) \tag{4.14}$$

where L is a linear operator. The determination of the impedance field needs to solve (4.14). For this purpose, let g(x, x') be the Green function of operator L. By definition, g is the solution of :

$$L\left|g(x, x')\right| = \delta(x - x') \tag{4.15}$$

when (4.15) is solved, the solution of (4.14) is :

$$\delta E(x) = \int_0^L \delta I(x')\, g(x, x')\, dx' \tag{4.16}$$

From this expression it is possible to deduce the voltage fluctuation $\delta V(L)$ as :

$$\delta V(L) = -\int_0^L \delta E(x)\, dx = -\int_0^L \int_0^L \delta I(x')\, g(x, x')\, dx\, dx' \tag{4.17}$$

As previously shown, the voltage fluctuation $\delta V(L)$ is related to the impedance field by :

$$\delta V(L) = \int_0^L \frac{\partial Z}{\partial x}\, \delta I(x)\, dx \tag{4.18}$$

Comparing (4.17) and (4.18) :

$$\frac{\partial Z(x')}{\partial x'} = -\int_0^L g(x, x')\, dx \tag{4.19}$$

4.2. Numerical determination of the impedance field

As shown before, the problem is to solve

$$L \left| \delta E(x) \right| = \delta I(x) \tag{4.20}$$

Numerically, this equation can be written : (Nougier 1985)

$$\sum_j a_{ij} \; \delta E_j = \delta I_i \tag{4.21}$$

The coefficient a_{ij} depend upon :

- the operator itself
- the discretization scheme
- the DC values

A numerical solution of equation (4.21) gives :

$$\delta E_k = \sum_j b_{kj} \; \delta I_j \tag{4.22}$$

the impedance field is then given by

$$\left. \frac{\partial Z}{\partial x} \right|_j = - \sum_k b_{kj} \; \Delta x_k / \Delta x_j \tag{4.23}$$

All the device properties (ac, noise) can thus be obtained using that very general numerical method.

4.3. Analytical modeling using the local potential

In that case, the non linear relation describing the device behavior is of the following form :

$$f(V, x) = I \tag{4.24}$$

This kind of equation is encountered, for instance,in the study of FET's (Van der Ziel 1962, Van Vliet 1979).

Introducing $I = Io + \delta I \exp(j\omega t)$, the first order development of (4.33) yields :

$$L \left| \delta V(x) \right| = \delta I(x) \tag{4.25}$$

Let G(x, x') be the Green function of L. G(x, x') is the solution of

$$L\left[G(x,x')\right] = \delta(x-x') \tag{4.26}$$

The solution of (4.34) can thus be written

$$\delta V(x) = \int_0^L G(x,x')\ \delta I(x')\ dx' \tag{4.27}$$

The comparison between (4.4) and (4.27) with x = L yields :

$$\frac{\partial Z}{\partial x'} = G(L,x') \tag{4.28}$$

In that case, the impedance field is simply related to the Green function of the linearized equation (4.24). An application of this method to the noise behavior of FET's can be found in the litterature (Nougier 1987).

4.4. One dimensional (1D) numerical modeling

The device physical description needs often to solve a complex system of non linear differential equations (Carnez 1981, Cappy 1985) describing, for instance, the conservation of charges, energy or momentum. Such a system can be :

$$\begin{aligned}
I &= q\,S(x)\,n(x)\,v(x)\\
\frac{d\varepsilon}{dx} &= qE - \frac{\varepsilon-\varepsilon_0}{v(x)\,\tau_\varepsilon(\varepsilon)}\\
v(x) &= \mu(\varepsilon)\,E(x) - \frac{v(x)\,\mu(x)}{q}\,\frac{d}{dx}\left(m^*(\varepsilon)\,v(x)\right)\\
\frac{dE(x)}{dx} &= \frac{q}{\varepsilon}\left(N_d - n(x)\right)
\end{aligned} \tag{4.29}$$

In that case it is preferable to directly calculate the impedance field according to the following method : a small current ΔI is added between 0 and x. The resulting voltage fluctuation is ΔV. By definition, the impedance field can be written :

$$\frac{\partial Z(x)}{\partial x} = \frac{\Delta V(x+\Delta x) - \Delta V(x)}{\Delta x\,.\,\Delta I} \tag{4.30}$$

This method can be used for the modelling of the high frequency noise performance of FET's (Cappy 1989).

4.5. Two dimensional (2D) numerical modeling

2D modelling can be divided into two groups

- hydrodynamic modelling
- Monte Carlo modelling

The former method is not well suited for noise study because of excessive computation times. In addition, some basic problems concerning the noise in 2D modelling are still unresolved.

On the other hand Monte Carlo modelling is very useful in the noise study for two main reasons :

* In the bulk material study, this method provides the D-E relationship.

* Basically this procedure provides the diffusion noise because the velocity of each carrier is known at each time. Therefore, this method yields, for a given electrode, the evolution of the current value versus time i(t). We can then deduce the quadratic current fluctuations.

$$\Delta i(t)^2 = \left| i(t) - \bar{i} \right|^2 \qquad (4.31)$$

By Fourrier transform (Moglestue 1985), the knowledge of Δi^2 yields the spectral intensity Si of the current fluctuations.

It is then possible to calculate directly the equivalent noise sources at each electrode without the use of local noise source and impedance field concepts. Unfortunatly, this method is restricted to the high frequencies. As a matter of fact, for an observation time T, the lower frequency studied is 1/T. In addition, in order to avoid numerical noise, long computation times are necessary.

5. Application

5.1. The high frequency noise in FET amplifiers

Conventional FET's and more complex related field effect devices (HEMT, pseudomorphic...) have demonstrated remarkably low noise figures in the centimeter and millimeter wave ranges. As shown before, the theoretical noise performance calculation needs to define two noise sources (gate and drain) as well as their correlation coefficient. Using a method equivalent to the impedance field method, it has been shown (Van der Ziel 1962) that the noise sources can be written :

$$\overline{id}^2 = 4\,kT\,gm\,P\,\Delta f$$
$$\overline{ig}^2 = 4\,kT\,\frac{Cgs^2\,\omega^2}{gm}\,R.\Delta f \qquad (5.1)$$

In these equations g_m is the transconductance, Cgs the gate-to-source capacitance, P and R are dimensionless coefficients depending upon technological parameters and biasing conditions.

Neglecting distributed effects, the correlation coefficient is purely imaginary and its value is close to j.0.8 (Cappy 1985, 1988). Figure 6 shows the evolution of id^2, ig^2, P, R and C in the case of a half micron gate length HEMT.

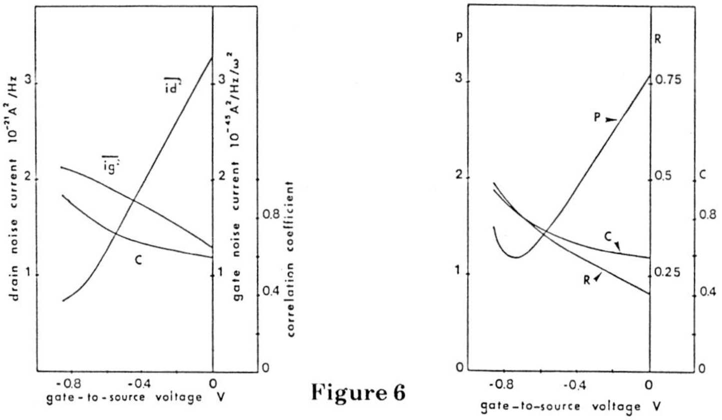

Figure 6

From P, R and C, the determination of the noise figure and the other noise parameters is straight forward (Pucel 1974). Introducing K_1 and K_2 defined by

$$K_1 = P + R - 2C\sqrt{RP}$$
$$K_2 = P\,.\,R\,(1 - C^2) \qquad (5.2)$$

we have :

$$F_{min} = 1 + 2\sqrt{K_1}\,.\,\frac{\omega Cgs}{gm}\,.\sqrt{gm\,(Rs + Rg) + \frac{K_2}{K_1}}$$

$$R_n = R_s + R_g + P/gm \qquad (5.3)$$

$$G_n = \frac{\omega^2\,Cgs^2}{gm^2}\left[\frac{(Rs + Rg)\,gm + K_2/K_1}{R_n}\right]$$

$$Y_{opt} = \left(\frac{G_n}{R_n}\right)^{\frac{1}{2}} + j\omega\,Cgs\,.\left[\frac{Co\sqrt{RP} - P}{g_m\,R_n}\right]$$

These expressions show different fundamental effects :

(i) The gate noise influences the noise figure <u>even at low frequency</u>

(ii) The noise figure (in linear scate, not in dB) keeps a linear variation versus frequency even if the gate noise is taken into account.

(iii) If Rs+Rg tends to zero, the noise figure is no longer close to unity and the gate noise is taken into account. The device can be characterized by an intrinsic noise figure

$$F_{int} = 1 + 2 \cdot \frac{\omega\, Cgs}{gm} \cdot \sqrt{PR(1 - C^2)} \qquad (5.4)$$

(iv) Due to <u>the correlation</u> between the drain and gate noise current sources, the gate noise is partially subtracted from the drain noise. This fundamental effect is expressed in the noise parameter expressions by the terms $P + R - 2C\sqrt{RP}$ and $PR(1-C^2)$. This reduction of drain noise is the basic reason why the field effect transistor is a low noise device. Therefore, the gate noise and the correlation coefficient have to be included in FET noise analysis.

In conclusion, the main problem in the noise modelling in FETs is to develop a modelling able to precisely provide not only the noise coefficient P and R but also the correlation coefficient C.

5.2. The spectral purity of oscillators

An equivalent circuit for an oscillator is shown in Fig. 7.a.

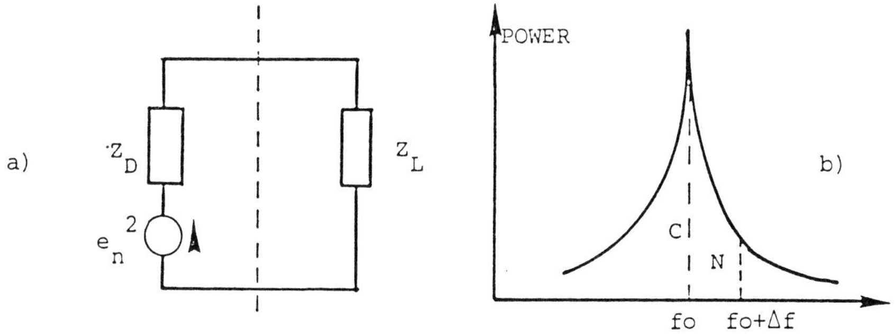

Figure 7

The active device can be a diode or an FET. The current i in this circuit is assumed to take the form :

$$i(t) = \left(I_0 + \Delta i(t)\right) \cos\left(\omega_0 t + \Delta\phi(t)\right) \tag{5.5}$$

The noise is then characterized in terms of amplitude modulation (AM) arising from $\delta i(t)$ and frequency modulation (FM) arising from $\delta\phi(t)$. In the noise oscillator theory (Edson 1960, Kurokawa 1969), it is assumed that the impedance Z_D of the equivalent active device is dependent on the amplitude Io and frequency fo. In theoretical analysis the following results for the noise to carrier ratio (N/C) in a single side band (Fig. 7b) arise.

FM case

$$\frac{N}{C}\left(f_0 + \Delta f\right) = \frac{Se(\Delta f)}{8R_L P_0}\left(\frac{f_0}{Q_0 \Delta f}\right)^2 \tag{5.6}$$

AM Case

$$\left(\frac{N}{C}\right)\left(f_0 + \Delta f\right) = \frac{Se(\Delta f)}{8R_L P_0} \cdot \frac{1}{\left(\frac{S}{2}\right)^2 + \left(\frac{Q_0 \Delta f}{f_0}\right)^2} \tag{5.7}$$

where Se(Δf) is the spectral intensity of the noise source e, Po the output power developed in R_L and S a dimensionless factor which characterizes the non linear behavior of R_D.

These results show that the device low frequency noise is mixed with the carrier signal via the non linearity of the device and up converted to microwave frequencies. As a consequence, the spectral purity of oscillators is mainly determined by the LF noise of the equivalent device noise source e.

6. Conclusion

In the case of 1D device modelling (diodes or 1D FET modelling) the very powerful impedance field method allows to calculate the equivalent macroscopic noise sources for any noise type (G-R, diffusion, 1/f ...). This method can be used even in an analytical or in a numerical form. When the noise sources are known, the derivation of the noise parameters of practical interest (noise figure, equivalent noise resistance ...) needs some circuit calculations only. However, it should be emphasized that the device noise properties have to be considered as first order properties. For this reason they greatly depend on the various assumptions used in the description of the zero order (DC) properties. Noise modelling has to be based on a modelling providing accurate DC predictions.

Lastly, it seems interesting to precis some important problems encoutered in noise modellings.

(i) experimental

Any device modelling has to be compared with experimental findings. Unfortunately, the noise source values cannot be measured accurately. Therefore it is very difficult to check the validity of a modelling using the comparison with experimental findings.

(ii) microscopic spatial correlation

The noise sources attwodifferent points of a device are always supposed to be uncorrelated. In fact, this is only an approximation. A mathematical derivation of the two points correlation function has been proposed (Nougier 1983) for some specific scatterings. The effects of this microscopic spatial correlation are not presently well known.

(iii) noise in quantized devices.

As the device characteristic length decreases, the DC value determination as well as the noise analysis become more and more difficult since the electron dynamics (v-E, ε-E, D-E characteristics) are less well known.

(iv) 2D modelling

Actually, there are only few reported works concerning 2D modelling including a noise analysis. In the case of diffusion noise, the Monte Carlo technique seems however to be well suited for the noise analysis including the microscopic spatial correlation.

Cappy A. et al (1985) Noise modeling in submicrometer gate two dimensional electron-gas field effect transistor. IEEE Trans. Electron Devices 32, n° 12 : 2787-2795.

Cappy A., Heinrich W. (1989). The high frequency FET noise performance : A new approach. IEEE Trans. Electron Devices 36, n° 1.

Cappy A. (1988). Noise modelling and measurement techniques. IEEE Trans. MTT 36, n° 1 : 1-10.

Carnez B. et al (1981). Noise modelling in submicrometer gate FET's. IEEE Trans. Electron Devices, Vol. 28, n° 7 : 784-789.

Edson W.A. (1960). Noise in oscillators. Proc. IRE, Vol. 48 : 1454-1466.

Kurokawa K. (1968). Noise in synchronized oscillators. IEEE Trans. MTT, Vol. 16, n° 4 : 234-240.

Moglestue C. (1985). A Monte Carlo particle study of the intrinsic noise figure in GaAs MESFET's. IEEE Trans. Electron Devices, Vol. ED-32, n° 10 : 2092-2096.

Nougier J.P. et al (1985). Numerical modelling of the noise of one dimensional devices. Physica B 134 : 260-263.

Nougier J.P. (1987). Bruit dans les composants. Rev. de Physique Appliquée 22.

Nougier J.P. et al (1983). Microscopic spatial correlations at thermal equilibrium in non polar semiconductors. In noise in Physical systems, Elsevier Science Publishers : 15-18.

Pucel R.A. et al (1974). Signal and noise properties of gallium arsenide field effect transistors advances in electronics and Electron physics, Vol. 38 : 195-265.

Rothe H. and Dahlke W. (1956). Theory of noisy fourpoles. Proc. IRE, Vol. 44 : 811-818.

Shockley W. et al (1966) in quantum theory of atoms, Molecules and the solid state (Academic, New York) : 537.

Van der Ziel A. (1954). Noise, Prentice Hall, Englewood Cliffs, N.J.

Van der Ziel A. (1962). Thermal noise in field effect transistor. Proc. IRE, Vol. 50 : 1808-1812.

Van der Ziel A. (1988). Unified presentation of 1/f noise in electronic devices : fundamental 1/f noise source. Proc. of the IEEE, vol. 76, n° 3 : 233-258.

Van Vliet K.M. et al (1975). Noise in single injection diodes : I a sinvey of methods. JAP, Vol. 46, n° 4, 1804-1813.

Van Vliet K.M. (1979). The transfer-impedance method for noise in field-effect transistors. Solid-state electronics, Vol. 22 : 233-236.

Monte Carlo Modelling Techniques

Mustafa Al-Mudares
University of Glasgow

1. INTRODUCTION

The fast rate of progress in semiconductor devices technology has led to a rapid increase in interest in developing various numerical models for these devices.

These models can be divided into two types. The first type is the macroscopic model and the second is the Monte Carlo model. It is well realised that the Monte Carlo model is the most accurate of the two because of many reasons that will discussed in due course. However, before discussing the basic advantages of the Monte Carlo method, we have to discuss the weaknesses and problems with the macroscopic models.

To do this we have to remember that the basic macroscopic models are based on the drift-diffusion approximation which solves the electron current transport and the continuity equations,

$$J = e\mu nF + eD\frac{dn}{dx}$$

$$\frac{dn}{dt} = -\frac{J}{e}$$

Here J is the current density, and the two terms on the right side of the first equation are the contributions to the current from drift and diffusion respectively. In the drift term, e is the electron charge, μ is the carrier mobility, n is the carrier density, and F is the electric field. In the diffusion term, D is the carrier diffusion coefficient. Both μ and D can be field dependent parameters.

The problem with this model is the assumption that the average carrier velocity v adjusts instantaneously to the electric field via the relation $v = \mu F$. No account is taken for the time delays associated with the scattering of carriers to higher energy states. In many energy-valley materials such as GaAs, electrons may exist in different energy valleys and then it is necessary to take into account the time delays associated with scattering between these valleys. Also, since this model excludes energy dependent terms, the drift-diffusion model will be valid as long as the distribution function is a drift Maxwellian with an effective carrier temperature equal to the lattice temperature. Therefore, this model can not be used in cases where the distribution function becomes distorted. A situation usually occurs when hot carrier effects are present.

To include these hot carrier effects, attention was paid to modify the drift-diffusion model by solving the first three moments of the Boltzmann's equation:

momentum conservation

$$\frac{dv}{dt} + v.\nabla v = e\frac{F}{m^*} - \nabla\frac{(nKT_e)}{m^*} + \left(\frac{dv}{dt}\right)_{coll}$$

energy conservation

$$\frac{dE}{dt} + v.\nabla E = ev.F - \frac{1}{n}\nabla(nvKT_e) + \left(\frac{dE}{dt}\right)_{coll}$$

carrier conservation

$$\frac{dn}{dt} + \frac{J}{e} = \left(\frac{dn}{dt}\right)_{coll}$$

where m^* is the effective carrier mass, T_e is the local carrier temperature, and E is the average carrier energy. In III-V compounds, the effective mass is valley dependent and to take this into account the above equations should be solved for electrons in all valleys. Such a solution leads to a rather complex model and to simplify this problem a single carrier gas model was suggested in which an average value for the effective mass is used which accounts for the proportion of carriers in each valley as

$$m^* = m_1^*\{1 - r(E)\} + r(E)m_2^*$$

where m_1 and m_2 are the effective masses for electrons in the lower and upper valleys respectively and r is the proportion of electrons in the upper valley as a function of the average electron energy.

The above equations are still intractable because the collision terms are quite complicated. To avoid this, relaxation time models were used for these collision terms as :

$$\left(\frac{dv}{dt}\right)_{coll} = \frac{v}{\tau_p(E)}$$

$$\left(\frac{dE}{dt}\right)_{coll} = \frac{E - E_O}{\tau_E(E)}$$

$$\left(\frac{dn}{dt}\right)_{(coll)} = 0$$

where $\tau_p(E)$ and $\tau_E(E)$ are respectively the momentum and energy relaxation times and E_O is the thermal energy corresponding to the lattice temperature.

These relaxation times are usually estimated from analysing the steady-state results of Monte Carlo simulation on the bulk properties of the concerned material when an electric field is switched on from a zero start i.e. assuming that all carriers are initially at thermal equilibrium. Therefore, this model will overestimate the nonstationary effects in devices such as the FETs which have nonuniform field and energy distributions. This is because it treats carrier dynamics anywhere in the device using relaxation times to the energy of carriers starting from equilibrium to the lattice temperature, while the nonstationary travelling carriers in a device are affected by the relaxation times corresponding to their local field and energy distributions. These relaxation times were also estimated when a single field component, the longitudinal electric field, exists in a device. However, Monte Carlo simulations performed on bulk materials subjected to both longitudinal and transverse electric fields do show that the steady-state results are much different from those obtained when only the material is subjected to a longitudinal electric field. Therefore, in devices such as FETs, where two electric field components control device's operation, one must allow for momentum conservation in the transverse direction, to consider the contribution of the transverse field to the total local energy and to include the effect of the transverse field on the relaxation times.

Furthermore, this model can not distinguish between each individual scattering mechanism since one can not expect to represent the complexities of carrier scattering by two relaxation times and therefore, this can affect both the static and dynamic properties of a device. Moreover, this approach is used under the assumption that the distribution function is symmetric about the mean momentum. This would be true if electron-electron scattering is very strong. In GaAs and other III-V compounds, electron-electron scattering is probably not strong enough to maintain a symmetric distribution function.

From this it seems that we still do not have a model that accounts for many important physical issues which can play an important roll in predicting the behaviour of submicron GaAs and other III-V semiconductor devices. For example, how scattering mechanisms and the distribution function profile affect device properties and what are the individual effect of degeneracy, screening, plasmon modes , hot phonons and electron-electron scattering on device properties.

We seem to have opened a Pandora's box. There are now more unanswered questions than when we have started, and to answer them it is necessary to find another physical model which can calculate all these effects. As we will see in the next sections, the method that can care for these various effects is the Monte Carlo particle method. This method provides an exact solution for the Boltzmann's equation as long as the band structure and the various forms of scattering processes are known. There are many ways to apply the Monte Carlo method for studying the physics of carrier transport in both materials and devices. Since this course is devoted to the modelling of practical semiconductor devices, we will concentrate on applying the Monte Carlo method for device modelling and the example we will consider first is on the two-dimensional modelling of GaAs MESFETs.

2 2. Two-Dimensional Monte Carlo Model FOR GaAs MESFETs

The basic idea of any Monte Carlo simulation is to follow, on a time step basis, the trajectories of all electrons in the device. These electrons are followed as they move under the influence of self-consistent fields and interact with scattering sources. Therefore, this technique reflects the the actual physics of a semiconductor device.

The model considered can simulate various MESFET structures, planar and recessed gate MESFETs. In this model we will track the motion of electrons in two-dimensional (2D) real space (x,y) and in three-dimensional crystal space (K_x , K_y , K_z). The term two-dimensional real space means that we will assume a uniform potential distribution in the third dimension, Z, which is usually known as the device width and therefore we will neglect electron movement in this direction.

Let us first assume that the MESFET structure has a cross sectional area A and it consists of a single layer beneath the contacts. This layer, normally called the active layer, has a uniform doping profile of a density N_D , a thickness d and a channel length L . Therefore, this layer consists of N_D A electrons per unit width. For any practical values of N_D and A, we can find that we have to follow the trajectories of very large number of electrons. As an example, if $N_D = 10^{17}$ cm^{-3} , d = 0.2 microns and L = 3 microns, the number of electrons per cm that should be followed is 800 million. By any means, this is not possible and to avoid this problem we will assume that each cloud of electrons will be represented by a particle. These particles, which also are commonly called super-particles, can be regarded as rod of electrons moving through a 2D plane .

From the point of view of the statistical fluctuations associated with the stochastical nature of the Monte Carlo method, a large number of particles should be used to represent the electrons in order to reduce these fluctuations. The upper number of these particles is dictated by the amount of available storage and computer processing time, since increasing particle number makes the processing time so great that the Monte Carlo method may become useless. When a decision is made on N, the total number of particles in the device, each particle will represent N_D A/N electrons per unit width.

Over the whole device, a two-dimensional mesh cell will be imposed. The area of each mesh cell is

$$A_c = \Delta_x \Delta_y$$

where Δ_x and Δ_y are the mesh cell widths in the x- and y-directions. Δ_x and Δ_y are assumed to be uniform across the device. These meshes will be the frame of reference in which the particles move under the influence of the electric fields.

However, when the MESFET is biased, the particles will accelerate in both real and crystal spaces under the influence of the self-consistent fields. As the particles are accelerated, they will collide with the crystal lattice, with impurity centres as well as with other particles and other scattering sources. In order to describe all these effects accurately, the model needs to have a detailed picture of the device and material considered. So we need to know things like the exact device geometry and dimensions, the band structure of the material used and other physical constants that are necessary for calculating the rates for the allowed scattering mechanisms. The conduction band structure that can be used in the Model can either be based on the effective mass approximation or calculated from pseudo-potential calculations , and for GaAs and other III-V compounds, it should consist of the three non-parabolic Γ, L and X valleys.

After setting all these parameters, the model starts with assuming charge neutrality everywhere in the device. As the device is divided into number of two-dimensional mesh cells, the initial number of the simulated particles N will be distributed equally among all these mesh cells to preserve a charge neutral mesh. So the number of particles in each cell will be

$$N_M = \frac{N}{M}$$

where M is the total number of mesh cells in the device. The position of the all N_M particles in a mesh cell is assigned by a random number generator with a flat distribution. Since the simulated device only consists of the active layer which is uniformly doped, the charge carried by each particle per unit width will be

$$C_P = \frac{-eN_D \Delta_x \Delta_y}{N_M}$$

Initially, all the particles in the device will occupy the lowest, Γ, valley of the conduction band and are assigned wave vector components corresponding to a Maxwellian distribution. A Maxwellian distribution is defined as one when the probability of finding a wave vector with magnitude K between K and K+dk is given by :

$$P(K)dK = \frac{\exp\left\{\frac{-K^2}{2K_{th}^2}\right\}}{\sqrt{2\pi K_{th}^2}} (1)$$

where K_{th} is the root-mean-square wave vector for the distribution. Since the initial distribution is usually taken at the lattice temperature T_o , the particle is released with an energy that is related to K_{th} by the energy-wave vector (E-K) relationship. For non-parabolic valley, the E-K relationship is usually given as

$$E(1 + \alpha E) = \frac{h^2 K^2}{2m^*}$$

where m_Γ^* and α_Γ are the electron effective mass and the non-parabolicity factor for the central valley respectively and K is the wave vector.

Therefore, for electrons released at lattice temperature, K_{th} can e expressed as

$$K_{th} = \frac{\sqrt{m_\Gamma^* K_B T_O (1 + \alpha_\Gamma K_B T_O)}}{h}$$

where K_B is Boltzmann's constant,

The three components of the wave vector K are selected again using a random number generator. However, in Monte Carlo simulations, one always faces the problem of selecting a set of parameters with a given uniform probability P(...) when there is only one random number generator giving a uniform distribution. The solution for this is to calculate the cumulative distribution function

$$c(K) = \int_0^K P(K)\,dK$$

Using equation (1), yields

$$c(K) = \frac{erf\left(\frac{K}{\sqrt{2}K_{th}}\right)}{2}$$

where erf is the error function defined as

$$erf(x) = \frac{2}{\sqrt{\pi}} \int_0^x \exp(-x^2)\,dx$$

By using the above two equations, we can find that

$$-\frac{1}{2} \le c(K) < \frac{1}{2}$$

so we can set

$$c(K) = \left(r - \frac{1}{2}\right)$$

where r is a random number in the range $0 < r < 1$. Then solving for the required wave vector, we obtain

$$K = \sqrt{2}K_{th} erf^{-1}(2r-1)$$

Given the magnitude of the wave vector K, the three components of K can be selected as

$$K_x = K\sin\theta\cos\phi$$

$$K_y = K\sin\theta\sin\phi$$

$$K_z = K\cos\theta$$

where θ is the angle between K and the z-axis and ϕ is the angle accounting for the variation of K about the z-axis and measured anti-clockwise from the x-axis . θ and ϕ
are generated from the random number r as

$$\theta = \cos^{-1}(r)$$

$$\phi = 2\pi r$$

3 Having set these initial conditions, the main calculation can be carried out on a time step basis. The time step ΔT should be chosen according to the stability requirements that will be discussed latter.
The main calculation usually starts with determining the two-dimensional potential distribution from a known distribution of the charge density $\rho(x,y)$ which is given as

$$\rho(x,y) = \frac{e\{n(x,y) - N_D\}}{\epsilon_0\epsilon_S}$$

where n(x,y) is the electron density. To determine the electron density in each mesh cell, we have to determine the number of charged particles within that mesh cell. There are many ways to assign a charged particle to a mesh cell. The simplest method is the nearest grid point (NGP) scheme in which the total charge found in a mesh cell is assigned at the midpoint of that cell. These are other charge assignment methods which can include the effect of spreading the influence of one particle over several mesh cells and smoothing the charge distribution and then results in a less noisy potential distribution particulary when heavy doping densities are used large electron densities are expected to occur. Among these methods is the cloud-in-cloud (CIC) scheme. In this scheme, particles are assigned to the mesh cells in a way that corresponds to the weighted surrounding occupation areas of particles. However, the NGP scheme has been adopted in most of the reported Monte Carlo models because it is computationally less costly than the CIC and other high order schemes because it does not interpolate the force over several mesh cells. But as device size becomes smaller, this assumption is no longer valid and schemes similar to the CIC haveto be used. However, to avoid any numerical instability that may arise from charge assignment, it is advisable to use the CIC scheme.

After determining the charge distribution in every mesh cell, the potential distribution V(x,y) can then be determined by solving Poisson's equation in two-dimensions:

$$\frac{d^2V(x,y)}{dx^2}+\frac{d^2V(x,y)}{dy^2}=\rho(x,y)$$

There are many methods for solving this equation in two-dimensions. These can be indirect or direct methods, but the latterare much more accurate and faster. For this reason, most of the reported Monte Carlo models do use the direct schemes for solving the two-dimensional Poisson's equation. Among the most general two-dimensional Poisson's equation solvers is the Fourier Analysis Cyclic Reduction (FACR) method which accounts for non-planar geometries, nonuniform mesh sizes, surface effects and mixed boundary conditions.

However, for the planar MESFET structures, Poisson's equation must be solved with assuming that the electric fields normal to all free surfaces should be zero. Also, at the ohmic Source and Drain contacts, the potential must be fixed to the applied bias i.e. at the source contact the potential is zero while at the drain contact, the potential is V_{DS} . At the Schottky barrier gate, the potential is fixed at $V_{GS} - V_B$, where V_B is the metal-semiconductor built-in-potential.

From the evaluation of the two-dimensional potential distribution we can determine the distribution of the longitudinal and transverse components of the electric field, F_x and F_y respectively as:

$$F_x(x,y) = -\frac{dV(x,y)}{dx}$$

$$F_y(x,y) = -\frac{dV(x,y)}{dy}$$

These two field components will accelerate each particle in both crystal and real spaces for a short time before being scattered. The acceleration is assumed to obey Newton's laws of motion. The new x- and y- components of the particle wave vector can then be obtained from

$$K_x^{t+\delta t} = K_x^t + \frac{eF_x(x,y)\delta t}{h}$$

$$K_y^{t+\delta t} = K_y^t + \frac{eF_y(x,y)\delta t}{h}$$

where δt is the free flight duration between two successive scattering events. The z- component of the wave vector should be kept constant during the free flight period since the z-component of the electric field is set to zero.

The new particle position in the x- nd y direction will then be

$$x^{t+\delta t} = x^t + \frac{hK_x^t\delta t}{m_l^*} + \frac{eF_x(x,y)\delta t^2}{2m_l^*}$$

$$y^{t+\delta t} = y^{t} + \frac{hK_y^t \delta t}{m_l^*} + \frac{eF_y(x,y)\delta t^2}{2m_l^*}$$

where m_l^* is the electron effective mass for the l-th valley.

During a free flight, there is a possibility that a particle may cross a boundary region. If this boundary is a free surface boundary, the particle should be reflected back into the device but with reversing its momentum component perpendicular to that surface. If a particle crosses an ohmic contact, it should be counted in order to estimate the contact currents. These particles can be injected back to maintain charge neutrality at the contact region and when it is injected, it must have wave vector components corresponding to a Maxwellian distribution at thermal equilibrium. On the otherhand, since the region under the gate contact is to be depleted from any carriers, any particle crossingthe gate contact will be removed from the simulation, but in order to estimate the gate current, the number of particles that cross the gate contact must be counted.

However, the selection of the free flight time t between two successive scattering events is usually determined as a function of the calculated scattering rates. If the total sum of all scattering rates for a particle with energy E is $\Gamma_T(E)$, and if $\Gamma_T(E)$ is constant across the enrgy range of interest, the probability that this particle will scatter in a time between t and t + dt is

$$P(t)dt = \Gamma_T(E)\exp\{-\Gamma_T(E)\}dt(2)$$

The general solution for P(t) is to use again the cumulative distribution function

$$c(t) = \int_0^t P(t^{)\ \mathrm{dt\ super}}$$

If we apply equation (2) assuming that the $\Gamma_T(E)$ is constant at all energies, we can obtain

$$c(t) = 1 - \exp(-\Gamma_T t)$$

At t = 0, c(t) = 0, while at $t = \infty$, c(t) = 1 , so
$0 <= c(t) < 1$

Then, the value of c(t) can be chosen from a random number r that is uniformly distributed between 0 and 1 . So

$$r = c(t) = 1 - \exp(-\Gamma_T t)$$

$$t = -\Gamma_T^{-1} \ln(r_1)$$

where r_1 is a new random number, r_1 = (1-r), which is also uniformly distributed between 0 and 1 . t is the free flight time which was introduced earlier by the sybmol δt .

However, the total sum of all scattering rates is dependent on the energy of the particle and consequently it varies with particle energy. Therefore, the above selection procedure for δt can not be valid. In order to determine δt when $\Gamma_T(E)$ is energy dependent, we must use the so called Rees self-scattering process in which neither the momentum nor the energy of a particle will change. By including self scattering, the new total sum of all scattering rate Γ_S is

$$\Gamma_S = \sum_{I=0}^{N} \lambda_I(E)$$

where λ_0 represents the self-scattering rate and λ_1 , λ_2 , ... , λ_N are the scattering rates for each individual real scattering process. The self-scattering rate at any energy should be chosen to be that rate necessary to bring Γ_T , the sum of all real scattering rates, to a chosen constant value Γ_S which should be larger that Γ_T for all considered energies.

If we now consider self-scattering as a process, the total

scattering rate will be constant for all energies and then the free flight time can be calculated using the conventional procedure i.e.

$$\delta t = -\Gamma_S^{-1} \ln(r_1)$$

If any particle has an energy E_A , it will undergo a large number of self scatterings if Γ_S is chosen much larger than $\Gamma_T(E_A)$. In this case, we will calculate several free flight trajectories where no change in particle energy and momentum will take place and therefore a considerable amount of computing time will be lost. To avoid this, many schemes based on the Rees's self-scattering have been proposed. The most common of those is the iterative gamma scheme in which self-scattering is reduced to a minimum. This scheme is based on choosing Γ_S so that:

$$\Gamma_S = \Gamma_T(E) + \Delta\Gamma$$

So when a particle has an energy E_A at a time t and within a free flight, its energy decreases to a lower value, $\Delta\Gamma = 0$. On the otherhand, if the particle energy increases within a free flight, we would not include all real scattering rates unless we add a $\Delta\Gamma$ to the total scattering rate. $\Delta\Gamma$ is chosen in an iterative way as an integral multiple of a predefined parameter. This iterative process converges in not more than 5 iterations.

At the end of each free flight, the particle will have a new kinetic energy E_N which is determined from the new value of the wave vector K_N

$$K_N = \sqrt{(K_x^{t+\delta t})^2 + (K_y^{t+\delta t})^2 + (K_z)^2}$$

$$E_N(1 + \alpha_l E_N) = \frac{h^2 K_N^2}{2m_l^*}$$

Each particle will then undergo a scattering with E_N being the initial particle energy before scattering. This scattering process will be chosen by means of a random number, r, uniformly distributed between 0 and $\Gamma_T(E_N)$. Assuming that the scattering rate are time and position independent, i.e. not affected by the local electron density, the scattering mechanism number J is chosen when

$$\sum_{n=1}^{J} \lambda_n(E_N) \leq r < \sum_{n=1}^{J+1} \lambda_n(E_N)$$

After each scattering, the particle's energy will be modified according to conservation laws of energy that are imposed by the selected scattering process. The selection of the new wave vector and its components after scattering involves the calculation of the new wave vector which is specified by the new particle energy as well as the calculation of the angles of scattering according to angular probability distribution for the selected scattering process. The angles θ and ϕ (see Figure 2) are selected by uniformly distributed random numbers. The angle ϕ is determined by using a random number r_a from :

$$r_a = \frac{\int_0^{\phi} d\phi}{\int_0^{2\pi} d\phi} = \frac{\phi}{2\pi}$$

So $\phi = 2\pi r_a$

The angle θ is obtained from the following relation using another random number r_b as :

$$r_b = \frac{\int_{\cos\theta}^{1} .M(K,K')..^2...d\cos\theta}{\int_{-1}^{1} .M(K,K')..^2...d\cos\theta}$$

where $M(K,K')^2$ is the matrix element for a transition from an initial state K to a final initial K'. The matrix element of some scattering mechanisms does not depend on the angle and therefore all wave vectors that satisfy the appropriate law of conversation of energy are equally probable. Such scattering mechanisms are called randomising mechanisms which means that the scattered particles have the same probability of being in any direction after scattering. In this case, we can find that

$$r_b = \frac{(1-\cos\theta)}{2}$$

On the otherhand, when the matrix element is a function of θ, one has to determine θ by integration. The resulted complexity depends on the type of scattering process and the methods that deal with these cases are discussed in App 1.

It should be noticed from App.1 that every scattering process can have different rate formulas. This depends on whether we consider effects such as degeneracy, screening, plasmons, hot phonons, and other effects. The first three effects are a function of the local electron density and therefore including these effects will make the scattering rates to be position and time dependent since the electron density in a device is a function of position as well as time because during the transient behaviour of the device, the electrons can change from a position to another. Therefore, allowing these effects to take place means that scattering processes at each mesh cell should also be calculated at each time step to account for the change in electron density with time. On the otherhand, if hot phonon effect is included, the scattering rates should be re-evaluated when the phonon occuption number differs with time which means that the scattering rates can also be time, as well as position dependent.

It is also possible that a particle is accelerated and scattered more than once during a time step. If the the sum of free flights in one time step exceeds ΔT, we must curtail the last free flight at the end of the time step, remembering how much of that last flight left for each particle. Then during the next time step, we will use the rest of the last free flight for that particle. This implies that the time step must be chosen so short that only a few free flights can take place. Experience yields that ΔT must be chosen as

$$\Delta T = \frac{\min(\Delta x, \Delta y)}{\bar{v}}$$

where $\bar{v}$ is the average velocity of particles.
However, it is obvious that all particles will have new states in both K and real spaces after a free flight and these new states will be different from their previous ones. Therefore, each mesh cell will have a new charge density and as result, Poisson's equation is no longer satisfied by the newly obtained charge density distribution.It is then necessary to evaluate the new potential distribution and then new values of potential are computed. The new potential distribution will result in a new distribution of the electric fields and the particles will be accelerated into new states in real and crystal spaces and then particles may experience new scattering selection rules.
This calculation loop must be repeated iteratively for many time steps until the steady-state flow of particles is reached.

3. Monte Carlo Modelling of Multi-layered MESFETs

Suppose that we want to simulate a MESFET that contains a number of layers which may be have different doping densities and doped uniformly or nonuniformly and may contain N- type as well as P-type layers. If each doped layer I has a doping density S_I and its cross sectional area is A_I , there will be S_I A_I carriers per unit width. If each layer is assigned N_I particles, then each particle in layer I will represent $\frac{S_I A_I}{N_I}$ carriers per unit width.

After setting the initial number of particles in each layer, we must set the requirement of charge neutrality everywhere in the device. Since each layer is divided into a number of two-dimensional mesh cells, the initial number of particles in each layer N_I must be distributed equally among all the mesh cells in that layer to preserve a charge neutral mesh. So we will have $N_{I,M}=\frac{N_I}{M_I}$ particles per mesh where M_I is the total number of mesh cells in layer I. Assuming that all meshes in the device have the same cross-sectional areas, then the charge carried by each particle will be

$$c_I=\frac{\mp e S_I \Delta_x \Delta_y}{N_{I,M}}$$

where - is used for electrons and + is used for holes.
This means that a particle in a layer may have a charge density that differs from that of a particle in another layer.
Therefore, the basic difference in modelling multi-layer structures is to distinguish between the charge density of the simulated particles. Also if N- and P- types layers are used, the charge sign carried by each particle will be different and this must also be considered. In this case , two Monte Carlo procedures must be allowed; one for electrons and the other for holes. It should be realised that when different carrier densities are used, different rates for scattering mechanisms, ionised impurity scattering in particular, must be allowed. Different dopings may also affect other scattering processes if screening and degeneracy are included. This means that the scattering rates must be identified for each layer.
Furthermore, if P-type layers are included in the simulations, the transport properties of holes must be allowed and therefore, the scattering rates for holes must be considered.
Differences in doping densities and doping types must also be accounted for when solving for the two-dimensional charge density distribution which, generally, is given as:

$$\rho(x,y)=\frac{e\{n(x,y)-N_D,I+N_A,I-p(x,y)\}}{\epsilon_0\epsilon_S}$$

where $N_{D,I}$ and $N_{A,I}$ are respectively the donor and acceptor densities for layer I and p(x,y) is the two-dimensional acceptor density.

4. Monte Carlo Modelling of Heterostructure Devices

In Monte Carlo modelling heterojunction devices, two more important conditions must be allowed:-

a. The difference between the materials assumed in the simulations,

b. Solving for any quantum effects that occur at the hetero-interface(s). These effects include carrier confinement , localisation, transmission and reflection, resonant tunelling and many more.

An example of simulating GaAs-AlGaAs Modulation Doped FETs (MODFETs) will be considered in the course.

4 Under the influence of an electric field, carriers can be accelerated as they gain some energy from this field. If the crystal is perfectly periodic, the acceleration will continue without any perturbation, but due to the existance of many mechanisms that perturb this periodicity, carrier acceleration will be accompanied by a number of collisions with the lattice and with other sources that alter either carrier momentum or energy or both. Therefore, these collisions, which are known as scattering mechanisms, are infrequent perturbations of the motion of carriers. These perturbations should take place at a given rate defined by first order time dependent perturbation theory :

$$\lambda(\underline{K}) = \frac{2\pi}{h} \int |< \underline{K}' | H | \underline{K} |^2 \{1 - f(\underline{K}')\} \delta(E' - E) dN'$$

where K and K' are the wavevectors related to the initial and final states respectively, H is the perturbation energy, f(K') is the probability of occupation of a final states and the integration is over all final states restricted by the delta function conserving energy. In non-degenerate systems, f(K') is very small and is usually neglected but if carrier population is very high, degenerate statistics must be used and therefore, f(K') has an appreciable value.

These scattering mechanisms can be classified into three main types. The first is due to lattice vibrations and called lattice scattering; the second is produced by ionised impurity atoms and called ionised impurity scattering, while the third is due to collisions between carriers themselves and called carrier-carrier scattering. In the discussion below, the first two type of scattering will be summarised and only scattering of electrons in III-V compounds will be considered.

Lattice Scattering

Lattice scattering is defined by the intrinsic physical properties of the crystal and occurs due to the vibrations of atoms about their equilibrium sites. These vibrations will vary the periodic potential in the crystal with time and then alter the electron sites in time. In a fashion similar to the association of photons with electromagnetic waves, lattice vibrations can be considered as phonons with an associated energy and momentum of hw_K and hK respectively, where w_K is the vibrational frequency and K is the wave vector of the vibration. The scattering of electrons due to the perturbations of the periodic potential arises from lattice vibrations can therefore be considered as an interaction of electrons with phonons.

However, the scattering of electrons with phonons depends on the nature of these phonons. There are four branches that are divided into four groups. For one group, w_K (w is the phonon frequency and K is phonon wave vector) increases with K almost linearly near the Brillouin zone centre and attains a maximum at the zone edge. The phonons corresponding to this group are called acoustic phonons and the two branches of this group correspond to the longitudinal (LA) and transverse (TA) vibrations. For the second group, w_K is fairly constant around the Brillouin zone centre but reaches a smaller value at the zone boundary. The phonons corresponding to this group are called optic phonons and the two branches of this group correspond to the Longitudinal (LO) and transverse (TO) vibrations. The acoustic phonon group causes the neighbouring atoms to vibrate in phase while the optic phonon group causes the neighbouring atoms to vibrate in opposite phase.

The scattering of electrons by acoustic phonons occurs in two ways. In the first, the acoustic phonon vibrations change the spacing bewteen the lattice atoms. Consequently, the energy band gap and the position of the conduction and valence band edges will vary from point to point because of the sensitivity of the band structure to the lattice spacing. The fluctuations introduced into these bands will produce potential discontinuities in both bands. This potential is known as the deformation potential since it is produced by the deformation of the crystal, and the resulting scattering mechanism is called deformation potential scattering.

The second type of electron scattering by acoustic phonons occurs when the displacements of the atoms by acoustic phonon vibrations produce a redistribution of the charges that produce some changes in the potential. The potential is known as the piezoelectric potential since it is produced by the piezoelectric effect, and the resulting scattering mechanism is referred to as piezoelectric scattering.

The scattering of electrons by optical phonons also occurs in two ways. The first is resulted from the electrical polarisation of the lattice produced by the optical vibrations due to the ionic charges assiciated with the atoms forming the crystal binding of GaAs. In turn, the dipolar electric field arising from the opposite displacement of the negatively and positively charged atoms provides a coupling force between the electrons and the lattice which is the potential associated with the scattering of electrons. This scattering type is called polar optic phonon scattering and is known as the most important scattering mechanism for electrons in direct gap semiconductors.

The second type of electron scattering by optical phonons is similar to the deformation potential scattering in which the deformation of the crystal due to optical phonon vibrations produces a perturbing potential. The resulting scattering mechanism is called the non-polar optic phonon scattering.

In phonon scattering, electrons can also be scattered by phonons in two other ways. For the first, the electron is scattered from its initial state in a certain valley to a final state in another valley which is equivalent in energy to the initial state. In such process, the wave vector of the electron changes by a large amount so that only phonons with large wave-vectors may take part in this scattering. The resultant scattering mechanism is called the equivalent intervalley scattering (or intravalley scattering) and the phonon involved is referred to as the equivalent intervalley phonon which may be either an acoustic or optic phonon. This scattering process occurs only in the L and X valleys of the conduction band for III-V materials since the Gamma valley has only a single equivalent valley.

For the second, the electrons will be scattered from an initial state in certain valley to a final state in a nonequivalent valley. For example, In III-V materials, this process occurs when an electron in the Gamma valley heat up and will be able to transfer either to the L or X valleys only when it have aquired energy equal to the respective band gap between the respective valleys. The electrons in the L valley can also transfer to the X valley when they have gained an energy equal to the energy gap E_{LX} . The electrons in a higher valley may also transfer into a lower valley when they lose energy equal to the energy gap between the respective valleys. This scattering process requires phonons with a large wave vectors and the theoretical selection rules show that only longitudinal optical phonons are allowed when the satellite valley minima are at the edge of the Brillioun zone. The resulting scattering mechanism is called the nonequivalent intervalley scattering and the phonon involved is referred to as the nonequivalent intervalley phonon.

In all the above scattering processes, the principles of energy and momentum conservation must be obeyed. If P_1 is the crystal momentum of an electron before a scattering and P_2 the electron momentum after the scattering, a momentum of an amount $(P_1 - P_2)$ must be supplied by the lattice. This can be achieved in two ways, either by absorbing a phonon of a momentum $(P_2 - P_1)$

or by emitting a phonon of a momentum $((P_1 - P_2)$. From a frequency-wave vector relationship for the phonons we can also obtain the energy of the phonon that corresponds to the change in electron energy after scattering. If E and E' are the electron energies before and after scattering, E' will be given as $(E + E_p)$ when a phonon is absorbed and as $(E - E_p)$ when a phonon is emitted. The change in electron energy by the phonon energy E_p depends on phonon type and in some phonon scattering mechanisms it is negligible when E_p is quite small.

Therefore, an electron with a given wave vector K can be scattered into a state with a wave vector K' by a phonon with a wave vector q if the following equations are satisfied :

$$K'^2 = K^2 + q^2 \pm 2Kq\cos\theta_q$$

$$\frac{h^2K'^2}{2m^*} = \frac{h^2K^2}{2m^*} \pm h\omega$$

where the upper sign refers to absorption, the lower to emission of a phonon and θ_q is the angle beween K and q. Elimination of K' gives

$$\cos\theta_q = \mp\frac{q}{2K} + \frac{m^*\omega}{hKq} \qquad (1)$$

Since $\cos\theta_q$ must be between -1 and +1, q must lie between two limits, q_{min} and q_{max}. The values of q_{min} and q_{max} depend on the type of phonon scattering. For deformation potential and piezoelectric scattering, we put $\omega = v_s q$

where v_s is the longitudinal velocity of sound.

Noting that $hk = m^*v$, we find that :

$$\cos\theta_q = \mp\frac{q}{2K} + \frac{v_s}{v}$$

Since v_s is less than 6000 m/sec in semiconductors and the electron velocity v at room temperature is 10^5 m/sec, we can assume that $v_s \ll v$, which yields

$$\cos\theta_q = \mp\frac{q}{2K}$$

Then

$$q_{\min} = 0.....when.....\theta_q = \frac{\pi}{2}$$

$$q_{\max} = -2K....when.....\theta_q = \pi$$

or

$$q_{\max} = 2K....when.....\theta_q = 0$$

The values for q_{min} and q_{max} for optical phonons can also be obtained by putting in equation (1) $\omega = \omega_o$, the frequency of the phonon involved in scattering with electrons. Then

$$\cos\theta_q = \mp\frac{q}{2K} + \frac{m^*\omega_o}{hKq}$$

Using equation (1) for phonon absorption yields

$$q_{\min} = K\left[\left\{1 + \frac{h\omega_o}{E(K)}\right\}^{\frac{1}{2}} - 1\right]....with....\theta_q = 0$$

$$q_{\max} = K\left[\left\{1 + \frac{h\omega_o}{E(K)}\right\}^{\frac{1}{2}} + 1\right]....with....\theta_q = \pi$$

while for phonon emission,

$$q_{\min} = K\left[1 - \left\{1 - \frac{h\omega_O}{E(K)}\right\}^{\frac{1}{2}}\right] \ldots. with \ldots. \theta_q = 0$$

$$q_{\min} = K\left[1 + \left\{1 - \frac{h\omega_O}{E(K)}\right\}^{\frac{1}{2}}\right] \ldots. with \ldots. \theta_q = \pi$$

The values for q in the above four equations define a limited range of the wave vector of the final state K' which then defines the limits of the integration over the final states in the expression for the scattering rate.

However, for the all electron-phonon scatterings, the first-order perturbation theory yields the following expression for the scattering rate:

$$\lambda(\underline{K}, E) = \frac{\pi}{NM}\int \omega^{-1} C^2(\underline{q}) I^2(\underline{K}, \underline{K}') M^2(\underline{K}, \underline{q}, \underline{K}')$$

$$\left[n(\omega) + \frac{1}{2} \mp \frac{1}{2}\right]\{1 - f(\underline{K}')\}\delta(E(\underline{K}') - E(\underline{K}) \mp h\omega)dN'$$

N is the number of unit cells in the volume of crystal and M is the appropriate effective mass which is the mass of a unit cell for acoustic modes and is the reduced mass of the two atoms in the unit cell for optical modes. θ_q **is the angular frequency of the mode and can be a function of the phonon wave vector q , and** $n(\omega)$ **is the population of phonons. Unless the phonons are hot,** $n(\omega)$ **can be taken to be its value at thermal equilibrium**

$$n(\omega) = \frac{1}{\exp\left(\frac{h\omega}{K_B T}\right) - 1}$$

where T is the absolute temperature. The upper sign is to be taken when the rate associated with phonon absorbtion is

calculated while the lowersign is for phonon emission.
The remaining terms $C^2(q)$, $M^2(K,q,K')$ and $I^2(K,K')$ form together the electronic part of the matrix element which is defined as:

$$|<\underline{K}'|H|\underline{K}>_e|^2 = C^2(\underline{q})I^2(\underline{K},\underline{K}')M^2(\underline{K}',\underline{q},\underline{K})$$

where $C^2(q)$ is the coupling strength, $M^2(K,q,K')$ is the parameter describing the conversation of crystal momentum and $I^2(K,K')$ is the overlap integral. The coupling strength is a parameter that depends on the type of phonon involved in scattering.

Impurity Scattering

This scattering process arises by the addition of impurities to the crystal to provide a type of free carrier of the required doping concentration. The substitution of an impurity atom on a lattice site will perturb the periodic potential and thus scatters an electron since the potential associated with the impurity atom is different from that of the atom in the host crystal.

The principles of momentum and energy conservation will also be obeyed in this scattering. The change of electron momentum on passing an impurity atom will be supplied as a momentum analagous to the momentum of the lattice waves, but since the mass of the impurity greatly exceeds that of an electron, this scattering is very close to being elastic and therefore, it does not alter the energy of the electron. Impurity scattering can be also classified into intravalley and intervalley scattering, but the latter is an exremely weak and rare process. We will assume that impurity scattering is always an intravalley scattering.

Quantum Transport Modelling

John R Barker
University of Glasgow

1. INTRODUCTION

The following lectures are an elementary introduction to the computer modelling of quantum transport and tunnelling in very small structures . The emphasis is on *quantum ballistic systems* for which collision processes are less significant than the free carrier motion in the strongly inhomogeneous (and possibly quantising) potential fields provided by the device structure. Quantum ballistic transport occurs when the carrier transit time is substantially shorter than the mean free time for inelastic collisions. Effects which may arise include size quantisation, localisation phenomena, low-dimensional effects, tunnelling, resonance and interference phenomena .

2. ONE-DIMENSIONAL METHODS

The most straightforward methods use a reduction of the one-dimensional Schrödinger equation to a three-point difference equation over a discrete spatial grid which may be solved by a variety of techniques. Powerful transmission matrix or S-matrix methods also exist for 1-D which are particularly useful for disordered systems, localisation phenomena, superlattice problems, tunnelling, propagation of arbitrary wavepackets and for problems involving multiply-connected paths such as Aharonov - Bohm effect structures. Self-consistent propagation in the presence of space charge effects can easily be incorporated by coupling the Schrödinger equation to Poisson's equation. The basic quantum ballistics are readily computed but difficulties may occur when relating the microscopic processes to macroscopic current-voltage characteristics.

For simplicity let us take a semiconductor heterostructure or n-i-p-i device to be described in one dimension by a electron Hamiltonian $H_e = \varepsilon(p) + V(z)$ ($p = -i\hbar\, \partial/\partial z$). Here $\varepsilon(p)$ is an effective-mass conduction-band structure and $V(z)$ describes the spatial modulation of the band-edge . Let us also assume $\varepsilon(p)$ is parabolic: $\varepsilon(p)= p^2/2m^*$, so that the envelope wave-functions satisfy a basic Schrödinger equation. $V(z)$ may include the potential of an applied electric field. Fig. 1 sketches the sort of "potential" profiles which occur in recent devices (usually from a vertical geometry heterostructure or modulation doped semiconductor)

Carriers traversing such systems may be supposed to be injected into travelling-wave plane-wave states, or into wave-packet pure states or as mixed states. It remains an open question as to what the correct form of initial state might be in quantum transport. Whatever the choice, the basic ballistic transport problem requires the separation of the system into a "free" component on which is superposed a "potential barrier" structure represented by $V(z)$. Without loss of generality we may take

V=constant outside a "barrier domain" $0< z< a$; $V=V_1$ $(z<0)$, $V=V_2$ $(z>a)$.Analytical expressions for the reflection and transmission coefficients of the barrier potential may be obtained in only a few over-simplified cases although the important step junction, finite superlattice, chirp structure and triangular barrier are included. It has been common practise to use WKB methods for more general potentials but although this sometimes gives a good "first pass" at the problem it is rarely valid and cannot cope with self-consistent potentials which arise from space-charge effects. A useful modular approach to complex structures exists as an analogy with transmission line theory (see section 4).

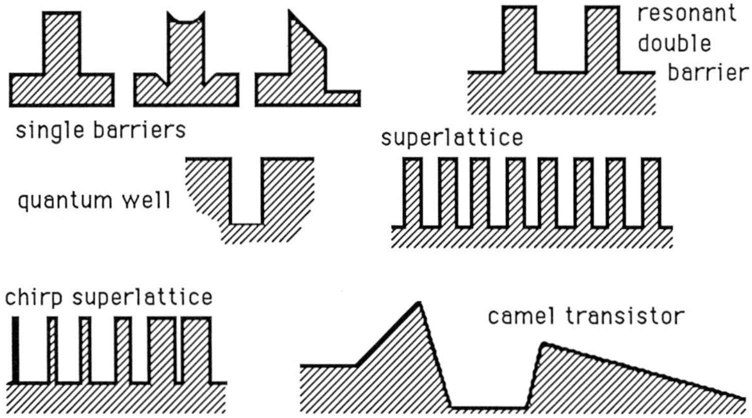

Fig.1 One-dimensional device profiles

3 THE DISCRETISED SCHRÖDINGER EQUATION : CONTINUOUS SPECTRA

3.1 DETERMINATION OF TRANSMISSION COEFFICIENTS

The *numerical problem* may be handled very easily by discretising the Schrödinger equation:

$$(-\hbar^2/2m^*\Delta^2)[\,\psi(z_{n+1})+\psi(z_{n-1})-2\psi(z_n)] + V(z_n)\psi(z_n) = E\psi\,(z_n) \tag{1}$$

where $\Delta = z_{n+1} - z_n$ defines the grid spacing and n=0 ,1,2..(P+1). Equation (1) is a three-point recurrence relation.The discretised quantum mechanical current follows as

$$J(n+1/2) = -2\hbar/m^*\ \mathrm{Im}\{\,S_n\,\} \tag{2}$$

$$S_n \equiv \psi(z_n)\,/\,\psi(z_{n+1}) \tag{3}$$

For a quadratic kinetic energy in the presence of a possible space-dependent effective mass and velocity dependent forces, the most general form for the time-independent Schrödinger equation may be written

$$L(n)\,\psi(z_{n+1}) + M(n)\,\psi(z_n) + N(n)\psi(z_{n-1}) = 0 \tag{4}$$

Suppose now that L,M,N are constant over a domain $[z_{n-1},z_{n+1}]$, the solution to (4) then has the form

$$\psi(z_n) = \chi^n \qquad (5)$$

which reduces (6) to a quadratic equation with the two solutions:

$$\chi = \chi^{\pm} = -M\pm[M^2-4LN]^{1/2}/2L \qquad (6)$$

From equation (2) it is clear that the solutions $\chi^{\pm}$ correspond to current flow in two different directions, they are therefore the discrete analogues of the plane waves exp(±ikx). Suppose a unit plane wave of energy E leaves the region $z>z_P=a$, and a superposition of forward and backward travelling plane waves exists to the right of the region $z=z_0=0$;

$$\psi(z)= Ae^{ikz} + Be^{-ikz} \quad (z<z_0) \qquad (7)$$

$$\psi(z)= e^{ik(z - Ì P)} \quad (z>z_P) \qquad (8)$$

where we assume $V_1=V_2$ for the present. The total reflection (transmission) coefficient R (T) is defined as the ratio of the reflected (transmitted) flux to incident flux:

$$R(k)= |B/A|^2 = |r(k)|^2 \quad , \quad T= 1/|A|^2=|t(k)|^2 \quad , \quad R+T=1 \qquad (9)$$

where we introduce the complex reflection amplitude r(k)=B/A and the complex transmission amplitude t(k)=1/A. The discrete analogue of equations (7-9) are

$$\psi(z_{-1}) = A(\chi^+) + B(\chi^-) \quad (z<z_0) \qquad (10)$$

$$\psi(z_{1+P})= (\chi'^+)^{1+P} \quad (z>z_P) \qquad (11)$$

R (and hence T) may be obtained as (Vigneron and Lambin,1980a)

$$R = | B(\chi^-)/ A(\chi^+) |^2 = | S_{-1} - \chi^- |^2 / | S_{-1} - \chi^+ |^2 \qquad (12)$$

where for the simple Hamiltonian

$$\chi^{(\pm)} = \beta/2 \pm i(1-\beta^2/4)^{1/2} \qquad (13)$$

$$\beta= 2 + \Delta^2[U_I - \epsilon] \qquad (14)$$

and U_I is the constant value of U(z) for z<0. $S_{-1} = \psi(z_{-1}) / \psi(z_0)$ may be obtained exactly by forward recursion (or continued fraction techniques, Vigneron and Lambin ,1980a) on the recurrence relations:

$$S(n-1) = -M(n)/N(n) - L(n)/[N(n)S(n)] \qquad (15)$$

subject to the boundary condition $S(n)= \chi'^{+}$ for n>P.

Good convergence requires $\Delta^2[U-\epsilon] << 1$.

3.2 COMPLEX TRANSMISSION AND REFLECTION AMPLITUDES

Collins and Barker have generalised the continued fraction method to extract the complex amplitudes r(k) and t(k) which are required to evaluate tunnelling times. The method also generalises to the case $V_1 \neq V_2$. Since t(k)=exp(ika)/A, r(k)=B/A, the ratio B/A = r is readily obtained from (12)

$$B/A = r = -[S_{-1} - \chi^-]/[S_{-1} - \chi^+)] \tag{16}$$

and we find from the recurrence relation

$$\psi(z_n) = S_n \, \psi(z_{n-1}) \; [\text{ initial condition } \; \psi(z_{P+1}) = 1] \tag{17}$$

$$A= -[S_{-1} - \chi^+] \prod_{n=0}^{P} S_n / [\chi^+ - \chi^-] \tag{18}$$

The routine used to calculate S_{-1} is thus available to compute A and hence t. Unfortunately the product of terms in (18) leads to an accumulation of numerical errors unless very high precision floating point arithmetic is used. An alternative approach (Barker and Collins) is to use a special property of one-dimensional potential problems. Let us write the phase angles of t and r as θ_L, θ_R and $\varnothing_L$, $\varnothing_R$ where the L and R subscripts denote flux incident from the left or right. A basic theorem in quantum mechanics (see Messiah, 1964) shows that

$$\theta_L = \theta_R \equiv \theta \qquad \text{and} \quad \varnothing_L - \theta_R = \pi - \varnothing_R + \theta_L \tag{19}$$

The calculation of |t| and |r| are found from R and T, $\varnothing_L$, $\varnothing_R$ are found from r thus θ and therefore t is obtained from (20). The values for $\varnothing_L$, $\varnothing_R$ are calculable from (20) using inverse tangent routines extended beyond the $-\pi/2,\pi/2$ range to ensure θ is a smooth function (Collins ,1986). Care is required for resonant tunnelling problems for which the reflection amplitude goes to zero. The phase angles are critical for interference phenomena, wave-packet propagation and the determination of traversal times. The solution of the generalised Schrodinger equation for non-parabolic bands is discussed by Barker, Collins, Lowe and Murray (1985).

Vigneron and Lambin (1980b) have used a similar numerical procedure based on continued fractions to obtain the E-k relationship for one-dimensional band structures. Barker, Collins and Pepin have recently used the technique for evaluating electron dynamics in semiconductor superlattice structures where appropriate corrections for different effective masses in different layers is required.

3.3 THE SINGLE BARRIER PARADIGM

The square barrier defined by : $V(z)= V_0$ ($z<0$); $V(z) = 0$ ($z>a$) is the archetypal heterostructure problem although practical structures (fig.1) have definite distortions (as well as spatially-varying effective mass).Typical parameters for GaAs=GaAlAs systems are : $V_0 = 0.2$-0.35 eV , $m^*/m_e=0.063$, $a = 2.5$ - 100 nm. It is useful for checking numerical procedures and more importantly it provides a basis for a numerical technique based on transmission matrix theory which works for arbitrary potentials.The transmission properties may be found exactly by standard wave-function matching across the two boundaries and the phenomena of *resonances and tunnelling* are easily demonstrated. The square barrier is realisable as an approximation in the conduction band edge of a heterostructure formed by sandwiching a thin layer of $Ga_{1-x}Al_xAs$ between two thick layers of GaAs (narrow bandgap). This raises the problem that the effective mass differs from material to material. If the carriers are moved perpendicular to the layers (no transverse energy component) then the elementary tunnelling/transmission problem follows the simple theory but with two modifications. First, the wave-vectors k,q are modified to $k=\sqrt{(2m_1E/\hbar^2)}$, $q=\sqrt{(2m_0[E-V_0]/\hbar^2)}$ to allow for the different effective masses m_1, m_0 outside and inside the barrier region. Second, the matching of current across the barrier discontinuities leads to matching $(1/m^*)(d\psi/dz)$ rather than the gradient of ψ. The modified transmission and reflection amplitudes are finally found by replacing k and q by the appropriate velocities $v(k)=\hbar k/m_1$, $v(q)=\hbar q/m_0$ *except* in the exponents. The resonances again are perfect and occur at $\sin(qa)=0$.

4 TRANSMISSION MATRIX FORMALISM

4.1 BASICS

A finite superlattice (N barriers) or infinite superlattice may be regarded as a chain of identical square barriers. Chirp super-lattice structures , which are useful for emulating continuously-graded potentials , are organised arrays of square barriers of identical height but different widths. In many other cases of interest , device configurations consist of a series of barrier or quantum-well structures separated by regions of constant potential. The question therefore arises : How do we construct the transmission /reflection properties of an array of elementary potentials? Heuristically one expects the coherence of the carrier wave-function to be an important factor. Current lore suggests that the critical coherence length is limited by Λ, the mean free path for *inelastic* scattering. In such a case it may be argued that if two barriers are separated by substantially more than Λ their combined transmission properties are determined by the incoherent superposition of their individual transmission properties. For the opposite extreme, infinite coherence lengths, interference between the reflected and transmitted waves within the barrier system will lead to coherent superposition. A new phenomenon may then occur: resonant tunnelling , for which perfect transmission may be set up at the resonance energy, even though the individual barriers are virtually opaque.

For one-dimensional systems there exists an elegant analytical technique for handling arrays of elementary potentials. An advanced group-theoretical treatment of the method is given by Peres(1981, 1983a,1983b) and applied to the determination of resistances, localisation length and chaotic band structures in 1-D disordered systems. To illustrate let us re-write the stationary Schrödinger equation as two coupled first-order equations for variables f and g defined as components of a spinor-like object:

$$\Psi = \begin{pmatrix} f \\ g \end{pmatrix} = 1/2 \begin{pmatrix} 1 & -i \\ 1 & i \end{pmatrix} \begin{pmatrix} \psi \\ \partial\psi/\partial kz \end{pmatrix} \tag{20}$$

where $k = (2mE)^{1/2}/\hbar$. For $V(z) = 0$, it easily shown that f and g correspond to forward and backward plane solutions. Using the Schrödinger equation we find the equation of motion for $\Psi(\tau)$ where $\tau = -kz$ is a dimensionless space variable :

$$i\,\partial\Psi/\partial\tau = \begin{pmatrix} 1-u/2 & -u/2 \\ u/2 & -1+u/2 \end{pmatrix} \Psi \equiv H\Psi \tag{21}$$

where $\mathbf{H} \equiv (1-u/2)\sigma_z - (u/2)\sigma_y$ is the effective "Hamiltonian" for the state Ψ, $u = V(t)/E$ and σ_z and σ_y are standard Pauli spin matrices. It is easy to demonstrate that the usual current is the conserved quantity

$$j = (\hbar k/m)\,\Psi^+ \sigma_z \Psi \tag{22}$$

If j is normalised to $(\hbar k/m)$ we see that σ_z acts as an indefinite metric for the normalisation of Ψ. The "Hamiltonian" **H** is pseudo-Hermitian in the sense that $\mathbf{H}^+ = \sigma_z \mathbf{H} \sigma_z$. **H** generates a pseudo-unitary transformation of Ψ such that the transfer or transmission matrix **T** belongs to the SU(1,1) group with the significance that $|T_{11}|^2 - |T_{12}|^2 = 1$:

$$\mathbf{T} = \begin{pmatrix} T_{11} & T_{12} \\ T_{12}^* & T_{11}^* \end{pmatrix} \tag{23}$$

$$\Psi(\tau) = \mathbf{T}(\tau)\ \Psi(0) \tag{24}$$

Numerically (or analytically, on a good day), one proceeds by integrating Ψ from a point $z = z_2$ ($>z_1$) back along the negative z-axis to z_1 (ie in direction of +ve τ). The transmission matrix for a null barrier ($V = 0$) of length L is given by $T_{11} = \exp(-ikL)$, $T_{12} = 0$. If $\mathbf{T}_1$, $\mathbf{T}_2$, $\mathbf{T}_3$,...are transmission matrices for successive regions (ordered left to right) the transmission of the total range is given by the product $\mathbf{T} = \mathbf{T}_1\mathbf{T}_2\mathbf{T}_3$...In terms of the usual transmission and eflection amplitudes for a plane wave incident from the left of a potential barrier it is trivial to prove that $t(k) = 1/T_{11}$ and $r(k) = T_{12}^*/T_{11}$. Advanced methods for parameterising the "equations of motion" are given by Peres(1983a). It is sometimes advantageous to deploy the scattering or S-matrix (a unitary matrix) which relates the outgoing waves to the incoming waves defined by:

$$\begin{pmatrix} g(\tau) \\ f(0) \end{pmatrix} = \mathbf{S} \begin{pmatrix} g(0) \\ f(\tau) \end{pmatrix} \tag{25}$$

4.2 NUMERICAL PROCEDURE

For an arbitrary potential one may divide the potential into small strips (usually corresponding to a subset of points on a discrete mesh) each approximated by a square barrier of height $V(z_n)$. The exact method of section 4.1 may be used to multiply-up the discrete sequence of transmission matrices to obtain the total transmission matrix for the problem. The thin barrier results are the best for small strip sizes and may be usefully used in conjunction with the fact that the transmission matrices are close to the unit matrix in that case. We have found it is often better to use the continued fraction method over subsets of the total potential and then to compound the results with the transmission matrix approach. Again if too many multiplications are involved one runs into floating point precision dificulties. Improved convergence has been found in a sub-matrix S-matrix formalism developed recently by Yukkei and Inkson (1988).

4.3 RESONANT TUNNELLING

A resonant tunnelling barrier has perfect resonances in the transmission coefficient for tunnelling. They arise in multiple simple barrier situations and have been extensively studied. Resonant tunnelling systems have technological importance because they have potential as fast switching devices and multi-state logic. Consider a single quantum well with n bound states ε_n. If the walls are now diminished to finite width d the bound state energies will now coincide with states in the continuum. If d is not too small the discrete states go over into quasi-bound states; an electron in such a state will gradually leak out of the double barrier system. Electrons incident from the outside at the energies ε_n will tunnel through without reflection even if the individual barriers are nearly opaque. The perfect transmission energies correspond to conduction bands when a long array of barriers or wells is considered.

Resonant tunnelling is easily analysed with the aid of transmission matrix theory . For example, consider a double barrier of spacing b and individual barrier height V_0 and width a respectively. Let the transmission amplitude for the single barrier be $t_1 = |t_1|e^{i\varphi}$ where φ is the phase angle. Using the product $\mathbf{T}_{(1)}.\mathbf{D}.\mathbf{T}_{(1)}$ where $\mathbf{D}$ describes the free propagation between barriers we obtain

$$t_2(k) = T_1 e^{-ik(a+b)}/\{R_1 + e^{i2\Theta}\} \qquad (26)$$

where $T_1 = |t_1|^2$, $R_1 = 1-T_1$ and $\Theta = \varphi - k[a+b]$ The transmission coefficient is

$$T_2 = T_1{}^2/\{ T_1{}^2 + 2(1+\cos 2\Theta)(1-T_1) \} \qquad (27)$$

Evidently the resonance energies are determined by $\varphi - k[a+b] = n\pi/2$ where n is an odd integer If b is large, kb varies rapidly, so that several resonances may be expected.

If inelastic scattering is present the coherence of the resonant tunnelling may be destroyed, but a sequential process may still lead to a resonance. In this incoherent alternative (which is more likely for thick barriers) the short coherence time only allows tunnelling into the quantum well between the barriers folowed incoherently by a second transition which escapes out of the second barrier. In the steady state there is little difference detectable between the these two processes so far as the current flow is concerned. Charge build up during resonance will distort the barrier potential and this must be included via Poisson's equation in any realistic modelling.

5. THE DISCRETISED SCHRÖDINGER EQUATION II : WAVE PROPAGATION

5.1 THE FORWARD CONSTRUCTION METHOD

If the stationary solutions are known, for example from the transmission or continued fraction techniques, then one may construct an initial wavepacket $\psi(z,t=0)$ in a free space region outside the potential according to

$$\psi(z,0)=\int dkG(z,k,0)=\int dkg(k)e^{ikz} \tag{28}$$

If the wavepacket is incident from a location deep to the left of the potential barrier regions we may expand it in terms of scattering states defined by the transmission and reflection amplitudes and propagate the components forward in time by the prescription:

$$\psi(z,t)=\int dkG(z,k,t)|t(k)|e^{i\varnothing(+)} \qquad (z>a) \tag{29}$$

$$\psi(z,t)=\int dkG(z,k,t)|\{1-|r(k)|e^{-i2kz}e^{i\varnothing(-)}\} \qquad (z<0) \tag{30}$$

where g(k) is non-zero for positive k values and $\Delta(\pm)$ are the phase angles for transmission and reflection and

$$G(z,k,t)=g(k)e^{ikz}e^{-i\varepsilon t/\hbar} \tag{31}$$

5.2 DIRECT SOLUTION OF THE TIME-DEPENDENT SCHRÖDINGER EQUATION

This method was popularised by Goldberg et al (1967) and there exist modified forms suitable for time-dependent potentials , non-parabolic bands and variable effective masses (Barker, 1985; Collins, 1986; Collins, Lowe and Barker, 1987,1988; Mains and Haddad, 1988). The method, which is based on the Crank-Nicholson algorithm, uses the fact that the Hamiltonian is the generator of infinitesimal time translations:

$$\psi(z,t+\delta t)=e^{-i\delta tH(t)/\hbar}\ \psi(z,t) \tag{32}$$

A direct linearisation in time leads to serious convergence problems. Instead the direct method uses a Cayley expansion of the infinitesimal evolution operator $e^{-i\delta tH(t)/\hbar}$ (equivalent to a low order Pade approximant), where we introduce $\tau=\delta t/2\hbar$

$$e^{-i\delta tH(t)/\hbar} \dashrightarrow (1+iH\tau)^{-1}(1-iH\tau) \tag{33}$$

If we introduce the uniform time-mesh $t=s\delta t$ and bring back the previous space mesh $z=n\Delta$ we find the discretised Schrödinger equation in the form

$$[1 + iH(s,n)\tau]\ \psi(n,s+1) = [1 - iH(s,n)\tau]\ \psi(n,s) \tag{34}$$

which for the simple parabolic time-independent Hamiltonian reduces to the form

$$a_n(s)\ \psi(n+1,s+1) + b_n(s)\ \psi(n,s+1) + c_n(s)\ \psi(n-1,s+1) = d_n\ (s) \tag{35}$$

where the a's,b's,c's,d's are known from the previous time step. The coefficients are

$$a_n(s) = c_n(s) = -i\hbar\delta t/(4m^*\Delta^2) \tag{36}$$

$$b_n(s) = 1 + i\hbar\delta t/(4m^*\Delta^2)\ [2+(2m^*\Delta^2/\ \hbar^2)\ V(n,s)] \tag{37}$$

$$d_n\ (s) = -a_n(s)\ \psi(n+1,s) + f_n(s)\ \psi(n,s) - c_n(s)\ \psi(n-1,s) \tag{38}$$

$$f_n\ (s) = 1 - i\hbar\delta t/(4m^*\Delta^2)\ [2+(2m^*\Delta^2/\ \hbar^2)\ V(n,s)] \tag{39}$$

Equation (40) may be solved exactly on a finite space mesh by the Thomas tri-diagonal algorithm. Conditions for convergence are given by Goldberg et al (1967). This technique is very fast; the algorithm is stable, unitary and accurate to second order in space and time. Computer movies are very easy to obtain. The method is useful for the rapid construction of the momentum wavefunction and the time-dependent Wigner function for wavepackets. Poisson's equation may be linked to the self-consistent potential without loss of generality.

5.3 INJECTION AND EXTRACTION OF WAVES

One practical problems with using time-dependent wave equation solvers is the handling of the boundary points. Usually one holds the wavefunction to zero at the endpoints of the space domain. This corresponds to perfectly reflecting boundaries which may cause unwanterd reflected waves in long time simulations. One way out is to damp out the outgoing waves at the boundaries by adding a complex potential i $|V|$ to the potential in the vicinity of the boundaries. This non-hermitian term absorbs incoming waves. Alternatively, one may use a delta function sink term in the Hamiltonian with an amplitude set to absorb incoming waves eg a term proportional to $i\hbar\delta(z-z_0)\partial\psi(z_{n+1})\partial z$. Similarly, monoenergetic plane wave injection may be achieved by a suitably started time-dependent delta function source term (a recent example is given by Mains and Haddad, 1988). The simulation of disspative processes such as optical phonon scattering may be simulated by using a time-dependent barrier potential.

5.4 TRAVERSAL TIMES AND TUNNELLING

Early numerical calculations of wavepacket propagation over tunnelling barriers brought confusion because different traversal times were found for different wavepacket shapes. We might expect that wavepackets with sharp momentum distributions which are a good approximation to plane wave trains would have comparable traversal times at the same energies. Indeed elementary tunnelling theory works quite well by assuming the states are plane wave outside the barrier region.

A recent investigation (Collins, Lowe and Barker, 1987,1988] has used extensive computer studies of a wide range of wave-packet versus barrier tunnelling problems where the space-time trajectory of the peak in the transmitted asymptotic wavepacket is tracked to evaluate the delay τ. The results have been compared with the various analytical models . The wavepackets were sized such that the transmission coefficient varies slowly across the momentum distribution . The numerical results were obtained in three independent ways: (i) directly solving the time-dependent Schrodinger equation; (ii) solving the Wigner equation ; (iii) constructing a superposition of stationary waves. The conclusion is that a generalisation of the *stationary phase time* is an excellent fit to the numerical curves under the above conditions and for narrow packets where velocity filtering is anticipated.

The phase time is easily obtained by approximate methods, but recently it has been shown to be an *exact* result (Barker, 1986), albeit in a more generalised form. The result is summarised as follows. In general an incident wavepacket will be limited in size by either the mean free path or by the dimensions of the electrode region. The momentum spread may thus be larger than the conditions we have discussed. In that case there is substantial velocity filtering upon tunnelling. The general analysis considers the case of an arbitrary wavepacket. Let us define τ_I as the average time spent by a particle within a domain D : { 0< z <a} associated with a tunnelling barrier of width less than or equal to a . For a wavepacket state $\psi(z,t)$ we may use the Born interpretation to find

$$\tau_I = \int_0^a dz \int_{-\infty}^{\infty} dt\, |\psi(z,t)|^2 \qquad (40)$$

$$\tau_I = 2\pi \int dk\, \{|g(k)|^2/v(k)\}\, \{a + |t(k)|^2\, (d/dk)\varnothing(+) + |r(k)|^2\, (d/dk)\varnothing(-)\} \qquad (41)$$

The interaction time evidently is an average over all the possible tunnelling (traversing) and reflection processes, so it does not give the tunnelling time directly. However, we can re-write (43) as the weighted sum of an effective tunnel/traversal time τ_b and a reflection time τ_r

$$\tau_I = R_{eff}\, \tau_r + T_{eff}\, \tau_b \qquad (41)$$

$$T_{eff} = 2\pi \int dk\, \{|g(k)|^2\, |t(k)|^2\}\ ;\quad R_{eff} = 2\pi \int dk\, \{|g(k)|^2\, |r(k)|^2\} \qquad (41)$$

$$\tau_r = 2\pi \int dk\, \{|g(k)|^2\, |r(k)|^2\, /v(k)\}\, \{a + (d/dk)\varnothing(-)\}/R_{eff} \qquad (43)$$

$$\tau_b = 2\pi \int dk\, \{|g(k)|^2\, |t(k)|^2\, /v(k)\}\, \{a + (d/dk)\varnothing(+)\}/T_{eff} \qquad (44)$$

It is the traversal time given by (44) which closely fits the numerical calculations. As such , (44) holds for arbitrary wavepackets provide they have very small negative k components in the initial state. If $|t(k)|^2$ varies slowly over the initial momentum distribution $|g(k)|^2$ which we presume is then peaked at $k=k_0$ we obtain the simple phase delay result

$$\tau_b = \{1/v(k_0)\}\, \{a + (d/dk)\varnothing(+)\} \qquad (45)$$

The essence of the analysis is contained in the concept of the time a particle spends in interaction with a scattering centre ; a similar three-dimensional analysis in terms of partial waves may be given.

The tunnelling time is somewhat slowed down in the vicinity of the resonance; it may be speeded up in the presence of bound states. There is considerable scope for calculating tunnel times analytically by this method in combination with numerical methods of transmission matrix theory.

6. I-V CHARACTERISTICS

So far we have considered the ballistics of individual electrons in simple potentials. Let us turn now to an elementary way of calculating the tunnelling current as a function of applied bias.The most widely used elementary theories of tunnelling *current response* are the stationary-state method and the equivalent transfer-Hamiltonian method . The generic tunnelling model involves a tunnelling barrier which is sandwiched between two degenerate three dimensional conducting materials , each electrode (1 and 2) being described by an equilibrium Fermi function $f_1 = f_0(E)$, $f_2=f_0(E+eV)$ where V is the applied bias across the barrier.

The net current density is asserted to be :

$$J=2e/(2\pi)^3\int\int\int d^2k_\perp dk_x v(k_x)T(E,V)\{f_1(1-f_2)-f_2(1-f_1)\}$$

$$=2e/((2\pi)^3\hbar)\int dE_x[f_0(E)-f_0(E+eV)\int d^2k_\perp T(E_x+E_\perp,V) \qquad (46)$$

where T is the current transmission probability (ratio of incident to transmitted probability current=tt*(1à2)v_2/v_1 where t is the particle transmission probability amplitude) *which depends on the bias V and energy E(k).* There is little hope of getting simple analytical expressions for T because it must be computed for each energy E and for each distorted barrier (which depends on V). This situtation is only pleasing to computer buffs and practioners of dark arts such as the WKB method!

For an ideal barrier of width a the barrier potential is: U(x,V)=U(x,0)-eV(x+a/2)/a.A staircase potential is obtained if U(x,0) is zero, and a classical staircase is obtained if T(E,V)= 1. For the true staircase we know that pronounced hot electron effects, particularly velocity and energy overshoot can occur if the drop V is large (few meV) over a short distance (few hundred Angstroms). Such a situation must be reconciled with the elementary tunnelling theory which incorporates no explicit details of scattering or hot electron effects (Barker, 1985).The elementary theory requires modifications for heterostructures where there is an effective mass variation from material to material which mixes the transverse and longitudinal wave motion. The non-parabolicity of the effective mass Hamiltonian also leads to modifications to T, but is manageable by a new technique described elsewhere. It has been argued that for sufficiently high conduction band discontinuities at GaAs-AlGaAs heterojunctions the commensurability between Γ-valley and satellite valley edges will lead to a failure of effective mass theory and deeper methods will be required for T.

A significant alternative due to Landauer (reviewed in Landauer, 1988) takes into account the self-consistent field due to applied bias. The finite temperature 1-D version gives the conductance as

$$G = (e^2/\pi\hbar)[\int dE(-df/dE)T(E)][\int dE(-df/dE)v^{-1}(E)]/[\int dE(-df/dE)R(E)v^{-1}(E)] \qquad (47)$$

A great deal remains to be done to establish how the microscopic processes relate to macro-currents.

7. INTRODUCTION TO ADVANCED METHODS

Let us now describe more advanced (and numerically more intenseve) techniques of computer modelling for multi-dimensional quantum transport and tunnelling in very small structures The devices are typified by fig. 2 which shows a set of lateral nanoelectronic devices which may be formed by patterning the gate of a HEMT heterostructure device; the devices range from quantum waveguides to two-dimensional superlattices.

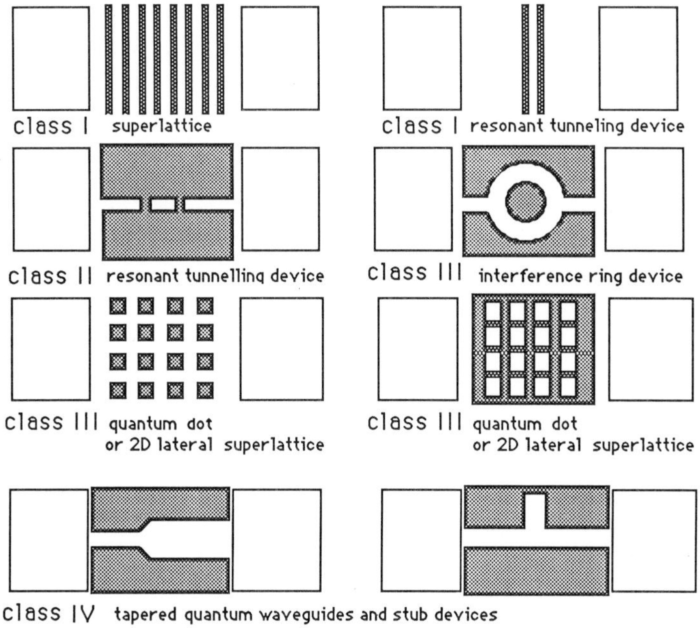

Figure 2. Nanoelectronic Devices

We first set up a formalism which can be used to approach the Boltzmann picture of transport in the classical limit. It is based on the Wigner distribution function which is intimately related to the finite temperature Green functions of field theory. Wigner distributions may be constructed directly or they may be determined from the wavefunction by the previously decribed methods.

8. QUANTUM PHASE SPACE DISTRIBUTIONS: WIGNER FUNCTIONS

Suppose a system is in a general state described by the density matrix ρ . The most direct recipe for computing the quantum-statistical expectation value of an observable Λ for that system involves:

$$<\Lambda> = \sum_{\mathbf{r}_1,\mathbf{r}_2} <\mathbf{r}_1|\Lambda|\mathbf{r}_2><\mathbf{r}_2|\Lambda|\mathbf{r}_1> \qquad (48)$$

The matrix elements of Λ and ρ are functions of two positions $\mathbf{r}_1$ and $\mathbf{r}_2$. From these we might construct a representative central position dependence on a vector **R** and a dependence on a central momentum vector **P**. A shift of origin to **R** on a line connecting the two fixed points could be used to locate $\mathbf{r}_1', \mathbf{r}_2'$ .It follows that $\mathbf{r}_1, \mathbf{r}_2$ may be expressed entirely in terms of the relative vector $\mathbf{r} = \mathbf{r}_1 - \mathbf{r}_2$ and the central locator **R**, where the parameter σ is in the range (0-1).

$$<\mathbf{r}_1|\Lambda|\mathbf{r}_2> = \Lambda(\mathbf{R},\mathbf{r},\sigma) = <\mathbf{R}+\sigma\mathbf{r}|\Lambda|\mathbf{R}-(1-\sigma)\mathbf{r}> \quad (49)$$

$$<\Lambda> = \sum_{\mathbf{R},\mathbf{r}} \Lambda(\mathbf{R},\mathbf{r},\sigma)\rho(\mathbf{R},-\mathbf{r},\sigma) \quad (50)$$

The relative vector **r** is a natural candidate for a transformation to the momentum representation, so let us define the phase-space representation of an operator Λ by

$$\Lambda(\mathbf{R},\mathbf{P},\sigma) = \sum_{\mathbf{r}} \Lambda(\mathbf{R},\mathbf{r},\sigma)\exp(-i\mathbf{P}.\mathbf{r}/\hbar) \quad (51)$$

$$\Lambda(\mathbf{R},\mathbf{r},\sigma) = (2\pi\hbar)^{-N} \sum_{\mathbf{P}} \Lambda(\mathbf{R},\mathbf{P},\sigma)\exp(i\mathbf{P}.\mathbf{r}/\hbar) \quad (52)$$

where N is the number of dimensions. Then (51) becomes:

$$<\Lambda> = \sum_{\mathbf{R},\mathbf{P}} \Lambda(\mathbf{R},\mathbf{P},\sigma)f(\mathbf{R},\mathbf{P},\sigma) \quad (53)$$

$$f(\mathbf{R},\mathbf{P},\sigma) \equiv \rho(\mathbf{R},\mathbf{P},\sigma)\,(2\pi\hbar)^{-N} \quad (54)$$

This form resembles a classical phase-space average over a distribution $f(\mathbf{R},\mathbf{P},\sigma)$. It includes (or can be used to construct) the majority of quantum distributions $f(\mathbf{R},\mathbf{P},\sigma)$ that have appeared in the literature (the cases σ=0 or 1 involve the product of momentum and direct space wave-functions).The case σ = 1/2 corresponds to the Wigner-Weyl transformation and we shall use $f(\mathbf{R},\mathbf{P}) \equiv f(\mathbf{R},\mathbf{P},\sigma=1/2)$ to represent the Wigner function from now on. For pure states $\rho = |\psi><\psi|$ or in terms of wavefunctions $\rho(R,r) = \psi^*(R-r/2)\psi(R+r/2)$. The density matrix and hence Wigner function is thus calculable if the wavefunction is known, e.g from one of the techniques described previously.

Of the possible functions (54) , only the Wigner function (σ =1/2) is real, although real distributions could be built out of linear combinations of functions with different ß parameters. The Wigner function is unique in another respect however; for a simple Hamiltonian H=T(p)+V(r) which is a general quadratic function of position and momentum the Wigner function obeys the corresponding classical Liouville equation. One should note some snags. First, the Wigner-Weyl transformation leads to Wigner functions which may assume negative as well as positive values, which rules out the interpretation of f(**R**,**P**) as a probability distribution. More seriously , f(**R**,**P**) can assume non-zero values at points **R** where the wave-function is zero. ie in non-physical regions (as indeed can the "Λ"-density $\Lambda(\mathbf{R},\mathbf{P})$); technically we say that f has non-compact support. Similarly the Wigner function may be non-zero in regions where the momentum wavefunction is zero.

9. QUANTUM BALLISTIC TRANSPORT

Consider the ballistic transport problem, which we define as the situation when the effects of dissipative collisions are vanishingly small. Switching to the centre of mass coordinates (and set $\hbar=1$ temporarily for readability) let us take the matrix elements $\langle \mathbf{R-r/2}|\quad|\mathbf{R+r/2}\rangle$ of the quantum Liouville equation

$$i\partial_t\,\rho = [p^2/2m + V,\rho] \tag{55}$$

$$[\, i\,\partial t + (1/m)\partial_{\mathbf{r}}\,\partial_{\mathbf{R}} - V(\mathbf{R} + \mathbf{r}/2,t) + V(\mathbf{R}\text{-}\mathbf{r}/2,t)]\rho(\mathbf{R},\mathbf{r},t)=0 \tag{56}$$

$$[\;\partial_t + (\mathbf{P}/m).\partial_{\mathbf{R}} \;+ \mathbf{L}]f(\mathbf{R},\mathbf{P},t)=0 \tag{57}$$

where Lf is the **"driving field term"**

$$\mathbf{L}f(\mathbf{R},\mathbf{P},t)= -\int d^3r\;e^{-i\mathbf{P}.\mathbf{r}}\{V(\mathbf{R} + \mathbf{r}/2,t) - V(\mathbf{R}\text{-}\mathbf{r}/2,t)\}\rho(\mathbf{R},\mathbf{r},t) \tag{58}$$

In the special case that V is a *constant quadratic function of position*

$$V(\mathbf{R},t)= V_0 + \mathbf{V_1}.\mathbf{R} + V_2\mathbf{R}.\mathbf{R} \tag{59}$$

the driving field term collapses to a familiar form :

$$\{V(\mathbf{R} + \mathbf{r}/2,t) - V_{eff}(\mathbf{R}\text{-}\mathbf{r}/2,t)\}=(\mathbf{V_1}+ 2V_2\mathbf{R}).\mathbf{r}= -\mathbf{F}.\mathbf{r} \tag{60}$$

where $\mathbf{F(R)}$ is the effective force. In that case equation (60) may be Wigner transformed to the form,

$$[\;\partial_t + (\mathbf{P}/m).\partial_{\mathbf{R}} \;+ \mathbf{F}.\;\partial_{\mathbf{P}}\;]f(\mathbf{R},\mathbf{P},t)=0 \tag{61}$$

which is just the classical Liouville equation .The general case may be put in a more elegant form ,

$$[\;\partial_t + (\mathbf{P}/m).\partial_{\mathbf{R}}\;]f(\mathbf{R},\mathbf{P},T)\;\; -i\int d^3\mathbf{r}\;e^{-i\mathbf{P}.\mathbf{r}}\{V(\mathbf{R} + \mathbf{r}/2,t) - V_{eff}(\mathbf{R}\text{-}\mathbf{r}/2,t)\}\rho(\mathbf{R},\mathbf{r},t) =0 \tag{62}$$

let us define

$$\mathbf{F}_{eff}(\mathbf{R},\mathbf{r},t)\;.\mathbf{r} \equiv \{V(\mathbf{R} + \mathbf{r}/2,t) - V(\mathbf{R}\text{-}\mathbf{r}/2,t)\} \tag{63}$$

then the second term in (62) becomes successively,

$$\begin{aligned} -i\int d^3\mathbf{r}\;e^{-i\mathbf{P}.\mathbf{r}}\;\mathbf{F}_{eff}(\mathbf{R},\mathbf{r},t)\;.\mathbf{r}\;\rho(\mathbf{R},\mathbf{r},t) &= \int d^3\mathbf{r}\;\mathbf{F}_{eff}(\mathbf{R},\mathbf{r},t)\;.\;[\partial_{\mathbf{P}}\;\;e^{-i\mathbf{P}.\mathbf{r}}]\rho(\mathbf{R},\mathbf{r},t) \\ &= \int d^3\mathbf{P}'\;\mathbf{F}_{eff}(\mathbf{R},\mathbf{P}',t).\partial_{\mathbf{P}}\;f(\mathbf{R},\mathbf{P}\text{-}\mathbf{P}',t) \end{aligned} \tag{64}$$

$$\mathbf{F}_{eff}(\mathbf{R},\mathbf{P},t) = \int d^3\mathbf{r}\;e^{-i\mathbf{P}.\mathbf{r}/\hbar}\;\mathbf{F}_{eff}(\mathbf{R},\mathbf{r},t)/(2\pi\hbar)^3 \tag{65}$$

The resulting equation is a non-local version of the collisionless Boltzmann equation

$$[\ \partial_t + (\mathbf{P}/m).\partial_{\mathbf{R}}\ ]\ f(\mathbf{R},\mathbf{P},t) + \int d^3\mathbf{P'}\ \mathbf{F}_{eff}(\mathbf{R},\mathbf{P'},t)\ .\ \partial_{\mathbf{P}}\ f(\mathbf{R},\mathbf{P}-\mathbf{P'},t)=0 \qquad (66)$$

Planck's constant is hidden away in the quantum force term (65) where we have re=exposed it. It is easy to check that if $V(\mathbf{R})$ is quadratic then $\mathbf{F}_{eff}(\mathbf{R},\mathbf{P},T) = \delta(\mathbf{P})(-\ \partial_R\ V(\mathbf{R})\ )$ and we recover the collisionless Bolzmann equation again. These equations may be formally solved by path variable techniques along the lines of the Chambers-Rees method but so far only the simplest of one-dimensional problems have been attempted. Convergence problems are very severe , a problem which may be traced to : (a) the breakdown of the area preserving mapping property of classical phase-space; (b) the non-compactness of f. A better approach is outlined in section 11.

10. STATIONARY WIGNER FUNCTIONS

The general motion of a particle in a potential field $V(\mathbf{R})$ is known to be complicated; we expect phenomema such as tunnelling, interference effects, resonances, Aharonov-Bohm effects, localisation phenomena to arise under particular conditions. So the general non-recovery of the collisionless Boltzmann equation is not surprising. There is one surprise however, the collisionless Bolzmann equation is recovered exactly for a quadratic potential such as the harmonic oscillator potential. The Wigner equation becomes purely classical. Evidently the familiar quantum structure of the harmonic oscillator must be contained in the *initial state chosen for f.* The initial state *must* be a linear superposition of *stationary* Wigner functions made up from the eigenstates of the Harmonic oscillator Hamiltonian. The Wigner equation for ballistic motion in a quadratic potential propagates an initial distribution classically. *This shows that the Wigner equation of motion gives an incomplete picture of physics- it cannot describe stationary states. A similar problem arises with the density matrix equation*

$$i\hbar\ \partial_t = [H,\rho] \qquad (67)$$

whose Wigner-Wey transform generates the Wigner equation of motion directly, it is not completely equivalent to the Schrödinger equation as can be seen by by trying to set up an eigenvalue equation. In fact the stationary states have to be obtained from an adjunct equation for the stationary density matrix,

$$\varepsilon\rho\ =\ \{H,\rho\}/2 \qquad (68)$$

which involves the anti-commutator { , }. The equivalent of equation (19) turns out to be

$$(\varepsilon_{mn}/2)f_{mn}(\mathbf{R},\mathbf{P}) = [\mathbf{P}^2/2m-(1/8m)\ \partial^2_{\mathbf{R}}\ ]f_{mn}(\mathbf{R},\mathbf{P})+$$

$$\int d^3\mathbf{r}\int d^3\mathbf{P'}\ e^{-i\mathbf{P'}.\mathbf{r}}\ \{V(\mathbf{R}+\mathbf{r}/2) + V(\mathbf{R}-\mathbf{r}/2)\}\ f_{mn}(\mathbf{R},\mathbf{P}-\mathbf{P'}/2(2\pi)^3 \qquad (69)$$

Here the $\varepsilon_{mn} = \varepsilon_m+\varepsilon_n$, where the ε_n are the usual eigenvalues of H. For m=n we obtain the usual stationary Wigner functions, they are real valued. The case $m\neq n$ gives *complex* functions which relate to the other eigenfunctions of the super-operator {H, }. The entire set of $f_{mn}(\mathbf{R},\mathbf{P})$ form a complete orthormal set for all Wigner functions, stationary or otherwise. *Any initial Wigner function should be projected onto this space if you want the correct boundary conditions for time-dependent transport.* All the special sum rules of f such as $|f| \leq 1/(\hbar)^3$ are then taken care of.

11. EXTENDED PHASE-SPACE REPRESENTATION

The time-dependent Wigner function for a Hamiltonian $H=\mathbf{p}^2/2m + V(\mathbf{r})$ may be written (following Barker and Murray, 1983):

$$\partial_t f + v.\partial_r f + (2\pi\hbar)^{-3}\int\int d^3Q d^3P \cos(Q.P/\hbar)\ F(r,Q).\partial_p f(r,p+P,t) = 0 \quad (70)$$

where F is derived from the relation $\mathbf{F(r,Q).Q}=V(\mathbf{r}\text{-}\mathbf{Q}/2)\text{-}V(\mathbf{r}+\mathbf{Q}/2)$. In the limit $\hbar \to 0$, F goes over into the classical force and (1) becomes the Liouville equation (the limit is approached non-uniformily however). Equation (21) is the fundamental starting point for quantum ballistic transport and the solutions are the Green functions for the more complex task of incorporating electron-phonon and other scattering processes. Equation (21) is non-local in momentum space (and in position space if the kinetic energy is non-quadratic). Ideally we would use this equation in analogy with the collisionless Bolzmann equation (or Vlasov equation) to describe short geometry structures. The effects of crystal structure can be taken into account through effective mass theory and the band structure.

The quantum ballistic transport equation cannot be emulated by an ensemble of point particles in phase-space except when the non-locality is removed. However, a classical-like picture does underly the representation. It has been shown (Barker and Murray, 1983) that f may be represented as the linear superposition of a set of distributions $f_c(\mathbf{r},\mathbf{p},\mathbf{Q},\mathbf{P},t)$ which obey *classical* equations:

$$f(\mathbf{r},\mathbf{p},t) = (2\pi\hbar)^{-3}\int\int d^3Q d^3P \cos(\mathbf{Q}.\mathbf{P}/\hbar)\ f_c(\mathbf{r},\mathbf{p},\mathbf{Q},\mathbf{P},t) \quad (71)$$

$$\partial_t f_c + \mathbf{v}.\partial_\mathbf{r} f_c + \mathbf{F}(\mathbf{r},\mathbf{Q}).\partial_\mathbf{p} f_c(\mathbf{r},\mathbf{p},\mathbf{Q},\mathbf{P},t) = 0 \quad (72)$$

The causal boundary conditions are $f_c(\mathbf{r},\mathbf{p},\mathbf{Q},\mathbf{P},0) = f(\mathbf{r},\mathbf{p}+\mathbf{P},0)$.

It follows that quantum distributions can be emulated by ensembles of classical particles in a family of phase-spaces and Monte Carlo techniques can be restored, but at extra computational cost. Phenomena like tunnelling can be readily pictured in this representation: For some values of the parameters **P** and **Q** the particle trajectories are at higher energies and the barrier potentials related to **F(r,Q)** are smeared out in comparison to the classical limit **Q=0, P=0**; penetration to forbidden regions is thus allowed off the classical shell.

12. MONTE CARLO EVALUATION OF FEYNMANN PATH INTEGRALS

In the last few years a number of attempts have been made to evaluate the Feynman path integral propagator for dissipative quantum systems (Fischetti, 1984; Fischetti and di Maria, 1985). So far these methods appear fraught with difficulties of a fundamental kind. The results are discouraging.

13. MONTE CARLO AND QUANTISING LONGITUDINAL MAGNETIC FIELDS

The quantising effect of a high uniform magnetic field has been taken into account by Barker and Al-Mudares et al (1985a,1985b) by representing the density matrix in Landau states rather than plane wave states. For slowly spatially varying electric fields parallel to the magnetic field one may construct a Wigner distribution f(R;N,k;t) where N is the Landau sub-band index and k is the wave-vector along the magnetic field. The equation of motion for f is Boltzmann-like except that the transition rates are evaluated between Landau states. The Monte Carlo method is thus applicable to this problem and has been used for very accurate determinations of magnetophonon data over a wide range of temperatures and hot electron conditions.

14. MONTE CARLO AND TUNNELLING PROBLEMS

Baba, Al-Mudares and Barker (1988) have used the ensemble Monte Carlo method phenomenologically for hot electron tunnelling processes in special device geometries where the carriers are treated classically up to the barrier potentials. At the barriers each carrier is split into a sub-ensemble of transmitted and reflected particles for which interaction /traversal times are computed from the methods described in section . A detailed account is given by Barker (1985). A faster algorithm due to Collins, Lowe and Barker(1988) has replaced the transmission and reflection coefficients by a continuous process of reflection and transmission which gives quite good results for the charge distributions in complicated processes such as resonant tunnelling. It cannot however describe effects such as the Aharononv-Bohm effect nor does it give a very accurate picture of tunnel times.

15. TWO-DIMENSIONAL QUANTUM TRANSPORT: FINITE DIFFERENCES

Recently Barker and Barker and Finch have devised a method for computing wavepacket propagation in two dimensional geometries including the quantising effect of an applied magnetic field which significantly extends the one-dimensional time-dependent method described earlier. The method again uses the fact that the Hamiltonian is the generator of infinitesimal time translations, but a variant on the Cayley expansion is used. Introducing the variable $\tau = \delta t/2\hbar$ we replace the infinitesimal evolution operator $e^{-i\delta tH(t)/\hbar}$ by

$$e^{-i\delta tH(t)/\hbar} \dashrightarrow (1+iH_y)^{-1}(1-iH_x)(1+iH_x)^{-1}(1-iH_y) \tag{73}$$

If we introduce the uniform time-mesh $t = s\delta t$ and the space mesh $x = n\Delta$, $y = m\Delta$ we find the discretised Schrödinger equation in the form

$$[1 + iH_x(s,nm)\tau]\ \psi(nm,s+1/2) = [1 - iH_y(s,nm)\tau\ \ \psi(nm,s) \tag{74}$$

$$[1 + iH_y(s,nm)\tau]\ \psi(nm,s+1) = [1 + iH_x(s,nm)\tau\ \ \psi(nm,s+1/2) \tag{75}$$

Here we write the total Hamiltonian as $H = H_x + H_y$ where H_x and H_y are the most symmetrical separation of the x and y dependent components. Two time steps are deployed. The method is analogous to the Alternating Direction Implicit method traditionally used for diffusion problems. Despite the recurrence relations the algorithm can be re-written into an easily vectorised form for computation on supercomputers.

16. MULTI-DIMENSIONAL QUANTUM TRANSPORT : LATTICE METHOD

Very recently a new approach has been developed by Barker and Pepin (1989-to be published) which represents the general multi-dimensional transport problem as a multiply-connected one dimensional flow on a lattice . The method gives very high accuracy numerically but also allows a semi-analytical attack on the problem using group theoretic methods developed for one dimensional potential problems. The discrete space mesh of the finite-differenced multi-dimensional Schrödinger equation by is replaced by a wavefunction defined on a finite, continuous 1-D network spanning the n-D structure (such as a quantum waveguide). On each branch the wavefunction is one dimensional and it is propagated by a 1-D S-matrix or T-matrix determined bertween nodes by 1-D algorithms. The wavefunctions along different branches are matched at the nodes by a unique S-matrix which preserves local physical continuity. The network equations are solved by a S or T-sub-matrix formalism which builds in the preservation of unitarity and pseudo-unitarity respectively. It can be shown that the method is intrinsically more accurate than finite-difference eqauiations at the same node spacing provided the energy is kept away from the band gap regions of the intrinsic lattice. It is only necessary to use a single k-vector along the network. This result appears surprising when we consider the normal modes in a rectangular guide for which a longitudinal and transverse wavevector must occur in the continuum problem. However, the discontinuities in the gradient of the wavefunction which occur at the nodes are sufficient to approximate the true modes in all directions and indeed can give exact results at the nodes themselves.

17. QUANTUM WAVEGUIDE MODELLING : COUPLED MODE EQUATIONS

A range of potentially exciting devices may be constructable as quantum electron waveguides: metal wires, GaAs wires, squeezed Q2DEG channels, ring structures, tapers.We may also include the problem of injecting electrons from a narrow channel into a wide contact. Until recently these have only been modelled as one-dimensional structures with a few numerical calculations based on the 2D Schrödinger equation. Frohne and Datta(1988) have described an approximate numerical technique based on wavefunction matching to calculate the scattering matrix for electron transfer between 2D channel regions with different confining potentials in the transverse direction. The problem has interesting analogies with electromagnetic tapered waveguide theory (Sporleder and Unger, 1979). Indeed, the analogy is exploited by Barker and Laughton (1989,to be published) in a new a formalism for one-electron 2 D quantum waveguides which generalises the one-dimensional transmission equations of section 4 to a set of one dimensional coupled-mode equations. Referring to fig. 4 for a tapered 2D channel, channel radius a=a(z), with an internal potential V= V(z) and perfectly reflecting walls, we represent the wavefunction at a location z,y as an amplitude factor multiplied by a local eigenstate $\varphi_m[a(z), y]\exp(ik_m z)$ of a uniform guide of width a(z) at the same energy (evanescent modes are included):

$$\psi = \Psi_m(z)\, \varphi_m[a(z), y]\exp(ik_m z). \tag{76}$$

where the amplitudes obey the coupled equations

$$\partial\Psi_m/\partial z = \Sigma \left\{A_{mn}\, \Pi_n + B_{mn}\, \Psi_n\right\} \tag{77}$$

$$\partial \Pi_m/\partial z = \Sigma \{A'_{mn}\,\Pi_n + B'_{mn}\,\Psi_n\}$$

The coupling coefficients A_{mn}, etc are known functions of the local guide eigenstates, the guide radius, the total energy and the internal potential V(z).

18. THE MADELUNG REPRESENTATION : THE FLUID ANALOGY

One is impressed from numerical studies of Schrödinger's equation that some concepts like velocity field v(r) and position density ρ(r) actually work quite well . This suggest that we should perhaps be using fluid models of matter to bridge the classical particle regime and the quantum wave regime for device modelling.The idea goes back to Schrödinger and in particular Madelung in the 1920s. It also illuminates the importance of phase in an intriguing way.Let us write the wavefunction for a pure state

$$\psi = \rho^{1/2} \exp(iS(\mathbf{r})/\hbar) \qquad (49)$$

where ρ and S are real functions of position. If we insert this expression into Schrödinger's equation using a quadratic kinetic energy for simplicity, and separate out the real and imaginary parts we get two equations

$$\partial S/\partial t + (\nabla S)^2/2m + V(r) = (\hbar^2/2m) \quad \rho^{-1/2}\,\nabla^2\,\rho^{1/2} \qquad (50)$$

$$m\partial\rho^{1/2}/\partial t + (\nabla S).\nabla\,\rho^{1/2} + (\rho^{1/2}/2)\,\nabla.(\nabla S) = 0 \qquad (51)$$

Now define the velocity field as in terms of the gradient of the phase $\mathbf{v}(\mathbf{r},t) = (\nabla S)/m$. We then find our two equations are just the continuity equation and an equation very similar to the Euler equation for the velocity field in classical fluid dynamics.

$$\partial\rho/\partial t + \nabla(\rho\mathbf{v}) = 0 \qquad (53)$$

$$\partial\mathbf{v}/\partial t + \mathbf{v}.\nabla\,\mathbf{v} + \nabla V(\mathbf{r})/m = (\hbar^2/2m^2)\quad \nabla\{\ \rho^{-1/2}\,\nabla^2\,\rho^{1/2}\} \qquad (54)$$

The RHS vanishes in the classical limit to give precisely an Euler equation - a perfectly valid construct to handle particles if we use delta functions to represent the position density and velocity field.If a vector potential is present we can use the form $\mathbf{v}(\mathbf{r},t) = (\nabla S - e\mathbf{A}/m)$ to pick up Lorentz force terms. An interesting point here is the significance of the classical fluid concept of circulation, defined as the loop integral of the velocity field: $\Gamma' = \oint \mathbf{v}.d\mathbf{r}$. From the definition of velocity this becomes: $\Gamma' = e \oint \mathbf{A}.d\mathbf{r}/m$ which is proportional to the non-integrable phase factor occurring in the Aharonov-Bohm effect. It should be remarked that the velocity field defined here is identical to what one gets from evaluating the velocity field from a Wigner function.

19. REFERENCES

Baba T, Mudares M, Barker J R (1989) Monte Carlo Simulation of Resonant Tunnelling Diodes. in the press

Barker J R, Mudares M Al, Snell BR, Guimaraes PSS , Taylor DC , Eaves L, Portal J C, Dmowski L and G Hill (1985) Hot electron magnetospectroscopy of sub-micron semiconductors and heterostructures. Physica 134 B: 17-21

Barker J R, Mudares M Al, Snell BR, Guimaraes PSS , Taylor DC , Eaves L, Hill G and Portal J C, (1985) Validation of magnetophonon spectroscopy as a tool for analysing hot electron effects in devices. App. Phys. Letters 47: 387

Barker J R (1985) Quantum theory of hot electron tunnelling in microstructures. Physica 134B: 22-31

Barker J R (1986),Quantum transport theory for small-geometry structures. in Physics and fabrication of microstructures and microdevices M Kelly and C. Weisbuch: editors, Springer-Verlag:London , 210-230

Collins S (1986) PhD Thesis, University of Warwick,

Collins S, Lowe D and J. R. Barker (1987) The quantum mechanical tunnelling problem re-visited J.Phys.C 20: 6213-6232

Collins S, Lowe D and J. R. Barker (1988) A dynamic analysis of resonant tunnelling J.Phys.C 20: 6233 - 6243.

Collins S, Lowe D and J. R. Barker (1988) Resonant tunnelling in heterostructures: numerical simulation and qualitative analysis of the current density J. Applied Physics 63:142 - 149

Fischetti M V (1984) Phys Rev Lett. 53: 1755

Fischetti M V and Maria D J (1985) Phys Rev Lett 55: 2475

Frohne R and Datta S (1988) Electron transfer between regions with different confining potentials. J.Appl. Phys. 64: 4086-4090

Goldberg A, Schey H M and J L Schwartz (1967) Am. J. Phys. 35: 177

Landauer R, (1988) Spatial variation of currents and fields due to localised scatterers in metallic conduction. IBM J. Res. Develop. 32: 306-316.

Mains R K, Haddad G I (1988) Time-dependent modelling of resonant tunnelling diodes from direct solution of the Schrödinger equation. J. Appl. Phys. 64: 3564-3569

Messiah A (1964) Quantum Mechanics, North Holland: Amsterdam

Peres A (1981) Calculation of localisation lengths in disordered chains. Phys.Rev. 24B:7463-7466

Peres A (1983a)Transfer matrices for one-dimensional potentials. J.Math.Phys.Rev. 24:1110-1119

Peres A (1983b) Chaotic band structure of almost periodic potentials Phys.Rev. 27B:6493-6494

Sporleder F,Unger H-G (1979) Waveguide tapers, transitions and couplers .Peregrinus Ltd-London

Vassell M, Johnson L and Lockwood H (1983) J.App.Phys. 54: 5206

Vigneron J P , Lambin Ph (1980) J. Phys A: Math.Gen. 13: 1135

Vigneron J P , Lambin Ph (1980) J. Phys A: Math.Gen. 12: 1961

Wolf E L (1985) Principles of electron tunnelling spectroscopy Oxford University Press:London

Computer Simulations

Trevor M Barton
University of Leeds

1. Introduction.

Models for semiconductor devices can be categorised into several different classes, and the major divisions are shown in figure 1. One of the most important distinctions is that between physical device models and circuit models. As the name suggests, circuit models employ circuit analogues for the device, often representing the device either as a "black box" there the transfer function is obtained from measurement of an actual device, or by an equivalent circuit consisting of ideal active and passive components. In the latter representation, the values of specific circuit elements are obtained either from measurement or by relating their values to specific physical parameters of a given device.

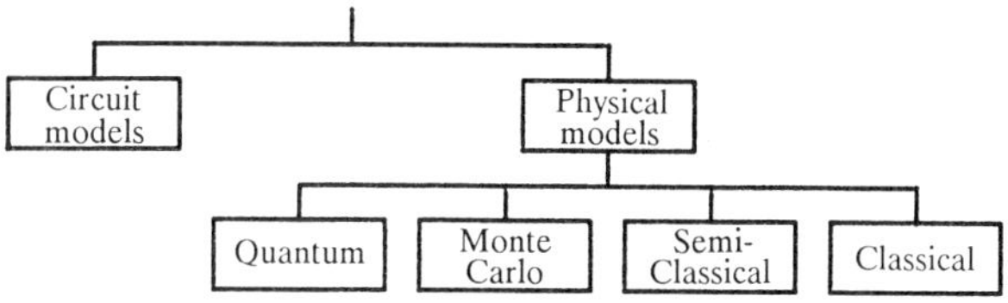

Figure 1: Classes of semiconductor device model.

Circuit models have the advantage that they are relatively easy to solve, and they therefore provide a computationally efficient means of modelling devices in applications such as large scale circuit simulation, where the use of more complex models would result in an unacceptable computing requirement. It is, however, often difficult to draw a direct relationship between the physical parameters and electrical behaviour of a real device and the elements of its circuit model. This can limit the predictive capability of this class of model, and they are most often used to model devices which are already well characterised.

The emphasis in this chapter is on the second class of model, the physical device model. This type of model utilises a more exact representation of the device, and the important physical parameters are incorporated directly into the model. This can result in a more accurate model, assuming, as always, that any simplifying assumptions used in its formulation are carefully considered. These models can provide a greater predictive capability, because the actual device is modelled based on the fundamental physical phenomena which govern its operation, rather than a circuit representation of the external electrical characteristics of the device.

Figure 1 shows several sub-classes of physical model, each employing different assumptions in the formulation of the carrier transport model, and having different ranges of application. The applicability of a physical model to a specific device is often determined by to the size of the important features of that device, for example, the length of the gate of a field-effect transistor. Physical models generally have a lower size limit placed on them by the physical assumptions used in their formulation, and an upper size limit determined by the complexity and computer requirements of the solution.

Quantum mechanical models provide solutions which are directly based on the underlying quantum mechanical principles which govern electron transport in semiconductors, and are required for very small devices in which quantum effects become significant. The complexity and computer resource requirements of this type of model tends to limit the size of device which can be modelled using these techniques to devices with dimensions less than a couple of hundred angstroms. These models are, however, valid for all of the smallest devices which are produced in the laboratory today.

Monte Carlo models are statistical models which provide a solution for the Boltzmann transport equation, and yield a somewhat less complex formulation to describe carrier transport. These models are most often used for simulating devices in which the particulate nature of the carriers is important, but in which the effects of very small scale phenomena, such as the Heisenberg uncertainty principle, become less significant. They are nevertheless computationally intensive, and are generally limited by the availability of computer resources to modelling moderate scale devices with dimensions of a few tenths of a micron, although they can be used on larger devices using further simplifying assumptions designed to decrease the total number of particles which are modelled. An important use of Monte Carlo models is for the calculation of material parameters, such as the velocity- and energy-field characteristic of carriers in bulk semiconductor.

Semi-classical models are based on an solutions for at least the first three moments of the Boltzmann transport equation, and neglect the particulate nature of carriers by treating electrons and holes as a continuum. This assumption requires that the device be large enough that there are sufficient carriers so that electrons and holes may be characterised by local average quantities, such as density, energy and momentum. This is often the case for devices with dimensions above a few tenths of a micron, and is true for a large number of the devices in common use today. As the dimensions of the simulated device become even larger and the magnitude and spatial derivatives of internal quantities such as electric field become small, further simplifications may be made to the semi-classical model, resulting in what is probably the simplest transport model of all, often referred to as the classical model, the phenomenological model, or the drift-diffusion approximation. It is these types of model which are most often used to simulate semiconductor devices at present, although it is likely Monte Carlo and even quantum mechanical models will become more and more common as device dimensions shrink.

In this chapter, the implementation of a fairly typical semi-classical simulator will be discussed, and illustrated using a simulator developed at the University of Leeds as an example. Examples of different types of simulator will be provided in order to demonstrate the range of results which can be obtained from numerical simulation of semiconductor devices.

2. Design and Implementation of a Semi-Classical Simulator.

The complexity of the numerical solution of the semiconductor equations means that the design and implementation of the software which is used for semiconductor device simulation must be considered carefully. Generally speaking, numerical simulation requires large amounts of computing resources, and in order for simulation to become an attractive proposition this resource must be used efficiently. In addition, the user interface, that is, the manner in which data is presented to the program, and, perhaps more importantly, the manner in which the results of the simulation are presented to the user, is an important issue which in itself can influence the overall efficiency of the simulation process.

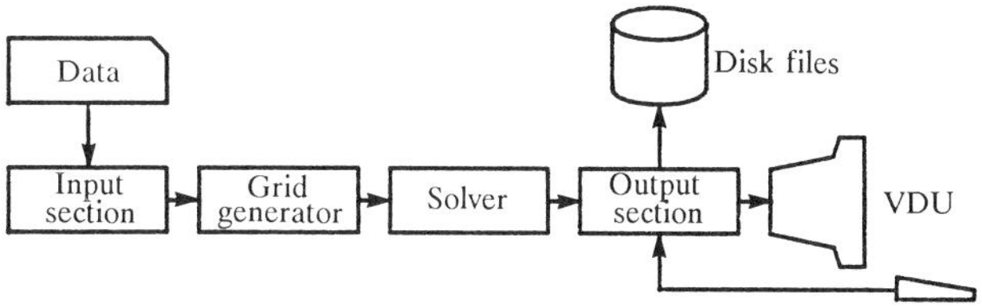

Figure 2: Sections of a typical simulation program.

A simulation program can be broadly divided into several distinct sections, and these are illustrated in figure 2. First, there is some means of reading and interpreting the data which specifies the characteristics of the simulation which is to be carried out. This data may contain information such as the geometry of the device, the doping profiles throughout the device, the voltages which are to be applied to the contacts, details of any transient or sinusoidal excitation of the contact voltages, and details of the numerical methods which are to be used to solve the semiconductor equations.

Second, many simulators determine the topology of the numerical mesh used for the solution based on the geometry of the device itself. In the simplest two-dimensional case, a rectangular geometry solved using finite difference techniques, the simulation domain would simply be mapped onto a regular grid of nodes represented by elements of a two-dimensional array. More complicated discretisations required by solution methods such as finite difference and finite element solutions on triangular or non-uniform rectangular meshes, finite boxes solutions and non-rectangular device geometries often require complex algorithms and data structures in order to describe the numerical mesh.

The next section is the numerical solver itself. It is generally the actual solution the semiconductor equations which requires the largest amount of computer resource of any part of the simulation process. The efficiency of this process is determined primarily by the method of solution, but this is often dictated by the type of solution required from the simulator, for example, whether steady state or transient solutions are required. Once the method of solution has been determined, however, the efficiency of the resulting code can depend critically on the details of the implementation, such as the exact form of the discretisation, the choice of data structures and variables, and the coding of the numerically intensive inner loops of the solver.

Finally the results of the simulation have to be communicated to the user in some fashion. The method of presentation is an important aspect of the simulation process, and it essential that this is done in the most efficient manner possible. In general, the results of simulations are graphic quantities, and the use of computer graphics is more or less required in order to maximise the amount of information which can be obtained from the simulation. It is much easier to obtain information from graphs, surface and contour diagrams than from columns of numbers, and without these features much of the important information from a simulation can be lost.

The implementation of these four parts of a typical simulator will be discussed in the following sections. As an example, the simulator FETSIM, developed at the University of Leeds is used [1,2]. This simulator is a general purpose two-dimensional unipolar GaAs device simulator (The FET in the name indicates it's most common use - that of simulating GaAs MESFETs). The geometry of the device is specified at run-time, and the device can be any shape, so any arbitrary device can be simulated. The solution is obtained using finite difference techniques on a rectangular, non-uniform, mesh which is generated at run-time based on the geometry of the device, and updated as part of the solution process in order to obtain a high degree of accuracy. The simulation solves a time-dependent form of the current continuity and, optionally, the energy conservation equation, and is therefore capable of providing transient as well as steady state results. The results of the simulation are dumped in binary files, which are read by a second program, FETOUT, an interactive graphical post processor which, along with several subsidiary programs, produces graphs, surface and contour plots of the results on a number of different types of video and hardcopy terminals.

2.1. The Input Section.

The task of the input section of the simulator is to read and process the data which determines the particular simulation to be carried out. Often a degree of input error checking is performed by this section; this can range from simple checks of the numerical range of data items to complicated checking of the input device geometry.

The form in which data is presented to the program can be an important consideration when designing the input section. Some simulators, especially those designed to model a particular type of device with a relatively fixed geometry, and therefore have much of the required information already built into them, take their input from a column of numbers in a data file. In this type of input, the position and order of the input data is important, and it is this which relates a particular data item to its associated internal variable. It becomes difficult, however, to check data in this format for errors, as missing data items may go undetected for several lines, or at least until the type of a data item conflicts with the type the program expects.

More complex simulators use keywords to associate data to variables. Typically, the input will consist of a keyword followed by one or more parameter. For example, the input

```
POISSON_RELAXATION 1.85
```

might be used to set the overrelaxation factor for the Poisson solver to 1.85. Figure 3 shows a data file for FETSIM which describes a simulation of a 1.0 μm gate length recessed gate MESFET, the geometry of which is illustrated in figure 4.

Figure 3 demonstrates the various types of keywords used as commands to the program. Simple commands, such as the POISSON_RELAXATION command mentioned previously, take a single parameter and assign the value to an internal variable. More complex commands require several parameters. For instance, the CONTACT_POTENTIAL command requires a parameter identifying the contact on which the potential is to be applied, and a second specifying the potential to apply. The input section searchs a list of contacts, which have already been defined in the geometry definition, until it finds a contact with the same name and applies the voltage to that

```
name 1.0um example MESFET.
surface_specification
    source  0.0      -0.1e-6   0.6e-6   -0.1e-6   13
    surf    0.6e-6   -0.1e-6   1.2e-6   -0.1e-6   10
    surf    1.2e-6   -0.1e-6   1.2e-6    0.0      10
    gate    1.2e-6    0.0      2.2e-6    0.0      17
    surf    2.2e-6    0.0      2.2e-6   -0.1e-6   10
    surf    2.2e-6   -0.1e-6   3.2e-6   -0.1e-6   10
    drain   3.2e-6   -0.1e-6   3.8e-6   -0.1e-6   13
    side    3.8e-6   -0.1e-6   3.8e-6    0.60e-6   7
    bott    3.8e-6    0.60e-6  0.0       0.60e-6   7
    side    0.0       0.60e-6  0.0      -0.1e-6    7
end
doping_density 1e23
x_doping_profile
    0.0    1.0
    3.8e-6 1.0
end
y_doping_profile
   -0.10e-6  1.0
    0.19e-6  1.0
    0.21e-6  0.00001
    0.60e-6  0.00001
end
poisson_relaxation 1.8
timestep 1.0e-14
max_delta_x 0.0501e-6
contact_potential source   0.0
contact_potential drain    0.0
contact_potential gate     0.0
energy_model no
newton
steady 500
dump rtaa0000 dumper
end
```

Figure 3: Data file used by FETSIM to describe a 1.0 μm recessed gate MESFET.

contact.

Finally, more complex commands require several lines of input. For example, the SURFACE_SPECIFICATION command is used to describe the geometry of the device, and hence the boundary of the simulation domain. Each line of input, up to the line starting with END, describes a single straight edge of the boundary. This description consists of an alphanumeric name, the coordinates of the start and end of the edge, and the integer type of the boundary which determines the boundary conditions which are applied to the edge.

These commands demonstrate some of the error checking that can be applied to the input data. For example, the POISSON_RELAXATION command checks that the relaxation factor is in the range $0.0<\omega<2.0$ prints an error message if this is not the case, and halts program execution if the program is not being run interactively. The CONTACT_POTENTIAL command checks that a contact with the specified name exists. Various checks are carried out on the SURFACE_SPECIFICATION command, including that the edges form a closed, non-overlapping polygon with each of its edges parallel to the axis of an orthogonal coordinate system. These checks increase the likelihood that the actual simulation, which is carried out in this case when the STEADY command is issued, will produce valid and meaningful results.

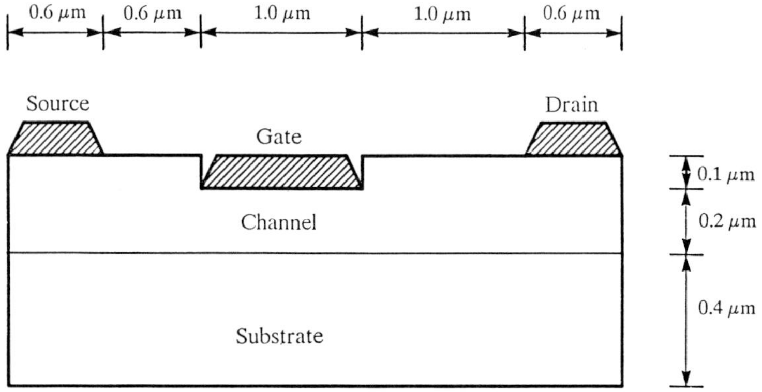

Figure 4: MESFET geometry described by the data file in figure 3.

2.2. The Grid Generator.

Grid generation schemes vary widely between simulators, and depend on various factors, such as the solution method, the form of the discretisation, and the complexity of the simulator. The simplest discretisations utilise a regular grid as shown schematically in figure 5(a), and these are often used to discretise simple one- and two-dimensional geometries. This is done be mapping the geometry onto the grid and storing the values associated with each node in an array. This can be a simple one-dimensional array for a one-dimensional simulation, and a two-dimensional, rectangular array in the case of a two-dimensional simulation requiring a rectangular simulation domain. Grids of this nature have the advantage that the discretised semiconductor equations have a simple form, which allows very efficient coding with a subsequent increase in the speed of the simulation. Unfortunately, these simple grids are often not able to describe the complex geometries and doping profiles of modern devices with sufficient accuracy, and a more complex discretisation must be used.

Perhaps the most common of these is the use of *non-uniform* discretisation of the simulation domain into rectangles or, in the case of a one-dimensional domain, line segments. An example of this type of discretisation is shown in figure 5(b). This results in a grid which is suitable for finite difference solutions of the semiconductor equations, and the rectangular nature of the grid still allows an efficient coding of the discretised equations. This form of discretisation often applies some limitation to the shape of the device in the two-dimensional case, in that the boundaries of the simulation domain have to be parallel to an orthogonal set of axis. There are, however, techniques available which allow a rectangular discretisation of devices having bevelled boundaries [3, 4]. A non-uniform discretisation has the advantage that areas of fine detail in the device can be modelled accurately, and that selective mesh refinement in areas of poor accuracy in the solution can increase the overall accuracy of the simulator.

One disadvantage of using a non-uniform grid of this nature is that grid lines must be carried through from one edge of the simulation domain to the other. Local refinement of the mesh by adding new mesh lines thus results in an unnecessary number of nodes in regions of the domain away from the area in which refinement was needed. This problem can be overcome, however, by use of a finite boxes discretisation, illustrated in figure 5(c) [3]. This employs a rectangular grid

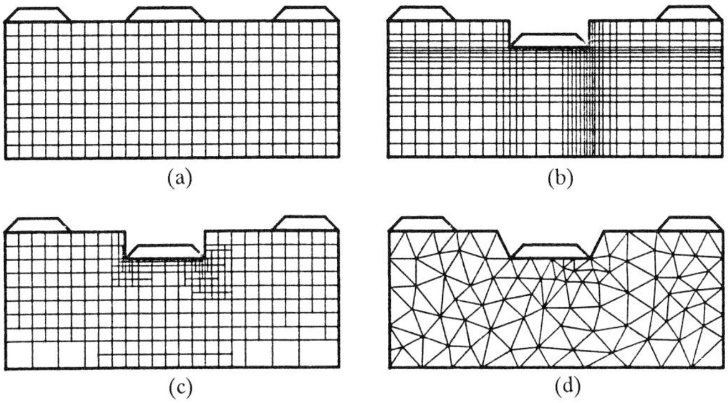

Figure 5: Schematic diagrams of various grid types. (*a*) uniform rectangular mesh, (*b*) non-uniform rectangular mesh, (*c*) finite boxes mesh and (*d*) triangulation.

with a special formulation of the finite-difference discretisation which allows mesh lines to terminate inside the simulation domain, and not just on the boundaries as is required by standard finite-difference discretisations. This has the advantage that selective mesh refinement can be carried out during the simulation process without the introduction of unwanted nodes in the domain, and can result in a simulator which combines the efficiency of coding found with a rectangular discretisation with the flexibility and accuracy provided by the ability to locally refine the mesh.

More general still are discretisations based on triangulation of the simulation domain, as illustrated in figure 5(*d*). Triangular meshes may be used with both finite difference and finite element solution methods, and provide a highly accurate means of describing the exact shape of the domain. While some triangular discretisations employ a regular triangular mesh, most modern simulations employing this discretisation use a non-uniform mesh with triangles of (almost) arbitrary shape [5,6,7,8]. This allows selective mesh refinement to be carried out with little difficulty, as a new node can easily be placed in a given triangle and joined into the mesh. Bevelled edges are easily implemented, and curves can be accurately approximated using short line sections. This allows a truly arbitrary device geometry to be described.

The choice of discretisation has a strong influence on the data structures used to store the information about each node in the mesh. Simple uniform and non-uniform rectangular grids are easily mapped onto rectangular arrays, and this is often the most efficient means of coding these meshes. Nodes are thus related to each other by their position in the array; adjacent nodes will be adjacent in the array. This method is easy to use when the geometry of the device is relatively fixed (such as in a special purpose simulator) and where no grid refinement is to be carried out. It becomes difficult, however, to map complex geometries onto these structures, and later mesh refinement is exceedingly difficult and time consuming.

More complicated meshes, such as those used in finite boxes solvers and for triangulations, often require complex data structures in order to retain an efficient coding. This is especially the case when there is no *a priori* knowledge of the device geometry or the final mesh pattern, which may, because of mesh refinement, be substantially different at the end of the simulation compared with the initial mesh. In these discretisations, nodes are not related to each other by their position in the array, but by a neighbour list which contains a list of all the nodes which are directly connected by the mesh to a given node. This allows new nodes to be inserted in the mesh by manipulating the neighbour list, breaking some links between nodes and adding new ones to the new node.

The data structures used to create the neighbour list depend heavily on the programming language which is used. In standard FORTRAN, the nodal information, including the neighbour list, is often stored in a series of one-dimensional arrays. For instance, in the case of a rectangular discretisation, there could be an array for the potential at each node, another for the electron concentration, and four arrays for the up, down, left and right neighbours. Neighbour lists for triangulations are complicated by the varying number of neighbours a node may have, but are nevertheless often implemented in the same manner.

Languages such as PASCAL and C, which allow complex structured types and the use of pointers to these types, simplify the coding to some degree. All the information about a given node can be stored in a single structure, which also includes fields for pointers to the node's neighbours, as shown diagrammatically in figure 6(a) along with the PASCAL code used to define the structure. This formulation allows a more efficient storage of the nodes, as the memory in which the node is stored can be allocated when the node is created (by the PASCAL *new*() statement), and may be later deallocated if the node is deleted during subsequent grid refinement. The use of pointers often results in faster access to neighbour nodes, as there is no need to calculate the address of array elements, and the actual address of the neighbour node is stored in the structure.

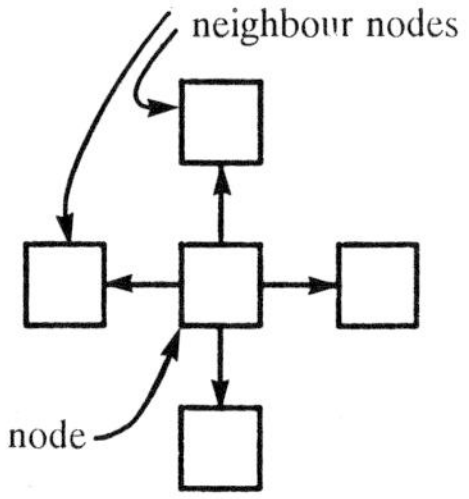


```
node_ptr = ^node;
node = RECORD
    phi:          real;
    en:           real;
    ...
    ...
    up_node:      node_ptr;
    down_node:    node_ptr;
    left_node:    node_ptr;
    right_node:   node_ptr;
END;
```

(a)

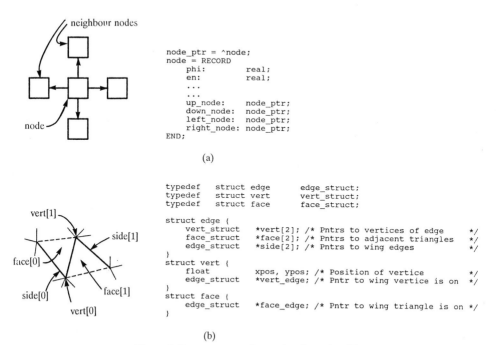

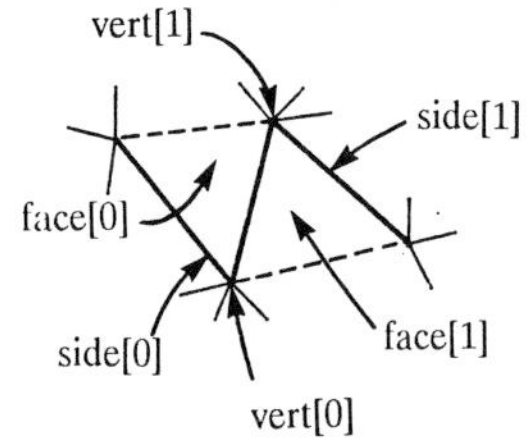


```
typedef    struct edge        edge_struct;
typedef    struct vert        vert_struct;
typedef    struct face        face_struct;

struct edge {
    vert_struct     *vert[2]; /* Pntrs to vertices of edge      */
    face_struct     *face[2]; /* Pntrs to adjacent triangles    */
    edge_struct     *side[2]; /* Pntrs to wing edges            */
}
struct vert {
    float           xpos, ypos; /* Position of vertice          */
    edge_struct     *vert_edge; /* Pntr to wing vertice is on   */
}
struct face {
    edge_struct     *face_edge; /* Pntr to wing triangle is on */
}
```

(b)

Figure 6: Data structures for storing dynamic grids.

Some grid generation schemes, such as Delaunay triangulation, require even more complex data structures. These grid generation schemes are often triangle, rather than node, based, and require a data structure which allows the relationship between nodes, edges, and triangles to be determined efficiently. This is in contrast to rectangular schemes, where it is generally sufficient to access adjacent nodes in the grid. One such structure is shown in figure 6(*b*), along with the C code with which it is described. This is based on a modified winged edge structure similar to that used by Davy [9] and De Floriani [10], and allows efficient manipulation of the grid during the generation process.

Grid refinement is often be treated as a special case of the generation procedure. Once again, the nature of the initial grid determines the ease with which new nodes can be inserted into the grid. Grids which are based on one- and two dimensional arrays are often more difficult to refine, as it becomes necessary to move whole blocks of the array in order to fit new nodes into the mesh. Pointer based meshes, and those using neighbour list arrays are generally fairly easy to refine, as new nodes or mesh lines can be added by manipulating the neighbour information. Grids based on triangulations can be refined by either placing a new node in the center of an existing triangle and joining it to the three vertices of the original triangle, or subdividing the three edges of a triangle and joining the new points to create a small inner triangle, and several new outer triangles.

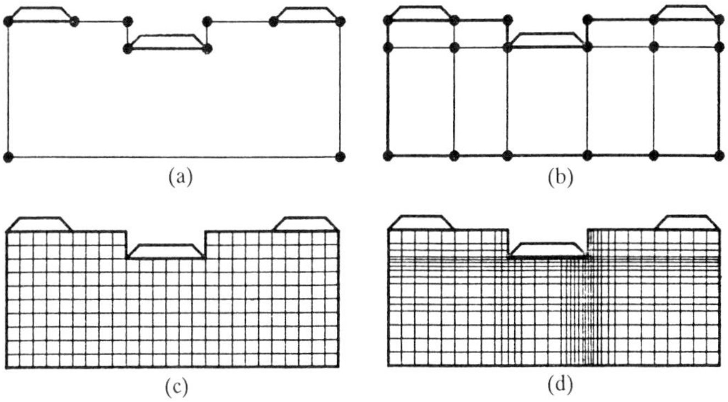

Figure 7: Steps in the FETSIM grid generation process.

The simulator FETSIM utilises a non-uniform rectangular mesh. Nodes are represented by structures containing pointers to adjacent nodes in the mesh, similar to that shown in figure 6(*a*). As was shown in the previous section, the geometry of the device is defined at run-time, and the mesh is generated from this information. At present, a standard finite-difference scheme is used, although it is intended that this will be extended to a finite boxes scheme with provision for bevelled edges as the program develops. The mesh generation scheme is described in detail below.

First, the edge data is read from the data file, and checked to ensure that all the edges of the simulation domain are parallel to the x or y axis, and that they form a closed polygon. Adjacent edges are then sewn together with a node, and all the nodes thus placed are joined together by making the pointer in the corresponding direction point to the next adjacent node on the boundary. Directions that point out of the device are initialised to point to a special node which indicates that

that direction is invalid. This is shown schematically in figure 7(a) for the device described by the data file in figure 3.

Second, all the uninitialised neighbour pointers are joined to the opposite side of the device, as shown in figure 7(b). This is done by creating a node on the opposite boundary, inserting it into the mesh by making it point to the two adjacent boundary nodes, initialising any outward pointers, and linking it with the original node. Nodes are also created in a similar fashion on any existing grid lines crossed as the line is extended across the device.

Finally, the mesh is completed by inserting sufficient nodes in the gaps between mesh lines so that the mode spacing is below a certain (default or operator specified) maximum. This results in a fairly regular initial mesh, which is shown schematically in figure 7(c), upon which a solution may be is obtained. Further refinement is carried out as the solution progresses, which results in a final mesh similar to that shown in figure 7(d).

2.3. The Numerical Solver.

The wide variety of solution methods for the semiconductor equations make it difficult to give a general description of the solution section of a simulation program. Some of these methods are described elsewhere in this text, and there are many publications dealing with solution methods in the literature [7,11,12,13,14,15]. There are, however, several considerations when implementing a numerical equation solver which are common to all methods.

The solver is generally the most time-consuming part of the simulation, typically using in excess of 90% of the CPU time requirement. For this reason, the coding of the solver should be considered carefully to ensure maximum efficiency. There are several means of increasing the efficiency of the solver, and these are discussed below.

Operation	Time (μs)
exponentiation (e^x)	5.400
square root	4.300
division	0.770
multiplication	0.160
subtraction	0.140
addition	0.072
absolute value	0.036

Table 1 Typical times of floating point operations on an AMDAHL 580 computer.

The first of these involves limiting the number and type of floating-point operations performed during the calculation. Careful formulation of the discretised semiconductor equations can result in a significant reduction in the number of operations. Complex operations, such as square roots and exponentiation should be avoided at all costs. Table 1 shows typical times of some sorts of double precision operations on an AMDAHL 580 computer, and is presented as a comparative guide rather than an absolute indication of the speed of these operations. It does show, however, that operations such as division and subtraction are relatively slow when compared to their complementary operations, multiplication and addition. Floating point division especially costly! Some language features such as the FORTRAN "raise to the power" operator (X**Y) should be used with care. For example, with many compilers, the operation X**2.0 (floating point exponent) will consume much more computer time than X**2 (integer exponent), which in turn is much more expensive than simple multiplication, that is, X*X.

It is often possible to reduce the number of operations performed during the calculation process by calculating terms which remain constant during the execution of a loop outside the loop. This is not so important for simple loops involving simple data types, as good optimising compilers are able to remove most constant calculations. Complicated loops, such as those found at the core of many solvers and those which involve more complex data types, cannot be resolved well by any compiler, and constants should be explicitly calculated before the loop. Often, terms are constant for each node rather than for the whole loop, and these can be calculated before the loop, and the results stored in a separate array, or in a special field within each node record.

The selection of the data structures that are used to describe the mesh can be important in determining the final efficiency of the code. The solution process requires many accesses to the data from the neighbour nodes of a given node in the mesh, and it is important that this information can be retrieved in the fastest possible manner. An example of this was found during the development of the FETSIM program. The first version of the program was written in FORTRAN, and used neighbour list arrays as described in the previous section. When the program was subsequently re-written in PASCAL these were replaced by dynamically allocated node records similar to those illustrated in figure 6(*a*), which include pointers to the records containing the neighbour nodes. This resulted in a speed increase of nearly 50%, which was almost entirely due to the more efficient way in which the address of the neighbour node was stored. The first method, which stored the index of the neighbour node in an array, required two array offset calculations, each involving costly integer multiplications, to determine the actual memory address of the neighbour node. The PASCAL implementation stored the address of each neighbour node directly in the node record, and thus only a single machine instruction is required to access a neighbour node.

It is often found that there is a definite trade-off between the CPU time and memory usage of a program. The CPU time can be reduced at the cost of increased memory requirements, and the memory usage can be decreased at the cost of increased CPU time. Sometimes language dependent tricks, such as careful allocation of variables to registers in C, can have a large benefit. Although this is often done automatically be the compiler, the choices that the compiler makes are not necessarily the best for a particular application. Multiple accesses into a given array element can be a source of inefficiency, as this involves a great deal of address calculation with poor compilers, and it is sometimes found that one read of the value into a local variable, which is subsequently used in all calculation, can result in a large improvement in the speed of the code. The use of pointers into an array where this is possible can have the same effect.

The solver used in FETSIM solves the time-dependent form of the current continuity and, if desired, the energy continuity equation, along with the Poisson equation. This allows the calculation of device response to fast transients, as well as the steady state behaviour, which is calculated by allowing the terminal currents to reach a steady state. Each individual equation is solved in turn using an iterative method combined with successive over-relaxation.

2.4. The Output Section.

Modern devices and simulators are so complex that, in order to gain the greatest possible amount of information, graphical presentation of the results of the simulation is often a necessity rather than simply a desirable option. This generally takes the form of graphs, contour plots and three dimensional surface representations of terminal currents and internal distributions. Unlike the simulation itself, the output of the results is often an interactive process, where the user can select which results are to be viewed after the simulation is complete. This generally requires a different program, often referred to as a post-processor, which can pick up the results from an intermediate dump file produced by the simulator proper.

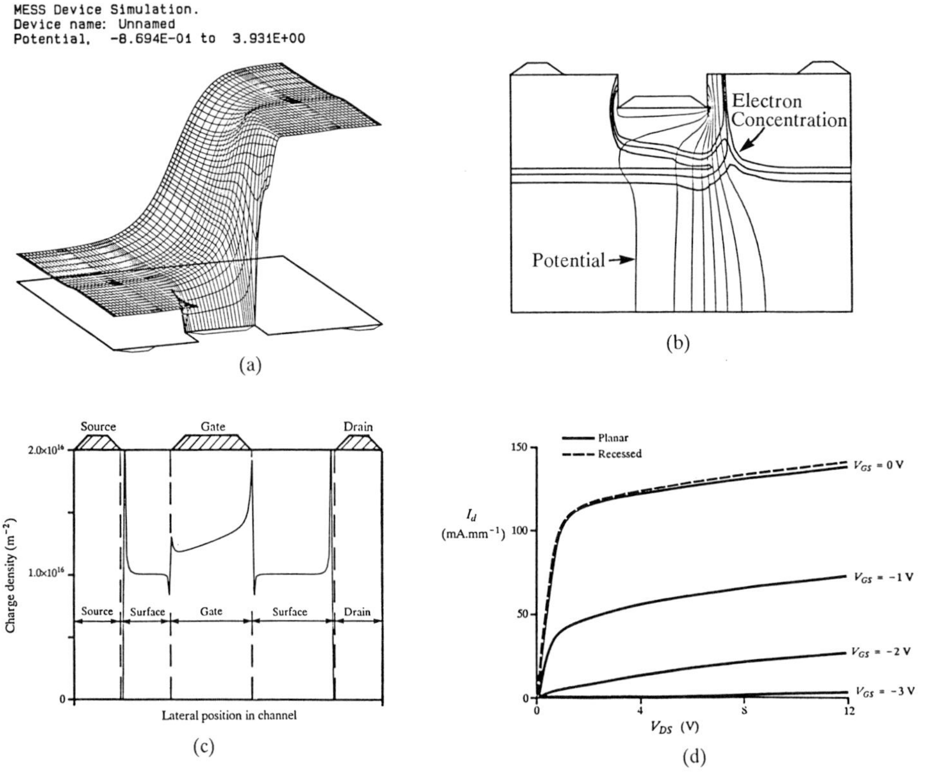

Figure 8: Some results from FETSIM. (*a*) potential surface, (*b*) contours of potential and electron concentration, (*c*) surface and contact charge distribution, and (*d*) output characteristics.

The post-processor used by FETSIM allows the user to interactively control the display of the results from the simulator. Input to the processor comes in the form of commands from the keyboard, and the graphical output can be directed to the same terminal if a graphics terminal is in use, or to a file which can be subsequently plotted on a plotter. This is achieved by using a hardware independent graphics package, in this case GINO, where the hardware device to be used for output is determined by the user when the program is run.

The program incorporates its own contouring routines, and is capable of overlaying contours of various functions from either the same device or different devices. This allows easy comparison of the results from several different devices, or the same device at different operating points. Three-dimensional surface plots of functions can be produced interactively using the same program, and can be viewed from any rotation and elevation. The surface plot can be passed to a separate program which performs hidden surface removal and colouring of the top and bottom surface. Graphs of transient terminal currents, surface charge and gate electric field distributions can be displayed by a graph drawing routine, and can similarly be passed to a separate program for further annotation, colouring, or the addition of data from other sources. It is also possible to graph sectional views of internal functions such as potential, current density, and electron concentration along grid lines within the device.

The post-processor reads the data from the simulation from a binary dump file which is produced by FETSIM. A binary format is used to store the results rather than a character file simply because of the amount of data which has to be saved. If a character file were to be used, the number of bytes that would be needed to store a double-precision floating-point number would be more than 20, compared with 8 for a binary file. The only limitation of this method of storage is that the restricts the transfer of result files between computers. For instance, it is not possible to transfer files between machines on which the internal representation of numbers is not the same, or between machines that do not have an 8 bit data path between them.

Figure 8 demonstrates some of the results that are available from the simulator for the device shown in figure 4. Figure 8(*a*) shows the potential surface inside the device for for V_{DS} = 4 V and V_{GS} = 0 V, and this is shown again in the contour plot in part (*b*), along with the electron concentration contours. The gate and surface charge distribution is shown in part (*c*) for V_{DS} = 1 V and V_{GS} = 0 V, and the output characteristics of the device in part (*d*). The diagrams are, of course, much more interesting in colour!

3. Further Examples of Device Simulators.

This section gives some examples of other simulators developed at the University of Leeds. These are, in the main, one- and two-dimensional classical or semi-classical simulators which are used to simulate a wide variety of high-speed III-V semiconductor devices, including MESFETs, HEMTs, TEDs, varactor diodes, and photodiodes. In addition, a set of Monte Carlo simulators has been developed in order to investigate both the electron and hole transport properties of GaAs, and to simulate high-speed GaAs photodiodes.

3.1. Monte Carlo Simulation.

Monte Carlo simulators are often used to calculate the transport properties of electrons and holes in semiconductor materials, and provide materials parameters which are of value for classical and semi-classical simulators. For example, there is relatively little information in the literature about the transport properties of holes in GaAs, this being a reflection of the fact that GaAs devices are generally fabricated on *n*-type material due to the superior electron transport properties of electrons compared with those of holes. Hole transport does, however, play an important role in some devices, and a good example of this is in GaAs photodiodes, where the current due to optical generation is comprised equally of both electrons and holes.

Monte Carlo simulators have been developed to model both electron and hole transport in GaAs. These are both single particle simulators, and are capable of providing steady state results. This is done by allowing a single particle to move in a uniform electric field, and averaging the velocity and energy of the particle over a sufficient length of time for the statistical uncertainty in these quantities to become small. In addition to the velocity and energy as a function of the electric field, the models also provide information on the distribution function as a function of field, and yield data on the relative importance of the various scattering processes in dissipating both energy and momentum. The electron transport model also provides information on the population ratios of the different valleys, and the hole model on the band population ratio.

The electron transport simulator models all three conduction band minima in GaAs, that is, the Γ, *L* and *X* valleys, along with the appropriate inter-valley scattering mechanisms. The hole transport simulator uses a two band model for the valence band (this is being extended to a three band model), and includes models for inter- and intra-band scattering. Other scattering processes include polar and non-polar optical phonon scattering, ionised impurity scattering, deformation potential and, in the electron model only, piezoelectric acoustic phonon scattering.

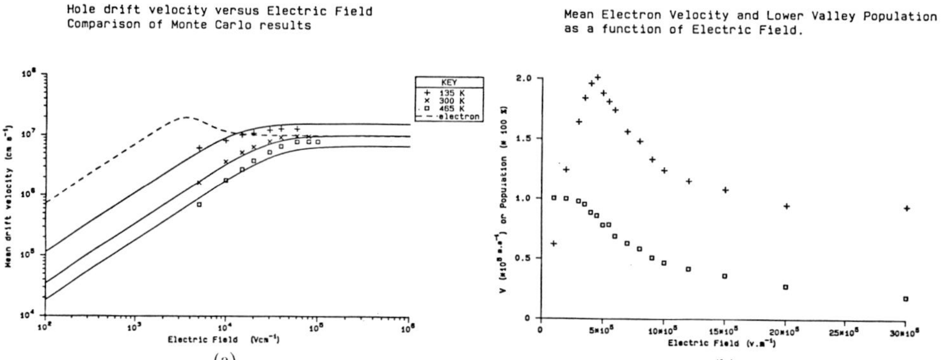

(a) (b)

Figure 9: Results of the single particle Monte Carlo simulators. (*a*) hole velocity-field characteristics and (*b*) electron velocity-field characteristics.

The programs are each comprised of 2000 lines of PASCAL, and run on both a VAX 8600 and an AMDAHL 580 computer. In order to calculate the transport properties at one value of electric field takes approximately 60 seconds on the VAX, and involves the calculation of 10,000 free flights, and a similar number of scattering processes. Figure 9 shows some of the results of the simulators. In part (*a*), the calculated static velocity-field relationship for holes in intrinsic GaAs is plotted for lattice temperatures of 135, 300 and 465 K (individual points), and is compared with an analytic curve fit for that data (solid lines) and the electron velocity-field characteristic at 300 K (broken line). Figure 9(*b*) shows the calculated velocity field characteristic for electrons in intrinsic material at 300 K (crosses, scaled by 10^{-5} m.s^{-1}), and the population ratio of the lower valley (boxes).

3.2. Monte Carlo and Classical Modelling of Photodiodes.

Both a Monte Carlo and a classical simulator for GaAs high-speed photodiodes have been developed as a basis for theoretical investigation of these devices [16]. These are both one-dimensional models of the device, the geometry of which is shown in figure 10. The classical model is based on the solution of the Poisson equation, combined with the time-dependent electron and hole current continuity equations. The continuity equations contain terms to account for both the photogeneration of carriers by the incident light wave and recombination processes, and, in order to account for absorption of the incident light, the photogeneration rate decreases exponentially from the surface of the device. The electron and hole mobilities are calculated using the single particle Monte Carlo models mentioned previously and are incorporated into the model using curve fits similar to those plotted in figure 9(*a*).

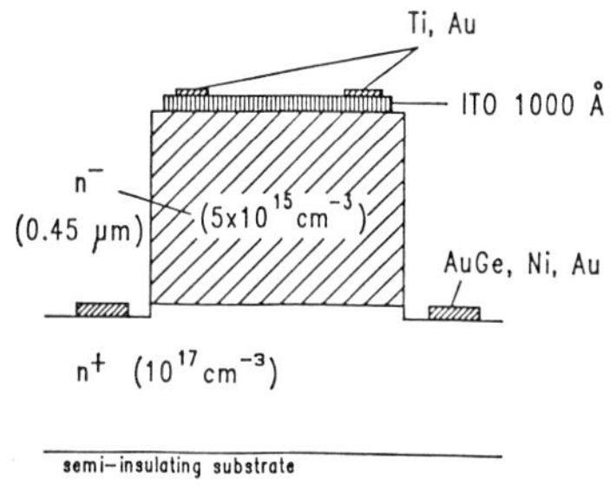

Figure 10: GaAs photodiode geometry.

The Monte Carlo photodiode model is a many particle model and is thus able to calculate the transient response of the photodiode to an incident light wave. The band structures and scattering processes used in this model are identical to those used in the single particle models, and these are combined with a Poisson solver used to calculate the electric field distribution throughout the device. This is determined by the instantaneous electron and hole distributions and the terminal voltages (boundary conditions). In order to keep the number of particles to a manageable level, superparticles are used to model both electrons and holes. These are particles which have the same transport properties as normal electrons or holes, but are assigned a multiple of the electronic charge when solving the Poisson equation. They thus represent the average electron in a region near the location of the superparticle. A typical simulation uses approximately 30,000 superparticles, each representing 1000 carriers.

Results from both the semi-classical and the Monte Carlo models of the photodiode have been compared, and show good agreement. Figure 11(*a*) shows the electron and hole concentrations inside the device under dark conditions, where the classical results are represented by broken lines and the Monte Carlo results by solid lines. Part (*b*) of that figure shows the transient response of the photodiode to the leading edge of an optical pulse, the broken line being a curve fit to the Monte Carlo data. The fluctuation in the concentration distributions at low concentrations in the Monte Carlo model is a consequence of the use of superparticles, as the average particle density is a sensitive function of the superparticle count.

The classical simulator incorporates an interesting animation package which allows the time evolution of the internal distributions, such as the electron concentration, photogeneration current and potential to be observed. This is performed by loading successive frames of the animation into picture segments on a graphics terminal, and then cycling the visibility of the picture segments on and off in sequence to give the illusion of motion. This display has been found to be very useful in determining the mechanisms which are important in the transport of photogenerated electrons and holes throughout the device. Unfortunately, it is difficult to illustrate this kind of output in this text!

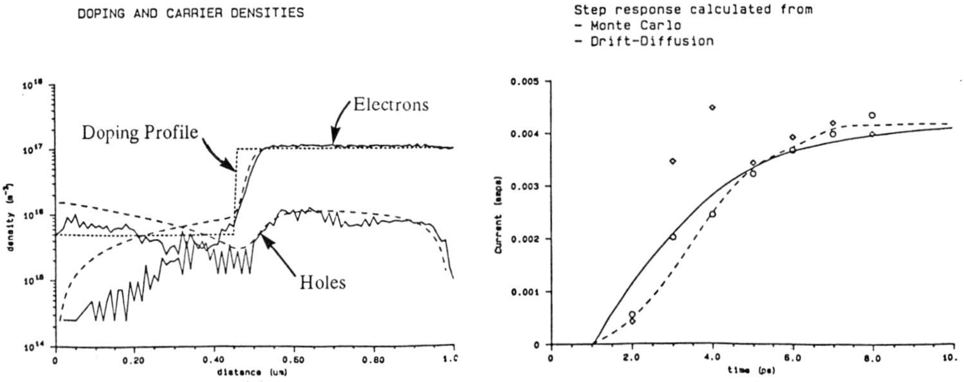

(a) (b)

Figure 11: Results of the photodiode simulation. (*a*) electron and hole distribution and (*b*) transient response, calculated using both the classical and Monte Carlo simulators.

3.3. Quasi-Two-Dimensional Simulation.

While fully two-dimensional simulators provide a highly accurate means of characterising devices with complex geometries, the computer resource required by these programs is often excessive compared to that available in the typical engineering environment. In some types of device, however, the current flow is predominantly in one dimension, and it is often possible to reduce the problem to a numerical solution in one dimension, and take account of the two-dimensional nature of a device using analytic means. A quasi-two-dimensional simulator of this nature has been developed at Leeds for simulating recessed gate and planar MESFETs, and forms part of a modelling package known as FETCAD [17].

This model uses a two-dimensional analytic solution for the shape of the depletion region near the Schottky gate, and numerically solves for the potential, carrier, energy and momentum distribution along the resulting channel profile. A quasi-two-dimensional description of the device is obtained by incorporating the width of the channel into the one-dimensional Poisson, current continuity, energy and momentum conservation equations. This results in a highly accurate two-dimensional description of the MESFET while retaining the simplicity and speed of a one-dimensional equation solver.

The simulator accounts for the non-uniform channel doping found in many devices, as well as effects such as surface depletion, substrate conduction, non-abrupt depletion region edges, avalanche breakdown, and gate conduction. The use of a mixed set of current and voltage boundary conditions avoids the need for iterative solution schemes, and results in an extremely rapid solution. A wide variety of results can be obtained from the model, including DC, transient, and RF results.

The program has been developed in PASCAL on an IBM AT, and is currently run on that type of machine, as well as IBM PS/2 model 80, VAXes, and an AMDAHL mainframe. An extensive graphical input and output system is utilised which allows graphical specification of the device geometry and doping profile, and plotting of results such as internal distributions, channel profiles, output characteristics, as well as circuit parameters such as the microwave S parameters of the device. In addition, the simulator incorporates a simple circuit analysis package which allows the user to embed the physical device model in an external circuit, and thus use the device model as an active element in a full non-linear circuit analysis.

FETCAD is most commonly run on IBM PS/2 based workstations, and on this hardware achieves a level of performance which could not be matched by a fully two-dimensional simulator, even on a mainframe. The simulation of a typical microwave MESFET requires approximately 100 seconds in order to calculate the DC output characteristics. Microwave circuit parameters, such as the S parameters requires less than 200 seconds. This is several orders of magnitude less than the time which would be required to characterise a similar device using a conventional two-dimensional simulator.

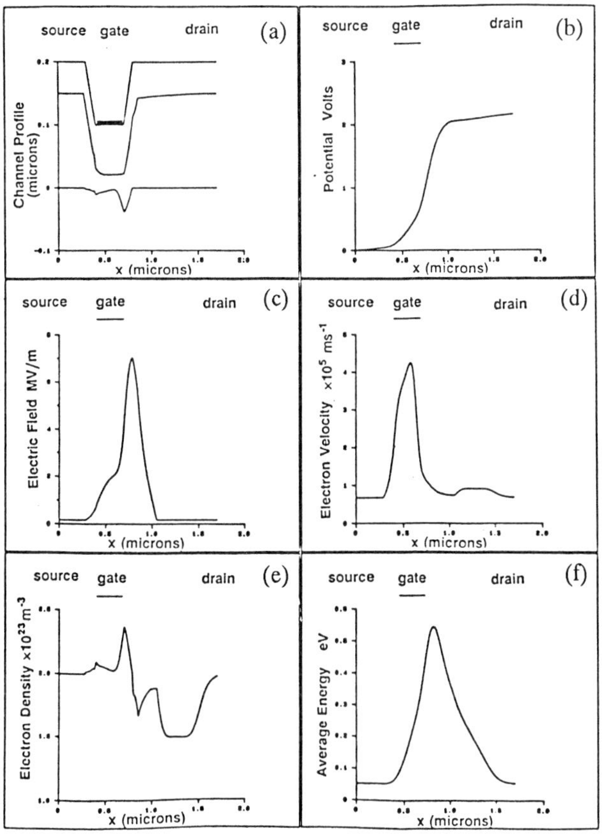

Figure 12: Internal distributions along the channel of an 0.3 μm MESFET. (*a*) channel profile, (*b*) potential, (*c*)electric field, (*d*) electron velocity, (*e*) density, and (*f*) energy.

Figure 12 shows the calculated channel profile, potential, electric field, electron velocity, electron density and average electron energy in the channel of a typical 0.3 μm gate length recessed gate MESFET. Calculated and measured S_{21} (gain) of an NEC9001 0.5 μm gate length device are compared in figure 13(*a*), and part (*b*) of that figure shows a comparison of the measured and simulated output characteristics of a 1.0 μm gate length device.

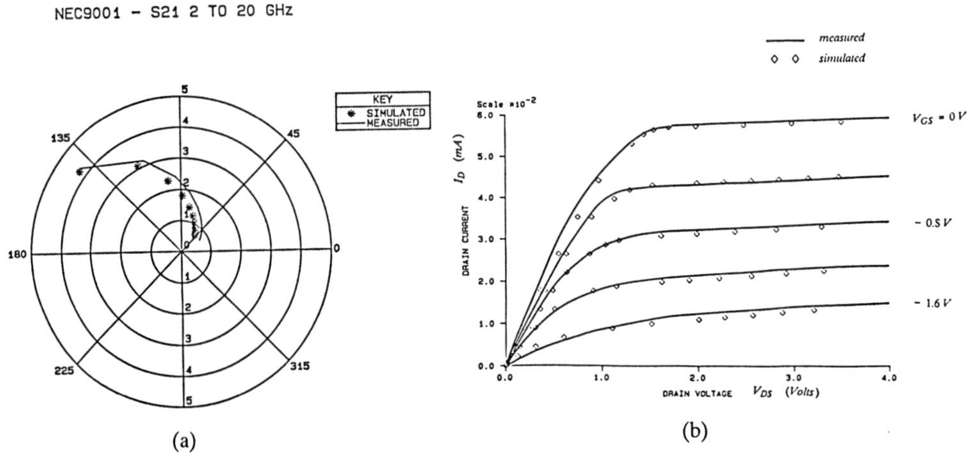

Figure 13: Calculated and measured S_{21} of an NEC9001 MESFET, and calculated and measured output characteristics of a 1.0 μm gate length device.

3.4. Heterojunction Bipolar Transistor Simulation.

Simulation of heterojunction devices is complicated by the change in the semiconductor band-gap which is found near the GaAs-AlGaAs interface. Electron transport in the region close to this interface is often best described using a quantum mechanical approach, and it is difficult to incorporate an interface model of this type into a classical or semi-classical simulator. Provided that the interface is not too abrupt, however, it is possible to re-formulate the current density equations in order to account for the conduction of both electrons and holes by the slope of the conduction and valence bands in the region close to the interface.

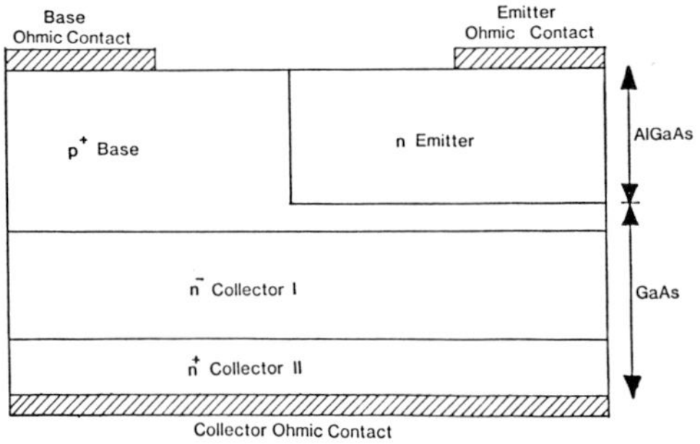

Figure 14: Geometry of simulated HBT.

A two-dimensional simulator of this type has been developed to perform transient and steady state simulation of heterojunction bipolar transistors (HBTs) [18,19]. These devices employ a graded heterojunction at the emitter-base junction in order to increase the emitter efficiency of the device. The effect of the heterojunction is to increase the conduction band potential barrier at the *p-n* interface, and thereby reduce the reverse current of the junction.

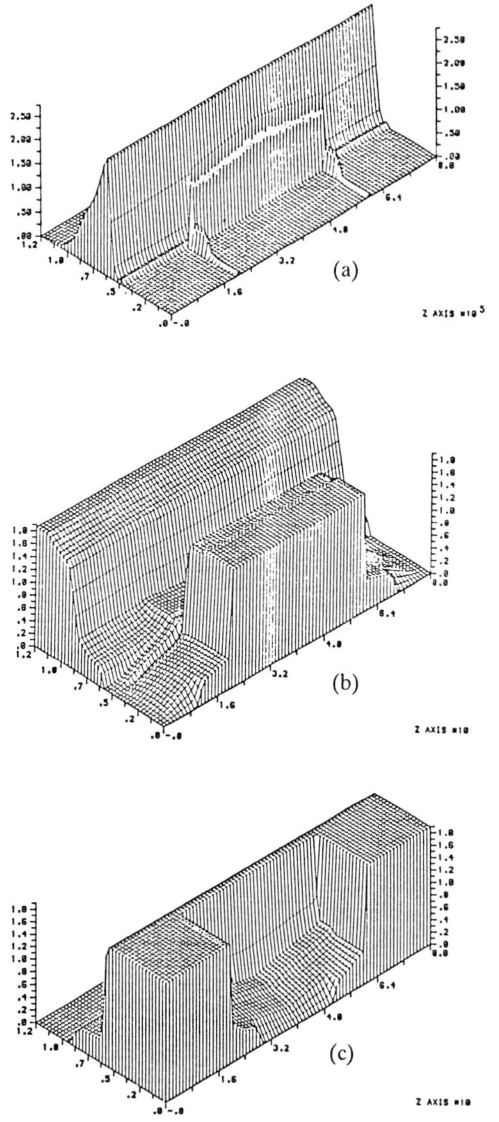

Figure 15: Electric field, electron concentration, and hole concentration in the HBT, V_{be} = 1 V, V_{ce} = 5 V.

The simulator solves the Poisson and modified time-dependent current continuity equations throughout the device to obtain the collector current as a function of the base current, and from these it is possible to calculate the microwave and DC characteristics of the device. The equations are solved using finite difference techniques on a non-uniform mesh, and the current continuity equation, which incorporates the modified current density equations, is discretised using a special formulation of the Scharfetter-Gummel technique [20]. In addition, the simulator uses a current boundary condition on the base ohmic contact to determine the potential of that contact, rather than the more usual Dirichlet boundary condition in voltage. This reflects the current controlled nature of the device, in contrast to an FET which is essentially a voltage controlled device.

Figure 14 shows the geometry of an HBT which has been simulated by the program. The device is symmetrical about the middle of the emitter contact, and it is therefore only necessary to simulate one side of the device, as the results for the other side may be obtained by reflection about the middle of the emitter. The emitter-base junction in this device is graded over a distance of 500 Å, and the Al ratio in the AlGaAs emitter is 0.3. Figure 15 shows a surface plot of the electric field, electron concentration, and hole concentration inside the device for V_{be} = 1 V, V_{ce} = 5 V. In these diagrams, the emitter occupies the middle 2.0 μm of the front long horizontal axis, the two base contacts the outer micron on the same edge, and the collector the entire back long edge.

References.

[1] Barton, T.M., Snowden, C.M., Richardson, J.R., *"Modelling of recessed gate MESFET structures"* in **"Simulation of Semiconductor Devices and Processes - Volume 2"**, Board K., and Owen, D.J.R. (eds), Pineridge Press, Swansea, U.K, 1986, pp 528-543.

[2] Barton. T.M., *"Characterisation of the physical behaviour of GaAs MESFETs"*, **PhD Thesis,** Leeds, UK, The University of Leeds, 1988.

[3] Franz, A.F., Franz, G.A., Selberherr, S., Ringhofer, C., and Markowich, P., *"Finite boxes - a generalisation of the finite-difference method suitable for semiconductor device simulation"*, **IEEE Trans. Electron Dev.,** Vol. ED-30, No. 9, Sept 1983, pp 1070-1082.

[4] Haigh, P.J., Miles, R.E., and Snowden, C.M., *"Simulation of non-planar GaAs devices"*, in **"Simulation of semiconductor devices and processes - Volume 3"**, Baccarani, G., and Rudan, M. (eds), Tecnoprint, Bologna, Italy, 1988, pp 291-304.

[5] Greenough, C., Hunt, C.J., Mount, R.D., and Fitzsimons, C.J., *"ESCAPADE: A flexible software system for device simulation"*, in **"Simulation of Semiconductor Devices and Processes - Volume 2"**, Board K., and Owen, D.J.R. (eds), Pineridge Press, Swansea, U.K, 1986, pp 639-652.

[6] Guerrieri, R., Ciampolini, P., Gnudi, A., Gibertoni, M., Rudan, M., and Baccarani, G., *"An investigation of polycrystalline-silicon MOSFET operation"* in **"Simulation of Semiconductor Devices and Processes - Volume 2"**, Board K., and Owen, D.J.R. (eds), Pineridge Press, Swansea, U.K, 1986, pp 468-479.

[7] Buturla, E.M., Cottrell, P.E., Grossman, B.M., and Salsburg, K.A., *"Finite-element analysis of semiconductor devices: The FIELDAY program"*, **IBM J. Res. Develop.,** Vol. 25, No. 4, July 1981, pp 218-231.

[8] Forghieri, A., Guerrieri, R., Ciampolini, P., Gnudi, A., Rudan, M., and Baccarani, G., *"A new discretisation strategy of the semiconductor equations comprising momentum and energy balance"*, **IEEE Trans. Computer-Aided Design,** Vol. 7, No. 2, February 1988, pp 231-242.

[9] J.R. Davy, Department of Mechanical Engineering and Computer Studies, The University of Leeds, Leeds, UK. **Private communication.**

[10] De Floriani, L., *"Surface representations based on triangular grids"*, **The Visual Computer,** Vol. 3, Springer-Verlag, 1987, pp 27-50.

[11] Gummel, H.K., *"A self-consistent iterative scheme for one-dimensional steady state transistor calculations"*, **IEEE Trans. Electron Dev.,** Vol. ED-11, No. 10, October 1964, pp 455-465.

[12] Bank, R E., Rose, D.J., Fichtner, W., *"Numerical methods for semiconductor device simulation"*, **IEEE Trans. Electron Dev.,** Vol. ED-30, No. 9 September 1983, pp 1031-1041.

[13] Polak, S.J., Den Heijer, C., Schilders, W.H.A., *"Semiconductor device modelling from a numerical point of view"*, **Int. J. Num. Meth. Eng.,** Vol. 24, 1987, pp 763-838.

[14] Hachtel, G.D., Mack, M.H., O'Brien, R.R., and Speelpenning, B., *"Semiconductor analysis using finite elements - Part 1: Computational elements"*, **IBM J. Res. Develop.,** Vol. 25, No. 4, July 1981, pp 232-245.

[15] Reiser, M., *"Large-scale numerical simulation in semiconductor device modelling"*, **Comp. Meth. Appl. Mech. Eng.,** Vol. 1, 1972, pp 17-38.

[16] Barry, D.M., Platt, S.P., Snowden, C.M., Howes, M.J., and Miles, R.E., *"Physical modelling of GaAs photodetectors"*, in **"Simulation of semiconductor devices and processes - Volume 3"**, Baccarani, G., and Rudan, M. (eds), Tecnoprint, Bologna, Italy, 1988, pp 31-42.

[17] Snowden, C.M., Submitted for publication in IEEE Trans. Electron Dev.

[18] Mawby, P.A., Snowden, C.M., and Morgan, D.V., *"Numerical modelling of GaAs/AlGaAs heterojunctions and GaAs/AlGaAs heterojunction bipolar transistors"*, in **"Simulation of Semiconductor Devices and Processes - Volume 2"**, Board K., and Owen, D.J.R. (eds), Pineridge Press, Swansea, U.K, 1986, pp 82-99.

[19] Mawby, P.A., *"Characterisation and fabrication of heterojunction bipolar transistors"*, **PhD Thesis,** The University of Leeds, Leeds, UK, 1988.

[20] Scharfetter, D.L., and Gummel, H.K., *"Large signal analysis of a silicon Read diode oscillator"*, **IEEE Trans. Electron Dev.,** Vol. ED-16, No. 1, January 1969, pp 64-77.

Practical Aspects of Device Modelling

Joseph A Barnard
Barnard Microsystems Limited

1. Introduction

We consider the practical ways in which an embedded device can be converted from a concept to artwork suitable for mask making. The masks can be used to realise the real device, or embedded device, which in turn can be tested to check the validity or otherwise of the device simulations.

GaS STATION is a layout software package that has been written to perform this task for microwave and optoelectronic devices and circuits. This software currently runs on IBM PC/ATs, Apollo and HP 9000 type computers.

As an example, we consider the realisation of a nonlinear transmission line consisting of a string of varactor diodes strung along a microstrip transmission line.

2. The device model

The varactor diode can be described as a variable capacitor with a series and parallel parasitic resistance. In a more exact model, the parasitic resistances will also vary as the capacitance is varied.

In an abrupt junction, uniformly doped pn junction diode, the capacitance varies as the inverse square root of the sum of the applied reverse bias voltage plus the inbuilt diode potential.

From a microwave simulation point of view, the equivalent circuit can be described in a nodal net list as follows:

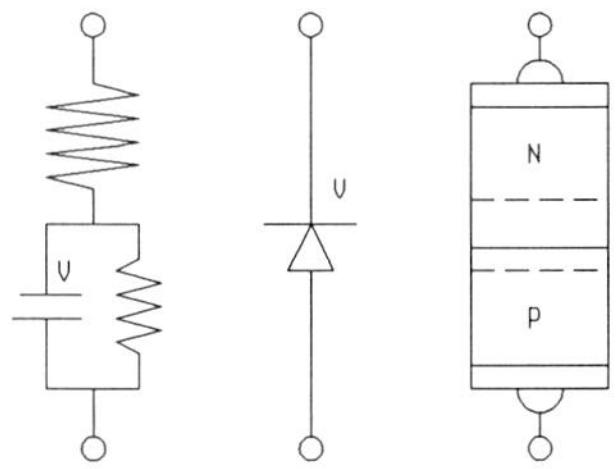


```
CV=12PF
BLK
    RES 1 2 R=5
    PRC 2 0 R=300 C=CV
    VAR: 1POR 1
END
```

where CV is a variable for the variable varactor capacitance that depends on the voltage applied accross the terminals of the diode. Ther is a 5 ohm resistor (RES) in series with a parallel RC (PRC) element, where the value of the resistor is 300 ohms, and the value of the capacitance is 12 picofarads. The entity is defined as the circuit subsection VAR.

The varactor characteristics could also be defined in terms of a microwave S parameter file derived by detailed computer simulation. In this case, we would use the following net list:

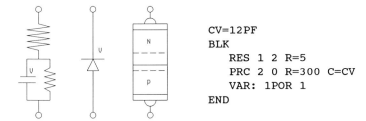

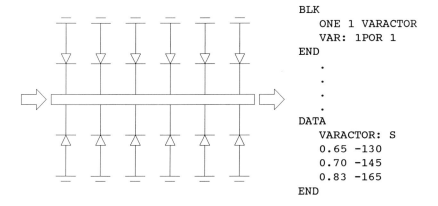

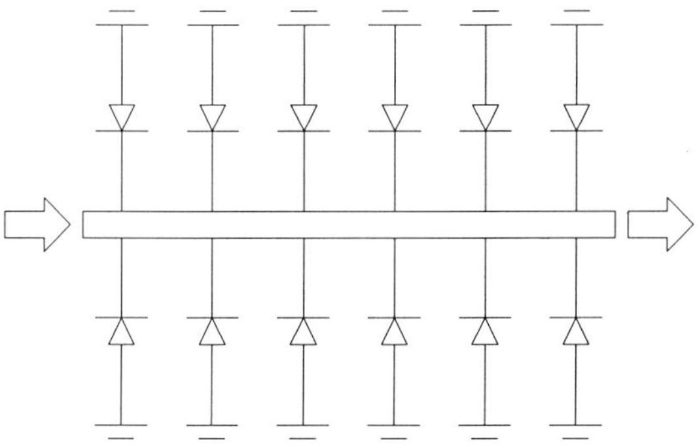

```
BLK
    ONE 1 VARACTOR
    VAR: 1POR 1
END
    .
    .
    .
    .
DATA
    VARACTOR: S
    0.65 -130
    0.70 -145
    0.83 -165
END
```

3. Relating the model to the device

The physical varactor was designed and stored in a graphics cell called VARAC. Essentially, the varactor is realised using a planar technology with a buried n+ layer. Mesa etching and via hole technology is used.

How can one relate the physical layout to the electrical simulation? The answer is to use a substitution command that is detected by the graphics software, but is ignored by the simulation software. In general, simulation software has

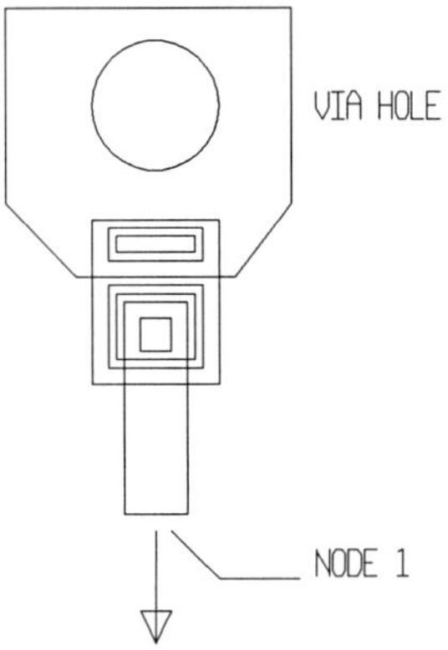

a comment symbol, eg. and asterisk (*) for Super Compact and an exclamation mark (!) for Touchstone. If the comment letter is followed by two dollar signs ($$), the graphics software will pick up information that follows while the electrical simulator will ignore it. Furthermore, GaS STATION uses the ASSUME command to substitute the physical cell for the equivalent circuit or S parameter file. In this way, the same net list can be used for electrical device and circuit simulation, and for an autoprocessed layout. The form of this substitution is:

```
*$$ ASSUME VAR VARAC
```

Now, whenever the net list calls for the element VAR, the graphics system will substitute the cell called VARAC.

4. Operation of the embedded device

Several varactors can be periodically located along either side of a microstrip transmission line to effectively load that line capacitively (and to a certain extent, resistively). As previously mentioned, the capacitance of the varactor varies as:

```
C = A/((Vbi + V)^0.5)
```

Now, the propagation velocity vp of an electromagnetic wave along a microstrip line is roughly as follows:

```
vp = 1/((LC)^0.5)
```

for an ideal transmission line. Consequently, the propagation velocity of the EM wave along a transmission line varies as

```
vp = B*(Vbi + V)^0.25
```

In other words, the EM wave travels faster as the amplitude of the wave is increased. This effect results in a compression of the leading edge of the wave, and an elongation of the trailing edge. With a suitable number of diodes, one can create solitons, which are short, sharp pulses.

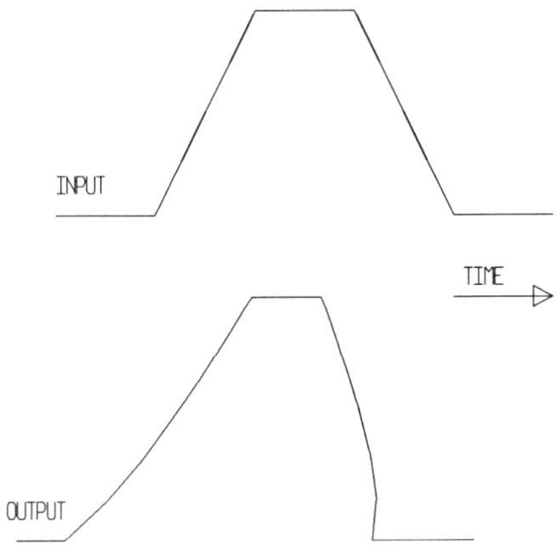

5. Creating a section layout

GaS STATION has been written to perform a direct mapping of a hierarchical circuit netlist to its hierarchical graphics equivalent. In this instance, we have a device (the varactor) and we will attach microstrip lines to a pair of these devices. The net list appears as follows:

```
BLK
   TRL 1 2 60 200 GAAS
   CROS 2 3 4 5 60 40 60 40 GAAS
   VAR 3
   VAR 5
   TRL 4 6 60 200 GAAS
   SECTION: 2POR 1 6
END
```

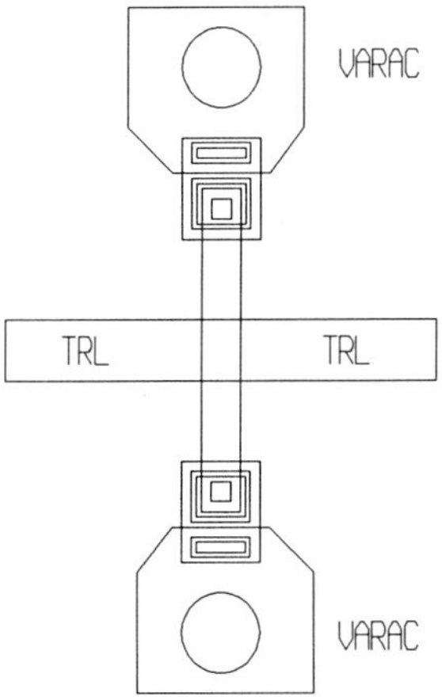

Here,

```
VAR  calls the cell called VARAC
TRL  is a system command
CROS is a system command
```

Note how the varactor (VAR) element is included in the netlist. How does GaS STATION handle this situation?

When autoprocessing a netlist, for each element, GaS STATION will first look for a cell with the same name as the element. If one is found, the system will attach the cell to the correct node, and with the correct orientation. If no cell is found, the system will search for a matching command in the system command library. If no match is found, the system will ignore the element. So, by opening the appropriate library before the autoprocessing begins, the user can control the layout.

```
BLK
  TRL 1 2 W=100 P=500 SUB
  CKT43: 2POR 1 2
END
```

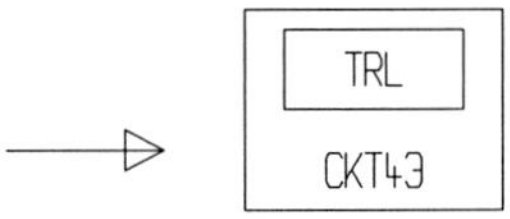


```
BLK
  CKT43 1 2
  CKT43 2 3
  CKT44: 2POR 1 3
END
```

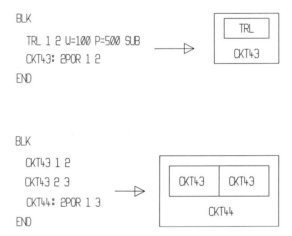

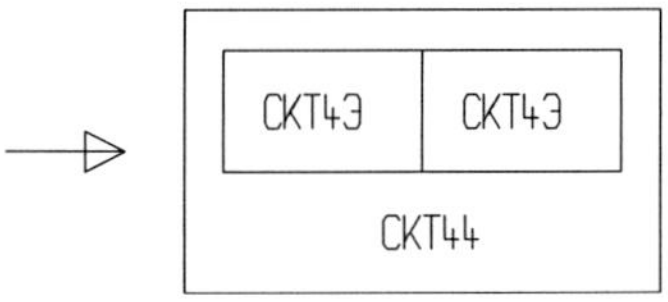

In this example, the library containing the cell called VARAC was accessable to the system prior to autoprocessing, so the physical varactor was located. VARAC was called instead of VAR due to the use of the ASSUME command to substitute VARAC for VAR in every instance after the ASSUME command was implemented. If the library contained schematic elements, a schematic representation of the netlist would be created. We have used Super Compact netlist notation. GaS STATION can just as easily handle Touchstone type netlists.

6. Hierarchical layouts

To create the travelling wave structure, we need to combine several of the SECTION circuit sections in a larger structure we will call STAGE. The netlist is as follows:

```
BLK
   SECTION 1 2
   SECTION 2 3
   SECTION 3 4
   SECTION 4 5
   SECTION 5 6
   SECTION 6 7
   STAGE: 2POR 1 7
END
```

Now, when GaS STATION creates a layout from the overall netlist, the cell called VARAC (the varactor layout) is contained in the cell called SECTION. SECTION contains a pair of varactor diodes and some connecting microstrip lines. The cell called STAGE contains six SECTION cells. There is an exact mapping of the hierarchical netlist file to a graphical hierarchy. This mapping is performed automatically, and guarantees a direct, and error free, relationship between the layout and the electrical simulation.

7. The overall net list

The overall netlist that can be used for both an electrical simulation and the net list driven layout is then as follows in Super Compact notation. As mentioned before, GaS STATION can interpret Touchstone format net lists in the same way.

```
CV=12PF
BLK
   RES 1 2 R=5
   PRC 2 0 R=300 C=CV
   VAR: 1POR 1
END
*$$ ASSUME VAR VARAC
BLK
   TRL 1 2 60 200 GAAS
   CROS 2 3 4 5 60 40 60 40 GAAS
   VAR 3
   VAR 5
   TRL 4 6 60 200 GAAS
   SECTION: 2POR 1 6
```

```
END
BLK
   SECTION 1 2
   SECTION 2 3
   SECTION 3 4
   SECTION 4 5
   SECTION 5 6
   SECTION 6 7
   STAGE: 2POR 1 7
END
```

The complete network layout is as shown below. This layout was created by autoprocessing the above net list file which made use of the varactor layout in the cell called VARAC.

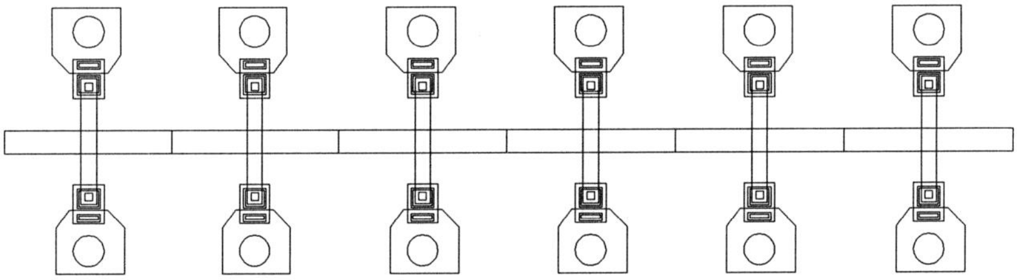

This precise link between the electrical net list used to simulate the performance of an embedded device and the physical network layout removes a source of potential error in the relationship between the measured and predicted network characteristics.

Subject Index